Production Methods and Workability of Concrete

BOOKS ON CONCRETE MATERIALS FROM E & FN SPON

Application of Admixtures in Concrete
    Edited by A.M. Paillere
Blended Cements in Construction
    Edited by R.N. Swamy
Cement-based Composites: Materials, Mechanical Properties and
    Perfomance. *A.M. Brandt*
Concrete in Hot Environments
    I. Soroka
Concrete in Marine Environments
    P.K. Mehta
Concrete Mix Design, Quality Control and Specification
    K.W. Day
Construction Methods and Planning
    J.R. Illingworth
Disposal and Recycling of Organic and Polymeric Construction Materials
    Edited by Y. Ohama
Durability of Concrete in Cold Climates
    M. Pigeon and R. Pleau
Euro-Cements: The Impact of ENV 197 on Concrete Construction
    Edited by R.K. Dhir and M.R. Jones
Ferrocement
    Edited by P.J. Nedwell and R.N. Swamy
Fibre Reinforced Cementitious Composites
    A. Bentur and S. Mindess
Fly Ash in Concrete: Properties and Performance
    Edited by K. Wesche
Hydration and Setting of Cements
    Edited by A. Nonat and J-C. Mutin
Interfacial Transition Zone in Concrete
    Edited by J.C. Maso
Manual of Ready-Mixed Concrete
    J.D. Dewar and R. Anderson
Performance Criteria for Concrete Durability
    Edited by J. Kropp and H.K. Hilsdorf
Special Concretes: Workability and Mixing
    Edited by P.J.M. Bartos
Structural Grouts
    Edited by P.L.J. Domone and S.A. Jefferis
Thermal Cracking in Concrete at Early Ages
    Edited by R. Springenschmid
Workability and Quality Control of Concrete
    G.H. Tattersall

For more details, contact the Promotions Department, E & FN Spon, 2-6 Boundary Row, London SE1 8HN, Tel: Intl +171-865 0066

Production Methods and Workability of Concrete

Proceedings of the
International RILEM Conference

Paisley, Scotland
June 3–5, 1996

EDITED BY

P.J.M. Bartos and D.L. Marrs

Advanced Concrete Technology Group,
Department of Civil, Structural and Environmental Enghineering,
University of Paisley, Scotland

and

D.J. Cleland

Department of Civil Engineering, The Queen's University, Belfast,
Northern Ireland

CRC Press
Taylor & Francis Group
Boca Raton London New York

CRC Press is an imprint of the
Taylor & Francis Group, an **informa** business

A CHAPMAN & HALL BOOK

CRC Press
Taylor & Francis Group
6000 Broken Sound Parkway NW, Suite 300
Boca Raton, FL 33487-2742

First issued in paperback 2019

ISBN-13: 978-0-419-22070-1(hbk)
ISBN-13: 978-0-367-44849-3 (pbk)

A catalogue record for this book is available from the British Library

Publisher's Note This book has been prepared from camera ready copy provided by
the individual contributors in order to make the book available for the Conference.

Visit the Taylor & Francis Web site at
http://www.taylorandfrancis.com

and the CRC Press Web site at
http://www.crcpress.com

Contents

Preface

Concrete production processes, particularly the mixing stage and the behaviour of ordinary or special mixes in their fresh state are fundamental to the achievement of a cost-effective but technically advanced concrete construction. To effect a genuine technology transfer, the advances in quality of ordinary concretes and properties of special concretes have to be matched by appropriate developments in the construction process itself.

Last decade has seen a very large expansion in the range of materials used in concrete construction, including many new high-performance, special concretes and innovative concrete construction methods. Practical trials showed that outstanding properties which were observed in research laboratories were obtainable in practice, however, the focus of the initial research and technical development had been firmly on the performance and properties of the final product, namely on the strength, durability and other key characteristics of hardened concrete.

The production process parameters and behaviour of the new high-performance, special concretes when fresh often differ very considerably from that of an 'ordinary' concrete. However, the production methods used during the research stage were often chosen arbitrarily and there was little practical guidance available for those who attempted to use the new high-performance, special concretes in full-scale, commercial projects. It soon began to be appreciated that the mixing processes had to be adapted and correct properties of the special concretes in their fresh state were the key quality control indicators critical for a successful execution of the construction process and a satisfactory performance of the material in its hardened form. The shortage of an **independent guidance** for design engineers, specifiers and contractors on a selection of a technically appropriate mixing plant and on cost effective construction based on the special, high performance mixes began to be felt strongly by the industry. This led to an establishment in 1992 of RILEM Committees TC 145-WSM on Workability of Special Concrete Mixes and TC 150-ECM on Efficiency of Concrete Mixes under chairmanships of Peter JM Bartos and Harald Beitzel respectively. The Committee 145 concerned with workability aims to produce practical guidelines for production of a range of special mixes while the Committee 150 will establish a comprehensive system for a rational assessment of performance of concrete mixers.

This book contains papers selected for publication and presentation at the International RILEM Conference on Production Methods and Workability of Concrete held in Glasgow on 3-5th June 1996. The Conference represented the culmination of work of the two RILEM Technical Committees and provided an international forum for a discussion of the proposed guidance on production and properties of fresh concretes. It was strongly focused on construction practice, offering a wealth of new information on the latest innovations, from entirely new types of concrete mixers, production and properties of many special concrete mixes to new, practical tests for key properties of fresh concretes such as those resistant to washout when placed underwater.

The organisation of this event relied heavily on a small group of dedicated and hard working staff of the University of Paisley. The administration was managed efficiently by Mrs Carol MacDonald, assisted by members of the Advanced Concrete Technology Group. Editing of this volume was a difficult task and I wish to thank my co-editors, Mr David Marrs (ACTG, Univ. of Paisley) and Dr David Cleland (Department of Civil Engineering, The Queen's University, Belfast) for their personal time and great effort without which the completion of the editing would not have been possible in the short time available. I also wish to thank Mrs Margaret Nochar for her untiring assistance with reprocessing of many of the texts and Mr S. Ainslie of the EDU of the University of Paisley for design and production of imaginative conference documentation. Editorial advice provided by Mr J.N. Clarke was very helpful.

Members of the RILEM TC 145-WSM and 150-ECM were involved in the selection and refereeing of the papers proposed and their assistance is very appreciated. Chairmen of Working Groups, namely Dr F. de Larrard (HSC), Prof C.D. Johnston (Fibre Concrete), Dr D. Beaupré (Sprayed Concrete), Mr. K. Juvas (Dry Mixes) and Mr Ö. Petersson (Flowing Concrete) made particularly valuable contributions. Prof. Dr-Ing Harald Beitzel and Mr. Y. Charonnat coordinated the contributions on concrete mixers and their assessment on behalf of the TC 150-ECM.

The RILEM conference coincided with Concrete Scotland 1996, an exhibition organised at the same venue by the Scottish Region of the UK Concrete Society, represented by Mr Andrew Sutherland. The technical co-sponsorship of the event by the UK Concrete Society and The International American Concrete Institute in addition to the RILEM itself was very valuable. Finally, it was the financial backing of the conference by the University of Paisley which made it possible to stage this event and which is gratefully acknowledged.

Peter JM Bartos
Chairman of RILEM TC 145-WSM
Conference Director

International Scientific Advisory Committee

PART ONE
PRODUCTION MIXERS AND MIXING PROCESSES

1 EFFICIENCY OF CONCRETE MIXERS

Y. CHARONNAT
Laboratoire Central des Ponts et Chaussées, Nantes, France

Abstract

A RILEM technical commission has been working for a few years to define the efficiency of concrete mixers. The commission is preparing a document which, it is expected, will become an international recommendation. This article describes the technical information that will constitute the recommendation.

The first part concerns definitions of the terms that characterise mixing: mode and principle. It is completed by a description of the parts of the mixer: container, paddles, discharge system, etc. This information should contribute to better understanding between the various parties to both economic and technical exchanges.

The second part defines the mixing cycle. In particular, it is stated what should be counted as mixing time. While highly conventional, this mixing time is of very great technical importance for the throughput of concrete mixing plant.

The third part describes qualification procedures and, in particular, fixes the parameters used to characterise a mixer, the testing conditions, and the performance criteria. Mixer efficiency is judged according to three performance levels: ordinary mixers, performance mixers, and high performance mixers.

Keywords: Qualification, mixer, homogeneous, concrete, RILEM, efficiency, mixing cycle.

Avertissement

Les informations à l'origine de cet article, proviennent des travaux de la commission technique de la RILEM - TC 150 - Efficiency of concrete mixers. Ce comité est présidé par le Professeur Dr Ing **Beitzel** de l'université de Trier (Allemagne). A ce comité participent également MM le Professeur **Bartos** de l'université de Paisley (Grande Bretagne), le Professeur **Legrand** de l'université de Toulouse (France), **Nebuloni** de Milan (Italie), **Pertersson** de Stockohlm (Suède), **Sikuler** de Haïfa

Production Methods and Workability of Concrete. Edited by P.J.M. Bartos, D.L. Marrs and D.J. Cleland. Published in 1996 by E & FN Spon, 2–6 Boundary Row, London SE1 8HN. ISBN 0 419 22070 4.

4 *Charonnat*

(Israël) **Steiner** de Widdegg (Suisse) et **Charonnat** de Nantes (France) qui en assure le secrétariat technique.

1 Introduction

Le malaxeur est l'élément essentiel d'une centrale à béton. Malgré cela et bien que beaucoup d'études aient été menées dans les différents pays et par les différents constructeurs pour mieux connaître ce matériel, aucune règle incontestée n'est publiée sur leurs conditions d'emploi et sur les résultats qu'on peut en attendre. La RILEM, en créant ce groupe, a voulu combler cette lacune. Le premier travail fut de rassembler les travaux faits dans les différents pays afin d'établir un lexique des termes utilisés pour décrire les malaxeurs. C'est l'objet de la première partie de cet article. Nous avons ensuite défini ce qu'était le cycle de malaxage. Sous ce vocable anodin se cache en fait un élément important qui régit très souvent les relations contractuelles entre l'acheteur et le vendeur du malaxeur ou entre le fabricant et l'utilisateur du béton fabriqué. Ce sera la deuxième partie de cet article. Enfin il fallait définir comment apprécier l'efficacité du malaxeur pour produire un béton de qualité. Le comité a donc défini les paramètres caractérisant le malaxeur, choisi les essais les plus discriminants pour faire la différence entre deux malaxeurs, déterminer les critères d'appréciation des malaxeurs et enfin préciser les conditions d'application. Ce sera l'objet des trois parties qui suivent. A l'issue de ce travail, chaque centre d'essais accrédité sera en mesure d'estimer les performances des malaxeurs et de les qualifier dans la classe d'efficacité correspondante.

2 Terminologie

Avant de rentrer dans le détail des éléments de brassage, il était important de distinguer les différentes familles de malaxeur. En premier lieu il a été analysé le mode de fabrication du béton. Il existe deux modes de fabrication du béton : le mode continu et le mode discontinu. Le premier est peu utilisé. En France par exemple ce mode de fabrication continu est essentiellement utilisé pour la production du béton des chaussées en béton de ciment. On conçoit bien que ce mode de fabrication convient principalement aux productions d'un béton de composition constante à fort débit, pendant de longues périodes. C'est le cas de la route mais également des barrages ou de grands ouvrages construit à partir d'un même béton. C'est la fabrication en discontinu, c'est à dire la fabrication par gâchée, qui est la plus utilisée dans tous les pays.
Un deuxième élément important, c'est le principe de fabrication. Longtemps en France on a séparé le malaxeurs de la bétonnière, ces dernières ayant toujours été sous considérées vis à vis de la qualité du béton. De nombreux essais réalisés ont montré que bien utilisées, comme il en est de tout matériel, ces bétonnières étaient capables de produire des bétons satisfaisant aux besoins. Dans cet article nous ne ferons donc pas de différence entre les termes bétonnière et malaxeur. Nous n'avons donc pas retenu ces termes comme qualifiant une famille de malaxeur et nous nous sommes orientés vers les principes de malaxage. Deux principes ont été retenus : les malaxeurs qui travaillent à partir de mouvements relatifs, c'est le brassage forcé, et ceux qui travaillent sans mouvement relatif, ce sont les premières bétonnières dans lesquels le

mélange se fait par relevage d'une partie du matériau suivi d'une chute dans la masse. Le brassage forcé a d'abord été rencontré dans les malaxeurs puis progressivement dans certaines bétonnières.

Nous avons ensuite distingué la position de travail. Celle ci est définie par la position de l'axe dans l'espace et nous avons retenu l'axe horizontal, l'axe vertical et l'axe incliné.

Enfin nous avons examiné la cuve du malaxeur, les outils de brassage et les dispositifs de vidange. Le tableau 1 récapitule les termes utilisés pour ces différents éléments.

Tableau 1 : Terminologie des éléments de brassage et d'évacuation du béton. Dans un malaxeur certains outils de brassage peuvent être associés.

Eléments	types existants	caractéristiques
cuve	annulaire	fixe
	cylindrique	fixe ou mobile
	auge	unique ou double
	tambour	cylindrique, tronc conique ...
	coquilles	à ouverture diamétrale
outils	pales	pleine, ajourée, peigne , à mouvement concentrique ...
	tourbillon	grande vitesse de rotation
	planétaire	mouvement épicycloïdal
	hélice	continue ou discontinue
	spire	périphérique ou intérieure
	vibreur	interne
vidange	trappe	latérale, de fond, basculante, coulissante
	basculement	avec ou sans inversion du sens de rotation
	inversion	du sens de rotation
	goulotte	basculante
	médiane	malaxeur à coquilles

2 Le cycle de malaxage

Les études sur le malaxage montrent que le remplissage du malaxeur a un rôle très important dans l'aptitude du malaxeur à produire un mélange homogène. En conséquence pour étudier l'efficacité d'un malaxeur il faut prendre en compte le remplissage du malaxeur et le temps de malaxage. Bien que moins importante la vidange du malaxeur peut être à l'origine de dégradation de la qualité acquise dans la cuve. Ces trois éléments forment le cycle de malaxage.

En général, c'est ce cycle qui détermine le débit de la centrale. Aussi cherche-t-on à le rendre optimal pour améliorer les rendements de production.

2.1 Le remplissage de la cuve

Certaines règles de base sont bien connues pour se placer dans les meilleurs conditions d'obtention d'un mélange homogène. Il y a les règles de conception et les règles d'utilisation [1]. Pour la conception du malaxeur, il faut chercher à faire arriver les matériaux solides dans la partie centrale de la cuve et éviter les angles dans lesquels le matériau stagnera et ne sera pas repris par les outils de malaxage. Les produits liquides doivent par contre être déversées en pluie sur toute la surface de la cuve. Toutes les positions de travail (orientation de l'axe) ne sont pas concernées de la même manière vis à vis du remplissage. Les constructeur le savent bien et positionnent les goulottes d'entrée en conséquence. Les modifications de ces dispositions, qui ne respecte pas les règles d'origine, entraînent souvent une perte importante d'efficacité.

2.2 Le temps de malaxage

On distingue plusieurs moments pendant lequel se fait le mélange. La figure 2 décrit un cycle théorique avec les différentes appellations. Le temps de malaxage est la durée qui s'écoule entre la fin de l'introduction du dernier constituant dans la cuve et le début de vidange de cette cuve. Bien entendu dans la réalité les introductions des solides et des liquides se superposent, ce qui peut changer le temps de prémalaxage. Mais, même dans ce cas, le temps de malaxage reste le même.

2.3 La vidange

C'est par un excès d'énergie qu'on peut constater une perte d'homogénéité (ségrégation). En conséquence on doit s'efforcer à maintenir un flot de matériau en masse en ouvrant progressivement le dispositif de vidange.

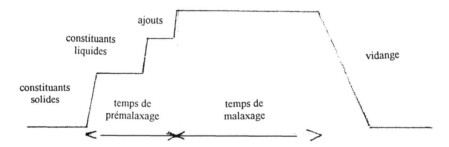

Figure 2 Exemple d'un cycle de malaxage. En pratique les introductions des solides et des liquides se chevauchent ce qui permet de réduire le temps de cycle et donc d'améliorer le rendement. Cela est cependant souvent au détriment de l'efficacité du malaxeur.

3 Parametres et essais caracterisant l'efficacite d'un malaxeur

Un malaxeur est efficace s'il distribue tous les constituants de façon homogène dans la cuve sans favoriser l'un d'entre eux. Dans la qualification d'un malaxeur, il y a donc lieu de caractériser les effets de cisaillement et de transport [2]et de vérifier qu'il ne provoque pas de ségrégation du fait d'une énergie trop grande (tableau 2). Pour effectuer ces mesures on réalise des prélèvements de béton. Le volume de ces prélèvements a toujours été un sujet de discussions sans fins.
En effet le prélèvement doit être :

- suffisamment petit pour que les mesures soient discriminantes ;
- suffisamment grand pour que l'effet de découpe n'influence pas le résultat ;
- suffisamment grand pour que les volumes soumis aux mesures soit en rapport avec la précision des essais.

Quelques études [3], [4] ont été menées sur ce sujet mais elles n'ont pas abouti à fixer des règles précises.

Les premiers paramètres retenus sont la teneur en eau et la teneur en éléments fins. Le seuil des éléments fins a été pris égal à 0,25 mm. L'objectif est la détermination du rapport E/C, rapport admis par tous les pays pour apprécier la durabilité d'un béton courant. Le volume du prélèvement dans lequel sont mesurés ces deux paramètres a été fixé à 3l ce qui correspond approximativement à 1,4 kg de apte. Vis à vis du malaxage on mesure la capacité du malaxeur à défloculer les éléments fins. On teste ainsi la fonction cisaillement du malaxeur. Le nombre de prélèvement est fixé à 15 pour les malaxeurs dont le volume nominal est inférieur ou égal à 1,5 m³.

Les paramètres suivants concernent la distribution granulaire du béton. Trois coupures sont retenues : 0,25 mm ; D/2 et D. Le volume des prélèvements est de 6 ou 8 l de béton. Vis à vis du malaxage on mesure la capacité du malaxeur à répartir indifféremment les éléments fins et les plus gros. On teste ainsi la fonction distribution du malaxeur.

Enfin le dernier paramètre concerne la teneur en air occlus (en absence de produit entraîneur d'air).Le volume du prélèvement est de 6 ou 8 l de béton. Le nombre de prélèvement a été fixé à 9. Par cet essai on apprécie également les facultés du malaxeur à cisailler le matériau.

Tableau 2 Ensemble des paramètres et des essais destinés à caractériser les malaxeurs.

Paramètres	essais	prélèvements	
		volume	nombre
teneur en eau teneur en éléments > 0,25 mm	séchage tamisage	3l	jusqu'à 1,5 m³ : 15 au delà 15 + 2(Cap - 1,5)
distribution granulaire 0,25 mm, D/2, D	tamisage	6 à 8l	jusqu'à 1,5 m³ : 15 au delà 15 + 2(Cap - 1,5)
teneur en air occlus	essai normalisé	6 à 8l	9

4 Criteres de qualification des malaxeurs et niveaux de performance

Il a été retenu quatre critères de qualification qui correspondent à des propriétés particulières du béton, propriétés recherchées dans les ouvrages. Pour chaque critère trois seuils ont été fixés (tableau 3). Ces seuils sont exprimés par référence au coefficient de variation (rapport de l'écart type à la moyenne) et pour la teneur en air la moyenne. Ces seuils (valeurs maximales à respecter) détermine trois niveaux de performances pour les malaxeurs :

- malaxeur ordinaire (OM) ;
- malaxeur performants (PM) ;
- malaxeur très performants (HPM).

Le premier critère est le rapport E/F qui caractérise la durabilité du béton (F étant assimilée à la quantité de ciment). Les trois seuils retenus sont 6%, 5% et 3 %.

Le second critère est la quantité d'éléments fins (> 0,25 mm). Elle figure la distribution du ciment dans la masse du béton. Les seuils retenus sont également 6%, 5% et 3 %.

Le troisième critère est la quantité du plus gros granulat (compris entre D/2 et D). Ce critère indique s'il y a ou non ségrégation. Les seuils sont 20%, 15% et 10%.

Le quatrième critère est la teneur en air occlus. Pour les OM il n'a pas été retenu de seuil. Pour les PM et les HPM il a été retenu pour la moyenne respectivement 2% et 1% et pour l'écart type 1% et 0,5%.

Tableau 3 Niveaux de performance des malaxeurs et seuils à respecter.

Critères		Niveaux de performances		
		OM	PM	HPM
E/F	avec F < 0,25 mm	s/m < 6%	s/m < 5%	s/m < 3%
teneur en F	avec F < 0,25 mm	s/m < 6%	s/m < 5%	s/m < 3%
teneur en D/2 - D		s/m < 20%	s/m < 15%	s/m < 10%
teneur en air		-	m < 2 %	m < 1 %
		-	s < 1%	s < 0,5%

5 Conditions d'application

Le comité technique a proposé différentes règles pour réaliser cette qualification des malaxeurs

5.1 Volume de la gâchée pris en compte

Compte tenu que c'est l'ensemble de la gâchée qui plus tard ira dans l'ouvrage c'est toute la gâchée qui doit être examinée. Cependant il a été admis que les début et fin de gâchée devaient être traités différemment.

La gâchée est divisée en trois parties début (les cent premiers litres), milieu et fin (les cent derniers litres). Sur le milieu on réalise les prélèvements prévus dans le tableau 2. Sur les parties début et fin on prélève trois échantillons sur lesquels on réalise tous les essais prévus au tableau 2.

Une gâchée sera réputée satisfaisante si :

- la population des résultats d'essais est gaussienne au seuil de confiance de 95%
- les moyennes des résultats d'essais effectués sur le début et sur la fin de la gâchée ne s'éloigne pas de plus de écarts types de la valeur moyenne des résultats d'essais effectués sur la partie centrale. L' écart type pris en référence pour l'essai concerné est celui calculé sur les résultats de la partie centrale.

Si ces deux conditions sont respectées pour chacun des essais, alors on analysera les résultats des essais effectués sur la partie centrale, conformément aux indications du tableau 2.

5.2 Prise en compte du volume du malaxeur

Comme il est indiqué dans le tableau 2 une distinction est faite pour les malaxeurs de plus de 1,5 m^3. Le nombre de prélèvement dans la partie centrale est supérieur en fonction de la capacité du malaxeur. Ainsi pour un malaxeur de 3 m^3 le nombre d'essais passe de 15 à 18.

Il a été également examiné le cas des petits malaxeurs qui sont le plus souvent utilisés dans les laboratoires. La limite a été fixée à 375 l de béton en place. Dans ce cas les prélèvements sont d'un volume inférieur respectivement 1 l au lieu de 3 et 5 l au lieu de 8. Enfin les extrémités ne seront plus distinguées, les prélèvements seront réalisés aléatoirement sur toute la gâchée.

5.3 Quels bétons tester ?

Comme les malaxeurs sont quelquefois étudiés pour des bétons particuliers, il était nécessaire de choisir des bétons représentatifs des fabrications courantes. La différenciation de ces bétons est faite par la dimension D des granulats et par la consistance du béton.

Il a été retenu deux granulométries et quatre consistances.

A partir de ces possibilités on considère couvrir l'ensemble des fabrications de béton existant sur le marché. Les bétons manufacturés sont représentés par le béton de caractéristiques D compris entre 8 et 10 mm et avec un affaissement compris entre 0 et 3 cm. Les bétons de chaussées ou de barrières de sécurité sont représentés par le béton de granulométrie comprise entre 20 et 32 mm avec un affaissement de 3 à 5 cm. Les bétons courants de bâtiment ou d'ouvrage d'art sont représentés par le béton de granulométrie comprise entre 20 et 32 mm avec un affaissement de 10 à 12 cm. Enfin les bétons fluidifiés sont représentés par le béton de granulométrie comprise entre 20 et 32 mm avec un affaissement supérieur à 18 cm.

Le tableau 4 récapitule les différentes possibilités. Bien entendu les combinaisons autres peuvent être également testées.

Tableau 4 Choix des bétons testés et exemples d'application.

Consistance (affaissement au cône)	Dimensions maximales des granulats	
	8 - 10 mm	20 - 30 mm
0 - 3 cm	bétons manufacturés	pistes d'aéroport
3 - 5 cm	bétons de sable	chaussées, barrières
10 - 12 cm	bétons fins	ouvrages courants
> 18 cm		bétons fluidifiés

6 Mise en place de la qualification

A partir de ce canevas les constructeurs de malaxeurs seront en mesure de faire reconnaître officiellement les performances des matériels qu'ils commercialisent. Bien entendu ce sera à ces constructeurs qu'il reviendra de fixer les paramètres suivants :

- Cycle de malaxage (ordre d'introduction des constituants et débit pour chacun d'eux) ;
- temps de malaxage (compté comme indiqué au paragraphe 2.2) ;
- choix des bétons à tester.

7 Conclusion

La connaissance des performances d'un béton commence par la connaissance des propriétés, des constituants et par la connaissance des matériels utilisés pour le fabriquer. Dans le domaine de la fabrication du béton deux postes sont importants: le dosage des constituants et le malaxage du béton. Jusqu'à aujourd'hui il n'était pas possible de donner les informations caractérisant la régularité de fabrication c'est à dire l'homogénéité du mélange.

Avec la mise en place d'une qualification des malaxeurs le concepteur de l'ouvrage pourra mieux prendre en compte les performances réelles du béton et concevoir les ouvrages en conséquence. Cette qualification aidera également les constructeurs de matériel en leur indiquant les points sur lesquels ils doivent progresser.

La qualité d'un produit, ici le malaxeur, peut toujours être améliorée quand on connaît ses caractéristiques réelles. Nous avons fait, avec la qualification des malaxeurs, un premier pas vers la qualité totale.

8 Bibliographie

1. Charonnat, Y. (1996) Quality control in concrete mixing, *International RILEM Conference on Production Methods and Workability of Concrete*, Glasgow, Scotland, June 3-5 1996.
2. Baroux, R., Charonnat, Y. and Rousselin, P.J. (1985) Application des traceurs pour l'étude du malaxage des matériaux de construction, *Bull Liaison Labo P. et Ch.*, 140 nov - déc 1985, Ref 3028.
3. Beitzel, H. (1981) Gesetzmässigkeiten zur optimierung von betonmischern, *Fakultät für Bauingenieur und Vermessungswesen der Universität (TH)*, Karlsruhe, 1 Juli 1981.
4. Robin, P. (1988) Contribution à la caractérisation de l'échantillonnage de mélanges granulaires en vus d'étudier les phénomènes de malaxage, *Université Blaise Pascal*, Clermont Ferrand II, 24 octobre 1988, No. d'ordre DI 213.

2 DEVELOPMENT OF A NEW-TYPE MIXER BASED ON A NEW MIXING SYSTEM

K. MAEDA
Maeda Corporation, Tokyo, Japan

Abstract

This paper reports about the initial steps towards the design of a practical mixer based on the new concept of kneading method. Not only the ordinary concrete, but materials like core-blend and clay-blend which are not possible to mix by conventional methods, can be mixed by the mixer developed from the new theory. The initial basic concept proposed of forming the 2^n layers has been advanced to form the 3^n, 4^n or even more, where n is the numbers of kneading; which effectively reduces the time of mixing and increases the uniformity of the mixed materials. It is also understood that other materials, e.g. rocks abandoned at dam-sites, can also be used as aggregate in the concrete with less possibilities of being broken up when mixing in the new-type mixer. This study has suggested that the mixer is practicable for these materials with the enhancement on its mechanical aspects.
Keywords: clay-blend, concrete, conventional mixer, core-blend, mixing, kneading, new-type mixer, soft aggregates.

1 Introduction

In the previous paper [1], presented at the International RILEM Seminar, CBI Stockholm, May 8-9, 1995, the mixing process of the new-type mixer, including the results obtained from the experiments using the prototype mixer and the bucket-roller

Production Methods and Workability of Concrete. Edited by P.J.M. Bartos, D.L. Marrs and D.J. Cleland. Published in 1996 by E & FN Spon, 2–6 Boundary Row, London SE1 8HN. ISBN 0 419 22070 4.

type mixer, were reported. In this paper, different stages of the development of mixer for practical use are reported.

With regard to the experimental results obtained from the new-type of mixing method, the following points can be noted as characteristics of this new-type mixer:

1. For the ordinary concrete, the similar results can be obtained as those from a conventional mixer, with kneading number of 10 [2].

Furthermore, the following materials which are not possible to be mixed in the conventional mixer can be mixed properly in the new-type mixer [3]:

2. Combination of materials; with extremely different specific gravities.
3. Lean concrete; aggregates of which need to be finely coated.
4. Highly viscous materials.

It is already understood that the new-type mixer has a very good efficiency for mixing the above-mentioned materials.

Since the application of this new mixing method is expected to be used in different fields, this paper reports about the manufacture and the experimental results of the mixer for mixing the core-blend of the rock-filled dam, the clay-blend and the concrete.

2 Reduction of the kneading numbers

Firstly, the basic concept of this mixer was based on the principle of the final mix containing 2^n layers of materials after lapping half of the materials to the other half and kneading to its original thickness with the total kneading numbers of n. However, it was desirable to reduce the total kneading numbers by forming the mixed materials in 3^n, 4^n layers or even more. According to the results from the various types of tests carried out, it was decided to roll the materials which consisted of two or more layers, after each kneading.

In the experiment, the following two cases were investigated.

2.1 Different types of clay-like materials with high viscosity
It was found that due to their high cohesiveness, such materials were not splitting during kneading and thus could be rolled into layers. However, rolling of the kneaded materials was possible only with a combination of a belt conveyor and a fixed roller, as shown in Fig. 1.

2.2 Soil-like materials of very low viscosity
In this case, rolled layer formation was not possible using the fixed roller, due to the splitting of the materials. However, rolled layers were formed when placing the materials on a belt conveyor consisting of semi-circle, as shown in Fig. 2.

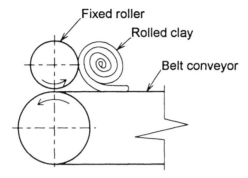

Fig. 1 Belt conveyor and fixed roller

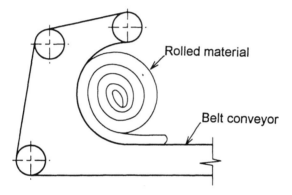

Fig. 2 Belt conveyor forming semi-circle

2.3 Avoidance of loss of water for concrete-like materials

As in the two cases mentioned above, it was not necessary to consider the water loss problem. However, sealing of the mixer was inevitably required to avoid the problem of water loss in the case of kneading of concrete mixes. Thus, the circular type mixer, as shown in Fig. 3, was designed and tested for kneading and lapping concrete materials.

3 Characteristics of the mixers

3.1 Clay kneading mixer

In a pottery production factory, it is generally very important to mix the clays which are taken from different sites with different colour and water content to produce the suitable raw pottery material. In this case, the clay with a very high viscosity is needed to avoid the problem of having different water content. For such clay, it is made possible to produce the roll with a correct combination of the roller's diameter and its revolving speed. The trial machine, photo shown in Fig. 4, produced the rolled layer of

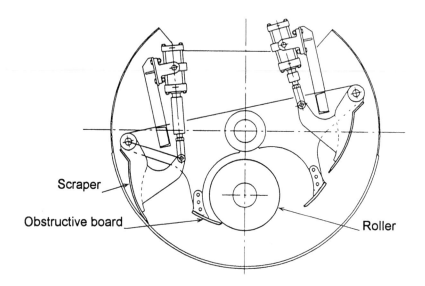

Fig. 3 Roller-rotary (RR) mixer

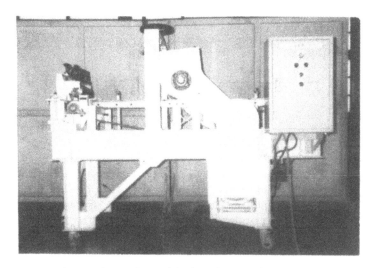

Fig. 4 Photo of the trial mixer for kneading the clay

clay, photo shown in Fig. 5. It can be seen from Fig. 5 that the central part and the external part of the layer are rolled in the form of 4^n and 2^n respectively. When this roll is kneaded and rolled again, the central part and the external part are rolled in the form of 2^n and 4^n respectively. Thus, by repeating this process the required number of times uniformly mixed materials can be obtained.

Fig. 5 Photo of the rolled clay

3.2 Mixer for core-blend materials

Since it was understood that the less viscous materials were impossible to be rolled by the clay kneading mixer, it was decided to use the belt conveyor to overcome the problem. However, it was very difficult to run the belt conveyor inside the mixer. Thus, as shown in Fig. 6, by fixing its two sides and the exact central part, it was made possible to run the belt forming a semi-circle shape. The shaft passing through the central part should have the vital role to interrupt the materials and to form the rolls. In reality, when carrying out this experiment in the machine, photo shown in Fig. 7, the following phenomena were observed (note that the thickness of the kneaded materials was kept equal to 50 cm in this case).

Firstly, the 50 cm-thick kneaded clay was mixed with sand and gravel. The mix was then placed on the semi-circled belt conveyor and rolled, photo shown in Fig. 8. It was found that the mix produced at least 3 layers or more, instead of only 2 layers with an average diameter of 150 cm. It was considered initially that it could have been kneaded with a thickness of about 50 cm, 100 cm being its original thickness. However in reality, when the belt's movement was reversed, the rolled clay was just congested with the belt at the lower part, due to the gravity effect, and thus as a result it

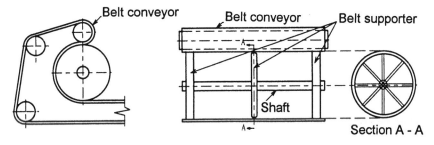

Fig. 6 Supporting method of the belt conveyor

Fig. 7 Photo of the mixer for kneading core-blend materials

Fig. 8 Photo showing the kneading condition of core-blend in kneading mixer

was collapsed in the extent of 60 cm. In this case its thickness would have been only 30 cm after kneading. If the mix was kneaded with maintaining its thickness in the scale of 30 cm each time, its efficiency would be decreased extremely and believed that it was not suitable in practical use.

The next case was considered to ascertain the rolled mix to return to the same diameter, i.e. 150 cm, even after reversing the operational direction. As shown in Fig. 9, the gradient of the belt conveyor was adjusted to avoid the accumulation of the left half portion of the rolled soil mix when reversing. It was judged that, due to the gravity effect, the accumulation of the mix in the belt conveyor would be avoided with an adjustment of gradient. However, the most suitable value of the gradient to be chosen

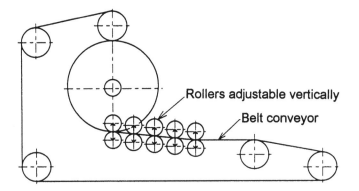

Rollers adjustable vertically

Belt conveyor

Fig. 9 Gradient adjustment of the belt conveyor

is still under trial.

Now, the application of the principle was used to mix the core-materials, sampled from the actual dam-site. In the actual site, three types of the core-materials, with the proportion of 1:1:0.5, are mixed in the sequence of three different layers consecutively until it reaches to the height of 10 m or so.

Then after about one year, these different layers are mixed together by cutting out by a bulldozer and loading into a truck, which carries them to the actual dumping site for placing. However, in this experiment, these three types of samples were originally collected separately from the site, and placed in three sequence layers. The layer was doubled and kneaded to its original thickness, i.e. in accordance to the principle of 2^n.

3.3 Test content

In order to determine the effect of the kneading numbers on the quality of core-materials using the experimental mixer, the following experiment was carried out.

In order to make the blend of the core-materials, three types of materials, namely BMD, BGD and BCL were taken from the Dam-site K. Materials were made in three different layers, as shown in Fig. 10.

Samples were taken from the top, middle and bottom layers of the mix each after kneading with the kneading numbers of 1, 3, 5 and 7. The grading test and water content test of each sample were carried out to evaluate the effect of kneading numbers on the extent of the uniformity of materials.

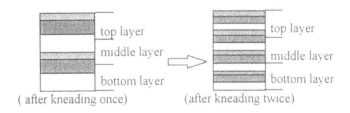

Fig. 10 Kneading method of core-materials

3.4 Test results and discussions

3.4.1 The variation on grading
The particle distribution of the materials, namely BMD, BGD and BCL, combined in the proportion of 1:1:0.5, which was obtained from the calculations, is shown in Fig. 11. The change in the particle ranges with the kneading numbers is shown in Figs. 12(a)~(d). It can be seen from the figures that when increasing the numbers of kneading from 1 to 3, 5 or more, the tendency for the uniformity on grading became more visible.

The mix proportion of the materials taken from the top, middle and the bottom layer, as calculated according to the theory, is shown in Fig. 13 and Table 1. When increasing the kneading numbers, it can be understood that the mix proportion of each layer goes on changing and finally the calculated mix proportion approaches a ratio to BMD:BGD:BCL=1:1:0.5=0.4:0.4:0.2. As shown in Fig. 12, the experimental value matches well with the theoretical value.

3.4.2 The change in grading index
The grading of the materials - represented by the indices as: (1) gravel content percentage; (2) sand content percentage; (3) the fine particle content percentage and (4) the average particle diameter; and (5) water content ratio - were plotted separately against the kneading numbers, as shown in Fig. 14. It can be observed that with increasing the numbers of kneading, a fixed value can be obtained and this concludes the uniformity of the materials with respect to any index.

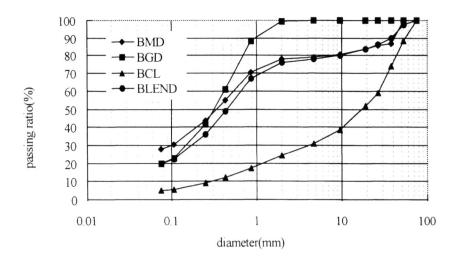

Fig. 11 Calculated particle distribution of three types of core-materials and their mix

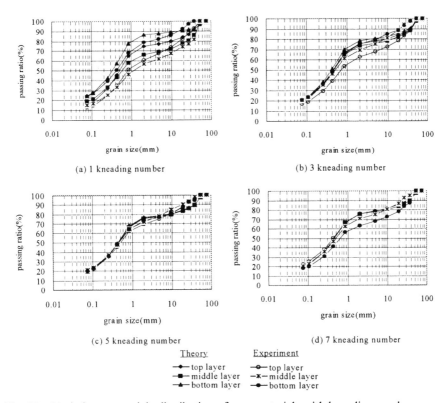

Fig. 12 Variation on particle distribution of core materials with kneading numbers

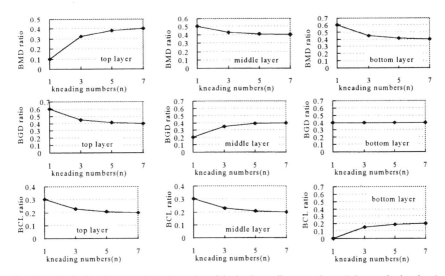

Fig. 13 Variation in material mix ratio with the kneading numbers (theoretical value)

Table 1   Variation of the mix proportion of three types of materials
in three layers with kneading numbers

Layers	Kneading numbers	BMD ratio	BGD ratio	BCL ratio
Top	1	0.100	0.600	0.300
	3	0.325	0.450	0.225
	5	0.381	0.413	0.206
	7	0.400	0.400	0.200
Middle	1	0.500	0.200	0.300
	3	0.425	0.350	0.225
	5	0.406	0.388	0.206
	7	0.400	0.400	0.200
Bottom	1	0.600	0.400	0.000
	3	0.450	0.400	0.150
	5	0.413	0.400	0.188
	7	0.400	0.400	0.200

4 RR-500 mixer

4.1 Development of the RR-mixer

4.1.1 Outline
With regard to the applicability of this new theory for concrete and other materials,
various experiments were carried out using a bucket-roller type experimental mixer
(BR-100). This experimental model mixer was manufactured to verify the applicability
of the proposed new mixing theory.

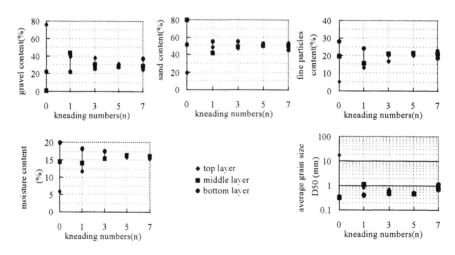

Fig. 14 Variation on the grading index with the kneading numbers

4.1.2 Design and manufacture of the practical mixer
The development of such a mixer was carried out step by step by conducting different experiments initially. In the designing process, various ideas were taken into consideration. However, when it was very difficult to judge the performance of the main parts from the drawings, models were manufactured and efficiency tests were carried out. In the meantime, a small mixer was manufactured and its actual performance was tested to verify the potential raised by this new theory.

The practical mixer has a typical circular shape. The internal parts of the mixer are composed of a roller and a pair of drum sides which serve to knead and return the materials respectively. The materials charged into the mixer are initially kneaded by the roller and then by tilting the drum; the first half of the returned materials is lapped into the other half. In order to confirm the process of kneading and lapping of the materials, as mentioned, an experimental mixer of 100 litre capacity (RR-100) was manufactured. The experimental mixing of concrete in this mixer raised some problems regarding the movement of the materials inside it.

These problems were considered to be due to the insufficient revolving speed of the drum, and since it was mechanically unreasonable to increase its speed the improvement work had to come from another standpoint.

The following points can be noted in the improved structure of the mixer, as shown in Fig. 3.

1. To increase the tilting speed of the drum, its method of driving should be modified.
2. To prevent the materials from sliding due to the insufficient speed, a baffle board should be fixed just near to the roller.
3. In order to ensure the lapping of the lifted material to the other half a movable scraper should be installed.

The improvements mentioned above solved the problems of the materials'movement. The modified mixer has 8 jacks for the following purposes:

For tilting the drum	2 jacks
For running the arm	4 jacks
For running the scraper	2 jacks

Altogether 8 jacks were need to improve the operation of the mixer. With these improvements, the results of experimental mixings were enhanced and finally the practical mixer of 0.5 m^3 was manufactured (Fig. 15). The photograph in Fig. 15 shows the completed version of the practical mixer (RR-500). In this mixer the required movement of materials was made possible with the adjustment of 3 jacks, two for the scraper and one for the arm inside the mixer.

4.2 Experimental application of the practical mixer to concrete containing soft rocks as aggregates

The quality of the aggregates on dam site in Japan is chosen according to standard. The quality of the abandoned aggregates ranges from a top soil to just a little short of the standard quality. From the viewpoint of dam's rationalisation and of the environmental safety an investigation on the use of these sub-standard aggregates for the dam

Fig. 15 Photo of RR-500 mixer

concrete has been started. To this aim a series of tests is necessary to be carried out before reaching the goal for the production of aggregates, supply system, testing and mix design method for aggregates, mixing method, casting method as well as the design method of dam itself.

For the production of concrete, i.e. the concrete mixing technique, it is important to establish such a mixing method, including mixer itself, in which the soft aggregates when mixed in the concrete would not be damaged so that the quality of the resulting concrete would not change greatly. Since the recently developed RR-500 mixer does not contain any rotating shaft and blades, as in a conventional mixer, it is thought that the probability of damage to the soft aggregates when mixing would be less than that in the conventional one. As a result, a better quality concrete would be obtained. With this consideration, a series of experiments using soft aggregates was planned to be carried out for a comparison study between the types of mixers. This paper describes only the initial part of this project.

4.2.2 Concrete mix proportion and aggregates
For using the abandoned rocks as the aggregates, as mentioned above, a series of investigations is needed to be carried out, from a design method to the construction method. Since the mix proportions of the concrete, designed according to the quality of the aggregates used, might be altered in the actual site conditions, a standard mix design could not be fixed. Therefore, two basic concrete mix proportions, with different slump values, were selected for the tests. Table 2 gives the details of the two different concrete mixes.

Three types of aggregates with their grain size ranging from 5~13 mm, 13~20 mm and 40~80 mm respectively were used for this experiment. The third type of aggregates were soft rocks taken from the dam site. The other two were aggregates used more for ordinary concrete. Furthermore, the purpose of making the grading gap, as from 20 mm to 40 mm, was to evaluate effectively the damage to the soft aggregates.

Table 2 Concrete mix proportion

Slump	Air	W/C	S/a	Unit Content (kg/m³)						Chemical admixture
(cm)	(%)	(%)	(%)	W	C	S	G_1	G_2	G_3	(SP-8N)
12	4.5	43.0	46.2	162	377	819	300	404	269	3.02
4	4.5	43.0	45.5	151	351	829	298	429	286	2.81

Note: G_3: (5~13 mm), G_2: (13~20 mm), G_1: (40~80 mm)

The abandoned aggregates (weathered granodiorite) taken from the dam site K were found to have small cracks inside them following a simple visual check.

4.2.3 Experimental method

The main purpose of this experiment was to compare the efficiency of the RR-500 mixer with a normal double-axis mixer for mixing the concrete containing soft aggregates. By mixing the same amount of concrete mix in both mixers, the damage to the soft aggregates was studied and the applicability of the new-type mixer for such a type of concrete was investigated. The mixing methods used in both mixers, double-axis and RR-500, are described below.

In the case of a double-axis mixer the concrete constituents, except soft aggregates, were charged first and mixed. After confirming the workability of concrete by a visual check weighed soft aggregates were then charged into the mixer and mixed again for 45 seconds. Then a decantation test for all the soft aggregates was carried out and the weight of the soft aggregates (>40 mm) was measured.

In the case of the RR-500 mixer the concrete was mixed in the same manner in the double-axis mixer. After a visual check the concrete was poured into the RR-500 mixer together with the soft aggregates. Then the mix was kneaded with a kneading number of 9 and the weight of all the soft aggregates was determined by the decantation test.

Note that the volume of the mix was fixed as 0.5 m³ for both cases.

4.3 Results of experiments

The damage to the soft aggregates mixed by the double-axis forced-action type mixer and the RR-500 mixer are shown in Table 3.

Here, the loss amount of the soft aggregates can be defined as the amount of the soft aggregates broken during the mixing.

For 12 cm-slump concrete, the percentage loss of soft aggregates was 9.7% when mixed in the double-axis mixer, whereas it was 6.5% in the case of the RR-500 mixer. Similarly, with the mix of 4 cm slump, the values were found to be 7.1% and 5.6% in the double-axis and RR-500 mixer, respectively.

4.4 Study of the result

The damage to the soft aggregates, in the case of the concrete with both wet and stiff consistency was found less when mixing in the RR-500 mixer, which used the new theory of mixing, than in the conventional mixer. However, since this result represents

Table 3 Damage caused to soft aggregates

12 cm-slump concrete

Type of mixer	Weight of soft aggregates (kg)		Loss in weight (kg)	Loss in percentage (%)
	Before mixing	After mixing		
Double-axis	150.0	135.5	14.5	9.7
RR-500	150.0	140.2	9.8	6.5

4 cm-slump concrete

Type of mixer	Weight of soft aggregates (kg)		Loss in weight (kg)	Loss in percentage (%)
	Before mixing	After mixing		
Double-axis	148.8	136.3	10.5	7.1
RR-500	148.5	140.2	8.3	5.6

only one particular case, other factors such as the range of the concrete mix and the quality of the soft aggregates, etc. will have to be taken into consideration before reaching a much more general conclusion about the applicability of this new mixing system for such low-quality abandoned aggregates. Nevertheless, this recently carried out experiment suggests a good potential for developing a practical new mixer using the new theory which overcomes the problem of damaging the softer aggregates encountered in the conventional mixer.

5   Conclusion

From the various experiments carried out for the practicability of the new-type mixer, the following points can be drawn:

1. The basic concept of forming the 2^n layers can be advanced to 3^n, 4^n or even more layers, with n the kneading numbers in order to shorten the mixing time but still producing a uniform mix.
2. The practicability of the new-type mixer has been verified for mixing the clay-blend and core-blend materials.
3. In the case of using abandoned low quality soft rock for a dam-concrete, the application of the new concept of mixing might lead to the development of an effective practical mixer.

However, further modifications will have to be carried out to improve new mixers, particularly concerning the enhancement of their mechanical aspects. It is expected that this continuing study will widen the applicability of these new mixers to various materials which are very difficult or not possible to be mixed efficiently by the conventional methods.

6 References

1. Maeda, K. (1995) *Development of New-type Mixer on the New Concept of Kneading Process,* presentation paper in RILEM seminar.
2. Maeda, K. (1994) *Study on Mixing Performance Test for New-type Mixer,* presentation report at AIT, Thailand.
3. Maeda, K. (1995) *Development of a New Mixer Based on the Kneading Method,* Doctoral dissertation presented in the University of Tokyo, Japan.

3 THE EFFECTS OF MIXING TECHNIQUE ON MICROSILICA CONCRETE

A.K. TAMIMI
Advanced Concrete Technology Group, Department of Civil, Structural and Environmental Engineering, University of Paisley, Paisley, Scotland, UK

Abstract

Concrete can be regarded as a chain comprising three links: cement paste, aggregate, and cement paste - aggregate interface. It has been found that microcracks occur at the cement paste-aggregate interface before the concrete is subjected to loading. This suggests that the bond is broken well below the ultimate load the concrete could sustain. Thus in order to improve the strength of concrete, the deficiencies in the cement paste-aggregate bond must be removed. Several methods were applied to produce high strength and durable concrete that can withstand extreme conditions of exposure while in service or during the period of construction. All these methods included either some additives to the concrete or use of complicated techniques such as vacuum processing, high-speed slurry mixing or revibrating. Despite the advantages of these methods, they are not widely used as they involve at least one additional step in the production of concrete, and hence increase the overall cost. The present investigations confirm the advantages of applying the two-stage mixing technique to improve the microsilica concrete with no extra cost. This was demonstrated with the mixing time being reduced to two minutes. The results showed improvement in the workability of the fresh concrete and overall improvement of the compressive and indirect tensile strength at early and late age.
Keyword: Two-stage mixing technique, Microsilica concrete, Cement paste-aggregate interface, Mixing, Workability, Compressive Strength, Indirect tensile strength.

Production Methods and Workability of Concrete. Edited by P.J.M. Bartos, D.L. Marrs and D.J. Cleland. Published in 1996 by E & FN Spon, 2–6 Boundary Row, London SE1 8HN. ISBN 0 419 22070 4.

1 Introduction

1.1 Cement paste - aggregate bond

The strength of the concrete depends on the strength of the paste, the strength of the coarse aggregate, and the strength of the paste-aggregate interface which is the weakest region of the concrete. The following phenomena that occur may explain this weakness

A- Development of higher porosity than in the bulk matrix.

B- Formation of larger crystal particles of the hydration products.

C- Deposition of calcium hydroxide crystals with a preferential orientation on the interface.

In general, bond failures occur before failure of either the paste or the aggregate. Bond forces are partly due to Vander Waal's forces, however, the shape and surface texture of the coarse aggregate are important, since there may be a considerable amount of mechanical interlocking between the mortar and the coarse aggregate. The bond region is weak because cracks invariably exist at the paste-coarse aggregate interface, even for continuously mois-cured concrete and before the application of any external load. These cracks are due to bleeding and segregation and to volume changes of the cement paste during setting and hydration. When ordinary curing takes place, which is accompanied by drying, the aggregate particles tend to restrain shrinkage because of their higher elastic modulus. This induces shear and tensile forces at the interface which increase with increasing particle size, and which will cause additional cracking if they exceed the bond strength. Under load, the difference in elastic moduli between the aggregate and the cement paste will lead to still more cracking.

1.2 Factors affecting the cement paste-aggregate bond

Bond strength seems to depend on many of the same factors that control other types of strength such as:

1.2.1 Water-cement ratio

The effect of the water-cement ratio on bond strength is similar to its effect on compressive strength, i.e. decrease in water-cement ratio increases the paste compressive strength as well as the strength of its bond to the aggregate. This effect of the water-cement ratio can only be demonstrated on paste and rock specimens because in concrete it affects both paste and bond strength. In any case, the fact that the strength of ordinary concrete is determined mainly by the strength of the paste and strength of its bond to the aggregate explains why the water-cement ratio is the most important factor in determining concrete strength. Earlier investigations[1] showed that for both mixing techniques (i.e. the two-stage mixing technique and the conventional mixing technique), the minimum and maximum paste hardness decreases with increasing water-cement ratio.

1.2.2 Segregation and bleeding

Segregation refers to a separation of the components of fresh concrete, resulting in a non uniform mix. In general, this means some separation of the coarse aggregate from the mortar. This separation can be of two type : either the settling of heavy particles to the bottom of the fresh concrete, or a separation of the coarse aggregate from the body of the concrete. This is generally due to improper placing or vibration.

Although there are no quantitative tests for segregation, it can be seen quiet clearly when it does occur. The factors that contribute to increased segregation have been listed by Popovics[2]as follows :

1- Larger maximum particle size (over 25mm) and proportion of the large particles.

2- A high specific gravity of the coarse aggregate compared to that of the fine aggregate.

3- A decreased amount of fines (sand or cement).

4- Changes in the particle shape away from smooth, well rounded particles to odd-shaped, rough particles.

5- Mixes that are either too wet or too dry.

Bleeding may be defined as the appearance of water on the surface of concrete after it has been consolidated but before it has set. Water, being the lightest component, segregates from the rest of the mix, and thus bleeding is a special form of segregation. This is the most common manifestation of bleeding, although the term may also be used to described the draining of water out of the fresh mix.

1.2.3 Mixing techniques
The object of mixing is to coat the surface of all aggregate with cement paste, and to blend all the ingredients of concrete into a uniform mass. There are many mixing techniques already adopted to improve the strength and other affiliated properties of concrete which can be summarised as follow :

A Revibration
On the basis of experimental results it appears that concrete can be successfully revibrated up to about 4 hours from the time of mixing. Revibration at 1 to 2 hours after placing was found to result in an increase in the 28-day compressive strength[3] (An increase in strength of approximately 14% has been reported). In general, the improvement in strength is more pronounced at earlier ages, and is greatest in concrete liable to high bleeding since the trapped water is expelled on revibration.

It is also possible that some of the improvement in strength is due to a relief of the plastic shrinkage stresses around aggregate particles.

B Vacuum processing
It is an extension of the water-cement ratio law, and is a mean of effecting compaction and early hardening of freshly placed concrete by the utilization of atmospheric pressure at 10 to 12 lb/sq.in[4]. By using this method the 3-day strength can be increased by 100% and the 5-day strength can be as high as the 28-days strength of the same concrete not vacuum processed. The 28-days strength can be increased by an average of 50% also the absorption of moisture can be reduced by up to 30%, but the impermeability of the concrete may not be improved unless an air-entraining agent is used and the process supplemented with vibration.

C High speed slurry mixing
The high speed slurry mixing process involves the advance preparation of a cement-water mixture which is then blended with aggregate to produce concrete. Higher

compressive strengths thus obtained are presumably attributed to more efficient hydration of the cement resulting from the most intimate contact between cement particles and water achieved in the vigorous blending of cement paste. Increases in strength are reportedly[5] in the area of 10% from concrete mixtures with high water-cement ratios. MacInnis et al[6] found that this technique was not effective in increasing the compressive strength at water-cement ratios in the range of 0.3 to 0.4, and a combination of high speed slurry mixing plus revibration produced increases which ranged from 3 to 9%.

D Vacuum mixing
Concrete mixed in a vacuum gives up most of the air contained in the mix, the ingredients are mixed dry, water is added and the vacuum produced in a special mixer. The mix is increased in unit weight by 6%[4], and the resulting concrete has greater strength (10 to 15%), greater strength and impermeability than the ordinary concrete.

E Two-stage mixing technique
Abram's water-cement ratio law has long dominated the strength of concrete.
    However, using the two-stage mixing concrete which is also called "Sand Enveloped with Cement" (SEC)[7], substantially higher strength concrete can be obtained using the same water cement ratio and materials. Hayakawa and Itoh[8] showed the premixing cement with sand and water of about 25% of the specified weight of cement was effective in producing very low bleeding mortar. Concrete produced by this method results in the aggregate surface being coated by the cement paste of a lower water-cement ratio. They showed that mortar produced by this mixing method has a 28-days compressive strength 5% higher than the ordinary mortar. Since the strength of the cement paste-aggregate bond depends on the water-cement ratio[9], such a technique would tend to increase the bond strength.
    In the two-stage mixing technique, saturated surface dry sand is pre mixed with water of only 0.25 of the weight of cement, in the first stage, for 30 seconds.
    Cement and coarse aggregate are then added and mixed for 90 seconds. In the second stage the remaining of the w/c ratio is added and mixed for further 60 seconds.
    In the present study, the mixing technique procedures are kept as before but the overall mixing time is reduced to only two minutes.

2 Mixing programme
Four types of mixes were examined. The first mix was the control mix which consists of the normal concrete mix with 0.50 w/c ratio. Round natural limestone aggregate with 20mm maximum size for the coarse aggregate and fine aggregate with 40% passing 600μ sieve were used. The concrete ingredients were mixed by the conventional method of mixing for two minutes. The second concrete was the two-stage mixing concrete which consists of the same materials and proportion and mixed in two stages. In the first stage the fine aggregate was mixed with portion of the mixing water equal to 0.25 of the weight of the cement.Cement and coarse aggregate were then added and mixed for 60 seconds. In the second stage the rest of the water was added and further mixed for 30 seconds. The complete procedure would take the same mixing time as the control concrete.

The third mix was concrete with 10% of microsilica by the weight of cement, mixed by the same procedures as the control mix. The fourth mix was the microsilica concrete which was mixed according to the two-stage mixing technique. For each mix two slump tests for workability were performed and the results were averaged. Six cubes and six cylinders were cast for each mix and average compressive strength and indirect tensile strength of three results were recorded at 3-days and 28-days.

3 Results and interpretations

Mix Ref.	Slump (mm)		Compressive Strength (N/mm^2)				Indirect Tensile Strength (N/mm^2)			
			3-days		28-days		3-days		28-days	
Control mix	75 70	ave. 70	20.0 19.5 18.5	ave. 19	38.5 39.5 40.0	ave. 39	2.60 2.40 2.50	ave. 2.50	3.75 3.90 3.80	ave. 3.80
SEC Mix	95 105	100	22.0 23.5 23.0	23	45.5 44.0 44.5	45	2.80 2.70 3.00	2.80	4.30 3.90 3.80	4.00
Microsilica Concrete	55 50	50	20.5 22.0 21.5	21	44.0 42.5 42.0	43	2.50 2.70 2.60	2.60	4.00 4.20 3.90	4.00
SEC Microsilica concrete	90 85	85	24.0 25.5 25.0	25	47.5 45.0 45.5	46	2.80 3.30 2.90	3.00	4.70 4.40 4.50	4.50

The results showed that the two-stage mixing procedures have resulted in a more workable fresh concrete in terms of the slump test and an increase of early compressive and indirect tensile strengths. There is an improvement of 21% at 3-days and 15% at 28-days in the compressive strength. This is accompanied by 12% increase at 3-days and 5% increase at 28-days in the indirect tensile strength. 10% microsilica of the weight of cement was added to the rest of the ingredients for both the conventional and the two-stage mixing concretes to study the effect of changing the mixing technique on the microsilica concrete. The results showed continuous improvement by applying the two-stage mixing technique. It indicated an increase of 19% of the compressive strength at early age and 7% increase at later age. There was also an increase of 15% at early age and 12.5% at later age in the indirect tensile strength. The workability of the fresh concrete was also improved.

Bleeding is generally caused by the fact that as the aggregate particles settle within the mass of fresh concrete, they are unable to hold all the mixing water. The upper layer of the concrete may become rich in cement paste which has a water/cement ratio that is too high. This leads to weakness, porosity and a lack of durability. Water pockets may form under aggregate particles leaving weak zones in concrete and reducing the bond. One reason of the improvement of the bond in the two-stage mixing concrete can be attributed to the reduction of bleeding rate which was demonstrated in the previous findings[1].

During the first stage of mixing, the aggregate is coated with a dense cement paste which was confined with low water/cement ratio of 25% of the weight of the cement. This dense shell cannot be easily washed away, even in the second stage of mixing. This may be caused by the build up of some hydrostatic pressure in the paste which enveloped the aggregate in this zone and reduce the bleeding. The result would also enhance the hydration process and producing more gel/space ratio and reducing the porosity in the bond. Evidence of more hydration products in the interfacial zone of the two-stage mixing concrete in the form of silica gel , portlandite and ettringite hydration was demonstrated by a previous research[10]. The coating of the aggregate with low w/c ratio during the first stage would act as a ball bearing to the aggregate in the mix and operate as a lubricant within the fresh concrete. This resulted in improvement of the mobility i.e. workability of the two-stage mixing concrete.

It can be also argued that the w/c ratio influences bond by affecting the amount of void space at the interface, i.e. the lower the w/c ratio the smaller the amount of void space presented in the interfacial film of the two-stage concrete specimen. The large surface area in the fine voids of the two-stage mixing concrete would need a greater quantity of energy to develop large well defined hexagonal portlandite crystals.

This is because at the first stage, the chemical hydration products around the aggregate became saturated earlier than in the conventional concrete specimens due to the precoating of the aggregate at this stage. It would lead to more filling of the fine voids in the two-stage concrete specimen with more hydration products. Although this will invite more investigations to study the porosity of the cement paste aggregate interface for both mixing techniques.

4   Conclusions

1.In the present investigations the duration of the mix for both methods of mixing technique was kept similar (i.e. two minutes). However, the two-stage mixing technique showed improvement in workability, compressive and indirect tensile strengths both at early and late age of hydration.
2.The same advantages of applying the two-stage mixing were shown when 10% microsilica was added to the concrete.
3.The advantages of the two-stage mixing technique are related mainly to the improvement of the cement paste-aggregate interface where more gel/space ratio was produced in this zone and raised the overall strength of the resulting concrete.
4.The improvement of workability could be attributed to the effect of ball bearing action of the paste between the aggregate particles which related to the build up of some hydrostatic pressure in this zone during the first stage of mixing.

5 Acknowledgement

Thanks are due to Mr Thomas McColgan for his assistance during the simulated industrial training.

6 References

1. Tamimi,A.K.(1994) The effects of a new mixing technique on the properties of the cement paste-aggregate interface,*Cement and Concrete Research*, Vol.24, No.7. pp.1299-1304.
2. Popovics,S.(1973) *Proceedings of a RILEM Seminar*, University of Leeds, U.K., Vol.3. pp.6.1-1 to 6.1-37.
3. Neville,A.M.(1973) *Properties of Concrete*, Pitman, London.
4. Taylor,W.H.(1977) *Concrete Technology and Practice*, McGraw Hill, London.
5. Saucier,K.L.(1980) High-strength concrete, past, present, future, *Concrete International*.
6. Macinnis,C. and Kosteniuk,P.W.(1979) High-speed slurry mixing for producing high-strength concrete, *ACI Journal*, Vol.76. pp. 1255-1265.
7. Higuchie,Y.(1980) Coated-sand technique produces high strength concrete, *Concrete International*.
8. Hayakawa,M. and Itoh,Y.(1982) A new concrete mixing method for improving bond mechanism. *Bond in Concrete*. Ed. by Bartos,P. Applied Science Publishers, London.
9. Carlis-Gibergues,A.,Grandet,J. and Ollivier,J.P.(1982) Contact zone between cement paste and aggregate. *Bond in Concrete*. Ed. by Bartos,P. Applied Science Publishers, London.
10. Tamimi,A.K. and Najeim,H.(1994) XRD and SEM investigations of the cement paste aggregate interface for concrete produced by both the conventional and a new mixing technique. *Proceedings of the 16th international conference on cement microscopy*, Virginia, U.S.A.

4 STRENGTH DEVELOPMENT OF MORTAR AND CONCRETE PREPARED BY THE 'SAND ENVELOPED WITH CEMENT' METHOD

P. ROUGERON and A. TAGNIT-HAMOU
University of Sherbrooke, Quebec, Canada
P. LAPLANTE
C.T.G. Research Centre of Italcementi-Ciments Français Group,
Guerville, France

Abstract
This paper introduces a new way of investigating the physico-chemical mechanisms which can explain the increased compressive strength generated by the application of a new mixing technique referred to as the "Sand Enveloped with Cement" (S.E.C.). In particular, the paper demonstrates that the cause of strength increase is not due to a mechanical effect of the densification of the paste-aggregate interface, but due to a thermal activation of the cement hydration.
Keywords: Mixing methods, S.E.C., experimental plans, cement paste aggregate interface, short-term compressive strength.

1 Introduction

In 1980, HIGUCHI [1] described a new mixing method based on a double introduction of mixing water (Fig. 1). Sand is first wetted with part of the mixing water and is then mixed with cement and the rest of the mixing water. According to HIGUCHI, the use of such mixing technique can result in increased short-term and long-term compressive strength and a decrease in bleeding.

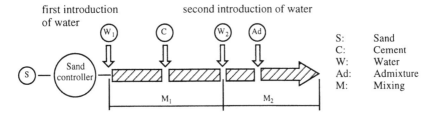

Fig. 1. Description of the mixing technique called "Sand Enveloped with Cement", after HIGUCHI [1]

TAMIMI [2] confirmed the effects of this mixing technique on compressive strength and attempted to offer an explanation to the causes of such strength increase.

Production Methods and Workability of Concrete. Edited by P.J.M. Bartos, D.L. Marrs and D.J. Cleland.
Published in 1996 by E & FN Spon, 2–6 Boundary Row, London SE1 8HN. ISBN 0 419 22070 4.

TAMIMI [2] explained that by wetting sand surfaces, cement grains can "stick" to the surface of the sand grains which generates a film having a relatively low water-cement ratio around the sand particles. The formation of such shell induces an increased compacity of cement particles around the sand grains. Thus, the interface between cement paste and aggregate can be densified. According to FARRAN [3], compressive strength is directly related to the density of such interface. The denser the cement paste-aggregate interface is, the stronger the concrete is. POPE [4] adds that the use of the S.E.C. mixing technique helps to avoid the formation of a thin film of water around sand grains.

The above authors agree that the use of the S.E.C mixing method produces a stronger concrete. They explain this phenomena with the modification of the spatial distribution of cement grains around sand grains.

2 Objective

In order to study the cause of strength increase in S.E.C. mixing technique, an experimental program was developped to first verify the effect of the technique on compressive strength of mortars and concretes. The origin of the observed strength increase was then investigated using the maturity concept in order to determine whether the source of strength is mechanical or chemical in nature. The last part of this study attempts to explain the mechanism of strength increase caused by the S.E.C. mixing technique.

3 Experimental program

Materials
Aggregate are first dried in an oven at 100°C during 24 h. Afterwards, sand is mixed with the amount of water needed to reach the desired level of humidity. The humidity of coarse aggregate is maintained at 4% by mass. All aggregate are kept in an air-sealed bag till the time of use.

Mixing procedure
Aggregates are first mixed with cement during 30 sec. at low speed. Afterwards, mixing water is added and mixed at low mixing speed during 30 sec. The mortar (or concrete) is then mixed for 4 min. at high speed.

Curing conditions
Mortar is cast in 4 × 4 × 16 cm moulds and kept at 20°C at 95% relative humidity until the time of testing.
Concrete is cast in 16 × 32 cm cylinders. After 24 h of hydration, the samples are removed from the moulds and kept in lime-saturated water at 20°C until the time of testing. A total of 6 and 8 samples were used to determine the mean strengths for mortar and concrete mixtures, respectively.

3.1   Confirmation of the effects of S.E.C. mixing technique on compressive strength of mortar
The effect of sand humidity on 24-h and 28-d compressive strength of mortar was investigated using a statistical experimental design approach (composite type). The two considered variables were the humidity of sand (amount of water added to sand in order to wet it) and the water-to-cement ratio. Table 1 summarises the composition of the experimental plan and the obtained compressive strength of 12 tested mortars.

Table 1. Compositions of tested mortars

Batch	Absolute water-cement ratio	Coded water-cement ratio	Absolute sand humidity	Coded sand humidity	f'_c 24 h (MPa)	f'_c 28 d (MPa)
1	0.45	-1	13.1	1	20.5	55.6
2	0.50	1	13.1	1	17.3	49.2
3	0.45	-1	3.9	-1	17.4	56.3
4	0.50	1	3.9	-1	12.1	50.1
5	0.475	0	8.5	0	19.2	55.1
6	0.475	0	8.5	0	18.7	54.4
7	0.44	-1.414	8.5	0	22.8	56.2
8	0.51	1.414	8.5	0	16.8	54.2
9	0.475	0	15	1.414	15.5	55.1
10	0.475	0	2	-1.414	20.5	53.8
11	0.475	0	8.5	0	19.4	55.4
12	0.475	0	8,5	0	18.6	53.6

From the results of the experimental plan, mathematical models for compressive strength at 24 h and 28 d were derived. These models are as follows:

$$f'_{c24h} = a_0 + a_1 \cdot \frac{W}{C} + a_2 \cdot \frac{W_1}{C} + a_{11} \cdot \left(\frac{W}{C}\right)^2 + a_{22} \cdot \left(\frac{W_1}{C}\right)^2 + a_{12} \cdot \left(\frac{W}{C} \cdot \frac{W_1}{C}\right)$$

where :W = total amount of water in mortar,
C = total amount of cement in mortar,
W_1 = amount of water added to the dried sand.

Results are presented as contour curves of measured strengths (Fig. 2).
By wetting the sand with some of the mixing water, the 24 h compressive strength is shown to increase by approximately 40%. This optimum is reached when sand humidity is approximately 12%. However, for 28 d compressive strength, the effect of initial sand humidity on compressive strength is not significant.

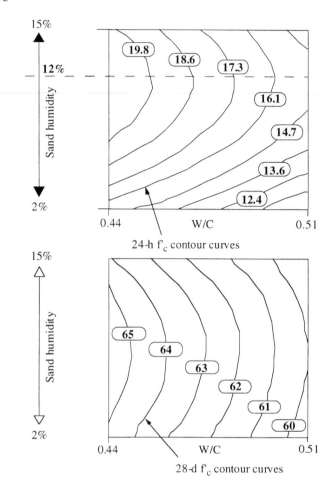

Fig. 2. Result for the models of 24-h and 28-d compressive strength

3.2 Confirmation of the effect of S.E.C. mixing technique on concrete properties
In the case of concrete, only the effect of sand humidity was investigated for concrete with a constant water-cement ratio. The humidity levels of large aggregate was fixed to 4%. The humidity levels of sand in reference concrete is fixed to 2%. Concrete composition is presented in Table 2.

Table 2 Concrete composition

CPA Cement	400 kg/m^3
Total Water	240 kg/m^3
Dry Sand	600 kg/m^3
Dry Coarse Aggregate	1050 kg/m^3

The external bleeding was measured using a concrete sample cast in a conical mould with a central vertical tube. External bleed water at the surface is then channelled through the tube down to a graduated cylinder to measure the cumulative bleeding.

The evolution of bleeding as a function of sand humidity shows the effect of the S.E.C. mixing technique: in wetting the sand, the kinetic of bleeding and total amount of bleeding are modified (Fig. 3).

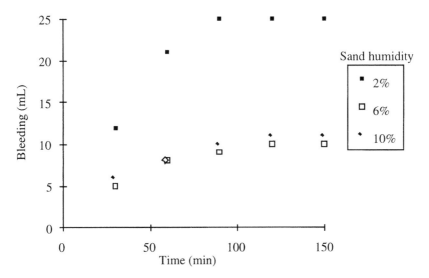

Fig. 3. Evolution of bleeding of concrete as a function of time and sand humidity.

The evolution of compressive strength of concrete with 2 and 10% of sand humidity was studied. At 24 h, 2 d, 7 d and 28 d of age, 8 samples were tested. Results are summarised in Table 3.

Table 3. Result of compressive strength of concrete at 24 h, 2 d, 7 d and 28 d

	24 h		2 d		7 d		28 d	
Sand humidity (%)	2	10	2	10	2	10	2	10
Mean strength (MPa)	8.7	12.6	16.5	22.3	40.1	43.2	55.3	56.1
Standard deviation	0.42 (14)*		0.56 (14)*		0.49 (14)*		1.1 (14)*	
Student test	**4.7**		**5.2**		**4.2**		0.64	

(* degree of freedom)

The strength results indicate that the effect of sand humidity on compressive strength is statistically significant until 7 days of hydration.

Based on these results and on mortar results, it appears that the effects of the S.E.C. mixing technique are effective on short-term compressive strength only. This observation confirms the results of TAMIMI [2]. Also, it seems that in the condition of this experiment there exists an optimum sand humidity. This value of the optimum (10% of water per mass of sand) corresponding to the formation of a thin layer of water around each sand grain. In fact, this film of water was shown by proton nuclear magnetic resonance (N.M.R.) of water (Fig. 4). N.M.R. were carried out on wet sand grains. These grains were wetted 24 h before analysis. Fig. 4 (calibration curve) shows that the total exited water is approximately equal to the humidity value. This means that the sand was well prepared for the study. This graph shows that up to 6% humidity (% of water per mass of sand), the link between water and sand was due to physical tensile forces, while beyond 6% humidity, the links were due to Van der Walls forces, which are weaker than physical tensile ones so this is illustrated in Fig. 4 by the dramatic increase in the long time of relaxation.

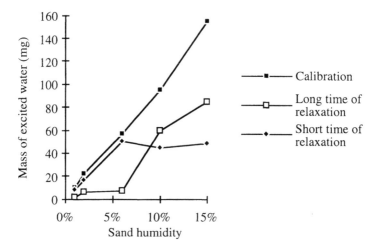

Fig. 4. Study of the link between water and sand as a function of sand humidity, using proton N.M.R. of water

4   Evolution of compressive strength as a function of calorimetric heat development

WIRQUIN [5] showed the advantages of studying the evolution of compressive strength as a function of calorimetric heat development. The amount of generated adiabatic heat is directly related to the degree of hydration of cement. So, the representation $f'(Q)$ (with f_c refers to compressive strength and Q to adiabatic heat) allows the comparison of compressive strength as a function of the amount of formed hydrates.

This technique allows to distinguish the chemical effect of an activation of hydration kinetic from the mechanical effect of the spatial arrangement of cement grains.

4.1 Description of the method

-    First, the evolution of temperature in two kinds of sample storage was recorded. Samples were kept at 20°C, under the same condition as samples used to evaluate compressive strength. The temperature evolution for a sample stored in a semi-adiabatic content (Langavant bottle) was also recorded. Following some corrections, it was possible to calculate the adiabatic heat evolution from the recorded temperature.

- Secondly, real time was transformed into equivalent time at 20°C for both strength and heat evolution. This transformation was based on the maturity law which allows to study the evolution of concrete characteristics taking into account the thermal history [6, 7].
- Thirdly, from f'_c (at equivalent time) and Q (at equivalent time), it was possible to determine graphically the compressive strength as a function of the adiabatic heat.

4.2 Representation of f'(Q) in case of studied mortars and concrete
Fig. 5 shows the evolution of compressive strength as a function of calorimetric heat development for concrete and mortar. In the Arrhénius function

$$f(\emptyset(t)) = A \times e - (E / (R \times \emptyset(t)))$$

mortar and concrete activation energies are expressed in terms of E/R or 5200 or 5000 J/Mol°C, respectively.

This figure shows that for concrete or mortar, the conclusion is the same. For a given state of value of adiabatic heat, compressive strength is shown not to be affected by mixing technique. This means that the mechanism increasing short-term compressive strength is a chemical one and is related to the activation of the hydration kinetic of cement.

Therefore, the effect explaining the variation of strength when using S.E.C. mixing method, seems not to be a mechanical consequence of the spatial arrangement of cement grains.

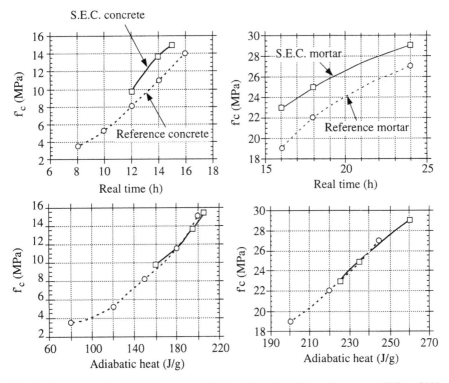

Fig. 5. Representation f'(Q) for mortar (E/R = 5200 J/Mol°C) and concrete (E/R = 5000 J/Mol°C) prepared by using S.E.C. mixing method and compared to reference prepared by using a traditional mixing sequence

If the hypothesis of the increase of the density of the cement-paste interface were valid, an effect of mixing method on f'(Q) representation should be observed. In fact, for a given hydration state, if the interface is improved, the strength should increase. Therefore, it is thought that the variations of short term strength is due to thermal activation of the hydration kinetic of cement.

5   Attempted explanation of the mechanism of activation of hydration kinetic in concrete prepared using the S.E.C. mixing technique

Using scanning electromicroscope S.E.M. observations of polished fresh mortar sample (Fig. 6), it was observed that S.E.C. mortar gets a special cement grain spatial structure. The density of cement grains around sand grains is higher than in the case of mortar prepared using a traditional mixing technique.

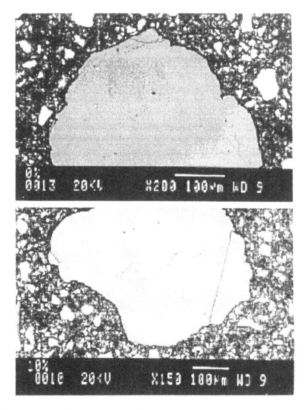

Fig. 6. Back scattered S.E.M. views on fresh mortar. On top, traditional mortar, at bottom mortar prepared using S.E.C. mixing method.

The use of the S.E.C. method seems to create a heterogeneous repartition of cement grains and of water-cement ratio. One can suppose that the cement paste with water-cement ratio (W/C) equal to 0.47, can be divided into two parts, one with water-cement ratio (W_2/C_2) equal to 0.35 (30% per volume of paste), and the other with water-cement ratio (W_1/C_1) equal to 0.53 (70% per volume of paste). Using this model, it was possible to study the evolution of temperature of different points of each part by mixing to pastes with different water-to-cement ratio as described in Fig. 7. This figure shows different points (T_1 to T_5) where temperature evolution was recorded. The ambient temperature (TEX) was maintained at 19°C.

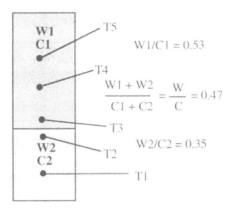

Fig. 7. Schematic representation of the model used to study temperature evolution of each part.

The results of temperature evolution in Fig. 8 show that the lower water-cement ratio zone (equal to 0.35), can heat the higher water-cement ratio zone (equal to 0.53). In effect, the temperature of the higher water-cement ratio zone was significantly lower than the temperature of the lower water-cement ratio zone (up to 40% more). This phenomenon can create a thermal activation of cement hydration and could explain the increase of short-term compressive strength.

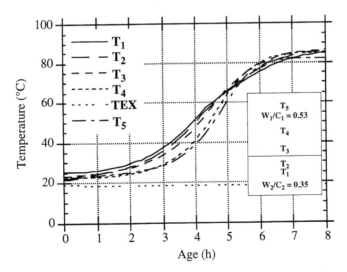

Fig. 8. Evolution of temperature recorded at different points of the model shown on Fig. 7

5 Conclusions

It appears that the S.E.C. method involves the formation of a shell of cement grains around sand grains. The creation of this zone seems to generate a thermal activation of hydration kinetic. This phenomenon could explain why the effect of the S.E.C. method are not sensible for long-term compressive strength.

Acknowledgements
This work would not have been possible without technical, scientific and financial support of C.T.G. (research centre of Italcementi-Ciments Français Group).

6 References

1.   **Higuchi, Y.**   (1980) Coated-sand technique produces high strength concrete. *Concrete International,* Vol. 2, No. 4, pp. 75-76.

2.   **Tamimi, A. K.**   (1989) The effect of two stage mixing technique on the properties and the microstructure of the concrete structure. *PhD thesis.* Paisley College, Paisley.

3.   **Farran, J.**   (1972) Existence d'une auréole de transition entre les granulats d'un mortier ou d'un béton et la masse de la pâte de ciment hydraté. Conséquences sur les propriétés mécaniques. *Compte Rendu de l'Académie des sciences,* Paris.

4.   **Pope, A. W. and Jennings, H. M.**   (1992) The influence of mixing on the microstructure of the cement paste aggregate interfacial zone and on the strength of mortar. *Journal of materials science*, No. 27, pp. 6452-6462

5.   **Wirquin, E.**   (1992) Étude de l'évolution des résistances du béton au très jeune âge en fonction de l'avancement de l'hydratation- Rôle joué par la microstructure à l'état frais. *PhD Thesis,* Toulouse.

6.   **Byfors,**   (1980) Plain concrete at early ages. CBI Research Report, n°3 80, Swedish cement and concrete research institute, Stockholm.

7.   **Regourd, M. and Gauthier, E.**   (1987) Comportement des ciments soumis au durcissement accéléré, *Annales de l'ITBTP*, n° 387

5 QUALITY CONTROL IN CONCRETE MIXING

Y. CHARONNAT
Laboratoire Central des Ponts et Chaussées, Nantes, France

Abstract

Managing quality means doing everything known to be necessary to attain the intended objective. In the mixing of concrete, the objective is to produce a homogeneous concrete and to guarantee this homogeneity each time a new batch of concrete is produced.

To attain this objective, it is first necessary to know the capabilities of the mixer. For this purpose, the article proposes a qualification procedure with reference to the recommendation of RILEM TC 150, "efficiency of concrete mixers".

It is then necessary to determine the mixing cycle that yields the expected performance, with the aim of optimizing the mixing time and plant wear. This is the reason for the initial test which is used to fix precise acceptable values for the various mixing parameters.

The evolution of homogenization in the course of the production cycle must then be tracked; this can be done by displaying the mixing energy.

The application of this concept, through the choice of a good mixer, a check of attainment of the objective, and assurance of permanent monitoring of the mixing operation, serves to ensure the transparency of the mix production operations. It gives to the customer the information he needs to accept his concrete with a minimum of risk. This makes it possible to reduce the number of concrete inspection tests while still ensuring that the concrete satisfies requirements reflecting the needs of the structure.

Keywords: Concrete, mixer, homogeneous, inspection, wattmeter, quality control, mixing cycle, RILEM.

Production Methods and Workability of Concrete. Edited by P.J.M. Bartos, D.L. Marrs and D.J. Cleland. Published in 1996 by E & FN Spon, 2–6 Boundary Row, London SE1 8HN. ISBN 0 419 22070 4.

1 Introduction

Maîtriser la qualité, c'est prendre les mesures reconnues nécessaires pour atteindre l'objectif visé. Dans le domaine du malaxage du béton, l'objectif est de fabriquer un produit homogène et de garantir qu'on mettra tout en oeuvre pour atteindre en permanence cette homogénéité. C'est en fait l'application de la norme EN ISO 9002 pour le malaxage du béton.
Cette définition implique que les mesures nécessaires pour atteindre l'homogénéité, définies au démarrage des travaux, soient conservées tout au long de la fabrication.
Lorsqu'un fabricant de béton pratique la "maîtrise de la qualité" et que, par des informations non contestables, il prouve qu'il a mis tout en oeuvre pour atteindre son objectif, alors il donne à son client l'assurance que le produit qui lui sera livré, répond aux exigences. La garantie n'étant cependant pas absolue, le client doit pouvoir effectuer des vérifications. Celles ci seront adaptées aux mesures prises par le fabricant, modulées selon la quantité et la véracité des informations transmises.
La maîtrise de la qualité sous-entend pour le fabricant de béton :

- l'utilisation de moyens performants ;
- une rigueur très stricte dans le traitement des anomalies ;
- l'information aussi complète que possible du client.

Nous ne reviendrons pas dans cet article sur la définition de l'homogénéité du béton et les règles de base à respecter pour sa mesure. Ils font l'objet d'autres articles [1] [2] et de normes d'application [3]. Nous n'aborderons pas non plus les règles de l'art permettant de respecter les dosages et les tolérances [4], nous considérerons qu'elles sont appliquées. L'objectif de cet article est d'examiner les moyens qui doivent être mis en oeuvre pour assurer à coup sûr la qualité du malaxage, c'est à dire l'homogénéité du béton et, pour un niveau de qualité équivalent, montrer ce qu'apporte l'application du concept de la maîtrise de la qualité au malaxage du béton.

2 L'homogénéité d'un béton

Le contrôle de l'homogénéité d'un béton est une opération longue et délicate c'est pourquoi il est rarement réalisé. On préfère en général se reporter au contrôle des performances du produit fabriqué.
Avec le développement des bétons hautes et très hautes performances, cette pratique n'est pas du tout réaliste car elle oblige à prendre des marges excessives compte tenu des niveaux visés (§ 6). Il faut donc en revenir à une prise en compte des moyens pour les associer à la mesure des performances du béton. L'homogénéité sera alors évaluée par les caractéristiques de distribution des constituants dans le mélange et par les propriétés de ce mélange.
Beaucoup de paramètres peuvent agir sur l'homogénéité du béton [5]. Tous ces paramètres sont indépendants (tableau 1) car, ils ne résultent pas des mêmes commandes ou des mêmes organes mécaniques. En conséquence, on peut considérer que les caractéristiques qui définissent l'homogénéité du béton et qui résultent des valeurs prises par les paramètres cités, suivent des lois gaussiennes. Ces lois peuvent être caractérisées par les deux grandeurs que sont la moyenne et l'écart type. Les règles qui régissent l'estimation de l'une et de l'autre impliquent un nombre d'essais très différent. Pour la moyenne on s'arrête le plus souvent à cinq essais voire quelquefois

Tableau 1 Paramètres maîtrisables agissant sur l'homogénéité du béton. Ce tableau récapitule les paramètres malaxage maîtrisables par le conducteur de la centrale de fabrication du béton. Si le béton est constitué de trois composants, il y a alors 8 paramètres qui peuvent être considérés comme indépendants. La distribution des propriétés caractérisant l'homogénéité est assimilable à une loi gaussienne.

phases du malaxage	paramètres	remarques
introduction des constituants	- moment d'origine - débit d'incorporation	pour chacun des constituants et en relation avec les autres
brassage des constituants	- temps de malaxage - réglage des pales - usure du matériel	pendant le remplissage et après introduction jusqu'à la vidange complète
vidange du malaxeur	- usure du matériel	vidange souvent incomplète

trois. Il n'en est pas de même pour l'écart type pour lequel le nombre de 15 essais est souvent jugé à la limite de la confiance qu'on peut lui accorder [6] [7].

L'analyse des causes de fluctuation des résultats montre que la valeur moyenne est en priorité liée aux performances propres des constituants et à leur proportion dans le mélange et relativement peu aux évolutions des paramètres de malaxage. Par contre l'écart type est en priorité influencé par les évolutions des paramètres de malaxage et en second lieu par les fluctuations des propriétés des constituants (tableau 2).

On comprend, à partir de ce tableau, tout l'intérêt qu'il y a à maîtriser les paramètres de fabrication car ils assurent la régularité des performances du béton. Les niveaux atteints restent principalement liés aux propriétés des constituants et à leur dosage.

Tableau 2 Priorité dans les paramètres influents pour l'homogénéité du béton. Cette description permet de déterminer les paramètres sur lesquels il faut agir si les résultats ne sont pas satisfaisants.

performances	grandeurs	paramètres fondamentaux	autres paramètres
résistance	moyenne	- classe du ciment - quantité de ciment - quantité d'eau	- temps de malaxage
	écart type	- répartition du ciment - répartition de l'eau - ségrégation des gros granulats	- variation des performances du ciment
consistance	moyenne	- dosage en eau - quantité d'éléments fins	- temps de malaxage
	écart type	- mode d'introduction de l'eau - temps de malaxage	- répartition des éléments fins

3 Les performances des malaxeurs

Connaître l'outil utilisé pour le malaxage des constituants est le premier pas vers la maîtrise de la qualité. Aussi anachronique que cela puisse paraître, le malaxeur est le matériel le moins connu et le moins bien caractérisé de toute la chaîne de production du béton. Ce point a été particulièrement bien mis en évidence avec le développement des bétons hautes et très hautes performances pour lesquels le meilleur cycle pour homogénéiser les fumées de silice par exemple, reste à découvrir pour chaque chantier.

Le malaxeur a pour rôle de distribuer, par des actions mécaniques, les différents constituants dans l'ensemble de la gâchée. Cette distribution doit être telle que la probabilité de trouver un constituant particulier en un point quelconque du mélange, est directement et uniquement fonction de la proportion de ce constituant dans le mélange [7].

Compte tenu que tout se fait dans le même malaxeur, celui ci doit être capable d'assurer cette répartition pour tous les constituants solides ou liquides, grossiers ou fins.

Or les études ont montré [8] qu'il ne faut pas la même énergie pour homogénéiser les gravillons, le sable, le ciment et l'eau, sans parler des adjuvants et des autres produits comme les fibres ou les fumées de silice. En voulant favoriser la répartition de certains des constituants, ciment notamment, on peut provoquer une ségrégation des autres par un excès d'énergie par exemple. Il est donc impératif, à partir du moment où tous les constituants sont mélangés en même temps et dans le même malaxeur, de faire des compromis afin que chaque distribution soit satisfaisante à défaut d'être optimale. Ce compromis n'est pas le même pour tous les bétons. On constate en effet que certains malaxeurs sont plus efficaces pour des mélanges secs ou des granulométries fines. il n'y a donc pas de lois générales.

Le travail exercé par les outils de malaxage [2] dépend du mode de remplissage du malaxeur [10]. Ce mode de remplissage doit être optimisé pour chaque malaxeur. On sait par exemple qu'un malaxeur à axe horizontal est beaucoup plus sensible au remplissage qu'un malaxeur à axe vertical et, en conséquence, nécessitera une analyse plus poussée pour que les résultats sur le mélange soit satisfaisant. Ce mode remplissage peut être caractérisé par la position des goulottes d'entrée des différents produits (position par rapport aux outils et angle d'inclinaison), par le moment où les produits entrent dans la cuve du malaxeur et par les débits de ces constituants. Le meilleur exemple ou, plus précisément, le plus visible est le mode d'incorporation de l'eau. La meilleure condition d'incorporation de l'eau est obtenue lorsque cette eau arrive en pluie sur l'ensemble des constituants solides et après le début d'incorporation de ceux ci. On a pu constater qu'un retard de l'arrivée d'eau de quelques secondes (8 s dans l'expérimentation menée) permettait d'obtenir une homogénéisation beaucoup plus rapide (gain de débit du malaxeur supérieur à 20 %) que lorsque l'eau était introduite en même temps que les autres constituants. On reproduit ainsi le malaxage à sec des constituants solides ce qui a toujours été considéré comme bénéfique pour l'homogénéité du mélange. Toutefois l'augmentation de l'usure des outils de malaxage et de la cuve qui en découle, n'incite pas les producteurs à suivre cette règle.

Ces quelques règles, souvent empiriques, montrent que la fonction malaxage est mal connue et qu'elle ne peut pas être mise sous la forme d'une équation. En conséquence on est toujours ramené à mesurer l'homogénéité du mélange à partir d'essais sur le béton (distribution des constituants ou propriétés du béton).

Le TC 150 de la RILEM Efficiency of concrete mixers a travaillé pour définir quels devaient être les paramètres à évaluer pour juger de l'aptitude d'un malaxeur [2].

A partir des résultats obtenus on pourra apprécier les possibilités des malaxeurs pour les différents types de béton en répétant les expériences pour différentes consistances et pour différentes granularités du béton.

4 Les essais initiaux

Malgré la connaissance que l'on a des malaxeurs et du fait essentiellement de l'absence de modélisation des actions de malaxage, il n'est pas possible, pour une nouvelle fabrication de béton, de prédire précisément le meilleur cycle de fabrication. Il est donc nécessaire de vérifier que le cycle retenu permet d'obtenir les performances attendues ou éventuellement d'affiner ce cycle au cas où il ne satisferait pas totalement.

Les paramètres qu'il y a lieu de déterminer sont le mode d'incorporation des constituants dans la cuve du malaxeur (moment et débit) et le temps de malaxage. A ce stade on considère que le réglage des pales est déjà réalisé et que celles ci sont en bon état.

Tableau 3 Exemple de cycle de remplissage du malaxeur. Les paramètres à déterminer sont t, v, x, y et z. Les valeurs prises sont primordiales pour la qualité du mélange et pour le débit du malaxeur. Ces deux objectifs ne sont pas contradictoire.

constituants	introduction dans la cuve		commentaires
granulats	0 _____ t	(en secondes)	les granulats tapissent la cuve
ciment	x _____ v		le ciment tombe sur les granulats
eau	y _____ z		l'eau est dispersée sur les constituants
adjuvant	(y+1) _____ (z-1)		les adjuvants sont incorporés dans l'eau

C'est pour fixer les conditions des essais que l'expérience est intéressante. Pour les bétons traditionnels, on sait que le ciment doit être légèrement retardé par rapport aux granulats. Un malaxage à sec est toujours bénéfique à l'efficacité du brassage, l'eau sera donc retardée par rapport aux constituants solides. Enfin il est toujours préférable d'incorporer les adjuvants avec l'eau. Le nombre de paramètres à tester est trop important pour qu'il soit retenu un plan d'expérience factoriel aussi on choisit au départ des conditions particulières reconnues valables pour l'efficacité du brassage des constituants.

En pratique on commencera par tester le temps de malaxage. Avant de lancer la série d'essais sur le mélange, l'inspection visuelle peut être d'un grand secours. L'aspect du béton dans la cuve lors de la fabrication et l'aspect de la gâchée de béton étalée sur le sol, permet le plus souvent de nous guider pour savoir où il est nécessaire d'intervenir.

Pour apprécier la qualité du malaxage et reconnaître sa qualité on réalise une série d'essais sur le mélange fabriqué.

Les essais retenus sont ceux qui peuvent être pratiqués sur chantier et les résultats sont immédiatement connus. Ces essais sont le plus souvent l'essai d'affaissement, la teneur en air occlus et la granulométrie. Pour cette dernière caractéristique on ne retient qu'un seul tamis qui devra être celui qui caractérise le mieux la granularité du béton. On choisit le plus souvent le tamis de 5 mm et on compare les valeurs des refus. C'est en effet sur cette classe que la teneur en eau aura le moins d'influence sur la précision du résultat (on ne sèche pas les granulats avant pesage).

Cette série d'essais est réalisée sur au moins trois prélèvements, le nombre de prélèvement étant lié à la garantie recherchée.

La norme [8] fixe les limites dans lesquelles les valeurs relatives à chaque paramètre doivent être comprises. Les essais sont répétés sur au moins trois gâchées consécutives afin de tester la reproductibilité du cycle de fabrication.

Dans le cas où les résultats sont considérés comme satisfaisants alors le cycle de fabrication, ordre et cadence d'introduction des différents constituants dans la cuve, et le temps de malaxage deviennent les conditions à respecter pendant tout le chantier.

5 Suivi du malaxage

La méthode est ancienne [11] mais n'a pas connu aujourd'hui le développement qu'elle méritait. Il s'agit du suivi de l'évolution de l'homogénéisation par le contrôle de l'enregistrement de la puissance de malaxage (figure 1).

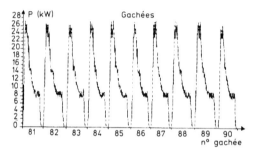

Figure 1 : Exemple d'enregistrement de la puissance de malaxage pour un lot de fabrication du béton. La similitude des différents graphiques successifs est la preuve de la régularité de la fabrication du béton

Avec le développement des moyens informatiques on est en mesure de rendre ce matériel tout à fait opérationnel et de permettre son emploi pour suivre la fabrication, rectifier en temps réel certaines anomalies et également pouvoir porter un jugement sur la qualité de la production tout entière et sur le fonctionnement du matériel. Il suffit pour cela de fixer les limites dans lesquelles les courbes de malaxage doivent se trouver, ces limites ayant été déterminées par les essais initiaux (§ 3).

Une étude très poussée [13] a permis de montrer que la courbe de malaxage était significative pour juger de la régularité du fonctionnement de l'automatisme de la centrale et pour juger de la consistance du béton en sortie du malaxeur. On a ainsi une preuve indiscutable du respect du cycle de fabrication du béton.

6 Apport de la maîtresse de la qualité pour le malaxage des bétons

L'apport qui est demandé, quand on propose un nouveau système, est le plus souvent la diminution immédiate des coûts avant même l'amélioration de la qualité. On oubli ainsi que l'une et l'autre sont liées mais avec une mise en évidence plus tardive pour l'économie financière directe.

Dans notre étude, nous proposons d'abord une amélioration de la qualité par l'assurance de fourniture d'un produit conforme. Il n'y.aura pas de diminution des coûts de contrôle et plutôt même une augmentation. Toutefois ces contrôles seront partagés (tableau 4) afin que ce soit celui qui utilise les résultats qui les réalise. C'est probablement ici que la première économie est réalisée : il n'y a plus de produits non conformes dans l'ouvrage ce qui devrait diminuer les coûts de réparation souvent constatés par la suite. Le volume global des essais ne sera pas diminué mais il sera plus en rapport vec la confiance qu'on attribue aux résultats, ce qui n'était pas le cas auparavant.

Une seconde économie concerne la prise en compte des performances réelles du mélange. Lorsque seuls des contrôles de performances du béton sont réalisés, comme le nombre d'échantillons ne peut pas être très élevé, on prend des marges importantes tant sur la prise en compte des performances de ce produit que pour les dimensions des ouvrages. Si cette procédure est acceptable pour les bétons courants, elle conduit, pour les autres bétons, à des niveaux de performance excessivement élevés et, en conséquence, devient de plus en plus coûteuse compte tenu de la progression des résistances mécaniques demandées aux bétons.

Tableau 4 : Description des actions selon le mode de contrôle. L'efficacité du contrôle et donc la qualité de l'ouvrage n'est pas garantie au même niveau selon les trois modes décrits.

Type de contrôle	Producteur	Client	Efficacité
performances	pas d'information	contrôle important sur le béton	fausse car le nombre d'essais n'est jamais suffisant
maîtrise de la qualité et contrôle extérieur	choix de moyens et utilisation de procédures reconnus et suivi permanent de la fabrication	réception des moyens, validation des méthodes et essais ponctuels sur le béton	**maximale**
moyens	déclaration des moyens et des méthodes	réception des moyens, validation des méthodes et surveillance permanente	efficace si rien ne change (usure, déréglages)

La maîtrise de la qualité dans la fabrication du béton permet de réduire les fluctuations des différents paramètres de malaxage. Ainsi on peut déterminer la valeur de l'écart type relatif à la distribution des valeurs de chaque caractéristiques du béton pour un mode de fabrication déterminé. L'expérience acquise par le producteur lui permettra de prévoir la dispersion liée à son outil de production. Il ne restera alors au producteur ou au client, qu'à déterminer la valeur moyenne de la caractéristique concernée, soit trois à cinq essais pour un lct de fabrication.

Si par contre il n'y a pas de garantie de régularité au niveau de la fabrication, il faudra pour chaque caractéristique du béton déterminer l'écart type et la moyenne soit au moins une quinzaine d'essais par lot.

Pour la fabrication du béton prêt à l'emploi en France, il a été jugé que, pour un même nombre d'essai et pour un B35/40, le surplus de résistance demandé au béton pour une fabrication ne bénéficiant pas de la maîtrise de la qualité était de plus de 10 % par rapport à un béton qui en bénéficiait. Inversement pour une même résistance et pour un même niveau de confiance, le nombre d'essais passe de 15 à 3 [9].

La maîtrise de la qualité est positive pour les deux parties car :

- en réceptionnant le béton à partir des informations "fabrication" le risque fournisseur (risque de se voir refuser une production conforme) est quasi nul ;

- du fait de l'élimination des productions non conformes au moment de leur production le risque du client (risque d'accepter une production non conforme) est en réalité, malgré le peu d'essai, très faible.

Ainsi il revient au demandeur de fixer les règles d'utilisation du matériel, les essais étant comme décrit précédemment.
Pour être qualifier dans une classe de performance particulière, il est nécessaire de satisfaire toutes les spécifications de la classe correspondante.
La qualification conduit alors à rédiger la fiche de certification comme indiquée ci après.

FICHE DE CERTIFICATION DES MALAXEURS

Description du malaxage

Mode de fonctionnement, principe de malaxage, description des outils, vitesse de rotation des outils, volume de la cuve...

Conseils d'utilisation

Mode d'introduction des constituants, réglage des pales en fond de cuve, réglage des pales latérales, dimension maximale des granulats, volume minimal de béton, consistance du béton, ...

Composition du béton	Cycle de malaxage

Caractéristiques du béton	Résultats des essais

Classification

Temps du malaxage :	8 - 10	20 - 32	Temps du malaxage :	8 - 10	20 - 32	Temps du malaxage :	8 - 10	20 - 32
0 - 3			0 - 3			0 - 3		
3 - 5			3 - 5			3 - 5		
10 - 12			10 - 12			10 - 12		
> 18			> 18			> 18		

Figure 2 : fiche de certification des malaxeurs à béton La fiche montre qu'il est possible, pour un même malaxeur, d'obtenir des classifications différentes selon les types de béton ou selon la durée de malaxage.

7 Conclusions

Le contrôle de la fabrication du béton par le seul client n'est plus envisageable aujourd'hui. Il doit être associé aux contrôles de suivi de la fabrication par le fabricant lui même. Cette association impose de fixer précisément les règles, que chacun les respecte et qu'une transparence totale accompagne les actions de chacun des partenaires.

Le concept de la maîtrise de la qualité est basé sur le principe suivat: les mêmes causes produisent les mêmes effets. Aussi, si on a vérifier en début de chantier que le cycle de malaxage conduisait à produire un béton satisfaisant, alors la répétition de ce cycle garantita la régularité recherchée.

L'économie qu'on peut espérer en tirer ne porte pas sur la suppression d'un poste de contrôle, poste qui, même s'il était considéré comme nécessaire, n'était pas toujours pourvu en conséquence. L'économie résulte en fait de la non mise en place de gâchées de béton qui ne satisferaient pas aux besoins de l'ouvrage. La barrière pour cela est double car le procédé de fabrication est conçu pour ne pas produire ces gâchées et son identification, au cas où elle franchirait ce premier obstacle, permettrait de la repérer et de l'éliminer.

Grâce à ces dispositions, le concepteur de l'ouvrage peut d'une part prendre en considération des performances du béton plus proche de la réalité (résistance mécanique) et d'autre part dimensionner les ouvrages plus justement (régularité de la masse volumique du béton).

Outre les gains de temps (attente des résultats des cntrôles) et l'amélioration des relations clients-fournisseur (transparence des actions du producteur), la maîtrise de la qualité est en fait une chasse au gaspillage et c'est là que se trouve la véritable économie.

8 Bibliographie

1. Gourdon, J.L., Charonnat, Y., and Robin, P. (1987) Contribution à la caractérisation de l'échantillonnage de mélanges granulaire en vue d'étudier les phénomènes de malaxage, in *From Materials Science to Construction Materials Engineering*, (ed J. C. Maso), Chapman and Hall, London, pp. 476-483.

2. Charonnat, Y. (1996) Efficiency of concrete mixers - towards qualification of mixers, *International RILEM Conference on Production Methods and Workability of Concrete*, Glasgow, Scotland, June 3-5 1996.

3. AFNOR (1992) Centrales de fabrication du béton de ciment - Définition des types de centrales et essais pour la vérification des réglages, *NF P 98-730*, Afnor 1992.

4. Charonnat, Y. (1982) Le dosage pondéral dans les centrales à béton, *Bull Liaison Labo P et Ch Spécial XII, Matériels de travaux publics*, juin 1982, pp. 59-70.

5. Jouandou, Th. (1982) Etude de l'influence des paramètres de malaxage sur l'homogénéité du béton, *Université Blaise Pascal*, Clermont Ferrand II, juin 1982.

6. Beitzel, H. (1981) Gesetzmässigkeiten zur optimierung von betonmischern,

Fakultät für Bauingenieur und Vermessungswesen der Universität (TH), Karlsruhe, 1 Juli 1981.

7.   Robin, P. (1988) Contribution à la caractérisation de l'échantillonnage de mélanges granulaires en vus d'étudier les phénomènes de malaxage, *Université Blaise Pascal,* Clermont Ferrand II, 24 octobre 1988, No. d'ordre DI 213.

8.   Bruneaud, S. and Charonnat, Y. (1989) Knowledge of concrete mixers: The mixer test station, in *ERMCO '89 The Norway to concrete,* (ed Finn Fluge, Odd Sverre Hunsbedt, Jörn Injar), FABEKO Norwegian Ready Mixed Concrete Association OSLO, pp. 172-181, Stavanger, June 7-9 1989.

9.   AFNOR (1994) Béton prêt à l'emploi, *P 18-305,* Afnor 1994.

10.   Bozarth, F.M. (1967) Influence of imbalances in charging of cement and water on mixing performance, *Research and Development Report,* US Department of Transportation, FHA Bureau of Public Roads, Washington, July 1967.

11.   Loring, Parrington (1928) The determination of the workability of concrete, *ASTM, Vol. 28, part II,* New Hampshire.

12.   Teillet, R., Bruneaud, S. and Charonnat, Y. (1991) Suivi et contrôle de la fabrication des mélanges - Une nouvelle jeunesse pour le wattmètre différentiel, *Bull liaison Labo P. et Ch., 174,* juil-août, pp. 5-16.

13.   Brachet, M., Charonnat, Y. and Ray, M. (1975) Vers un contrôle de qualité non conventionnel des bétons hydrauliques, *Annales de l'ITBTP, supplément au No. 336,* février 1976, p. 93-120.

PART TWO
PRODUCTION METHODS

6 SYSTEM-SPECIFIC EVALUATION OF FRESH CONCRETE TRANSPORT SYSTEMS IN THE CONSTRUCTION OF DAM WALLS

H. BEITZEL
Forschungstelle für Bauverfahrens und Umwelttechnik, FHRP,
Trier, Germany

Abstract
Manufacture, transport and placing of concrete on large dam sites requires a comprehensive technological and economical appraisal of alternative systems and local conditions. Performance criteria for different concreting stages are outlined and several types of plant and equipment systems are evaluated and compared. This approach is then applied to the case of the fresh concrete production and handling at the major Three-Gorges dam project in China.
Keywords: fresh concrete, transport, cranes, conveyor belts, concrete mixers, turret cranes, cable cranes, mix design, large aggregate.

1 Introduction

The process technology for the construction of dams is generally determined by the manufacture, transport and placing of mass concrete.

All the systems have in common the production of concrete in concrete-mixers.

At modern dam-building sites the following essential transport systems are used:

- Cable cranes - concrete buckets
- Transportation vehicles - cable cranes - concrete buckets
- Conveyor belt - turret cranes - conveyor belt
- Transportation vehicles - turret cranes - concrete buckets
- Other conveyors

The economic placing and compaction of large amounts of concrete in block-form is now carried out using modern bulldozers with integrated hydraulic vibrators.

Production Methods and Workability of Concrete. Edited by P.J.M. Bartos, D.L. Marrs and D.J. Cleland.
Published in 1996 by E & FN Spon, 2–6 Boundary Row, London SE1 8HN. ISBN 0 419 22070 4.

2   Concrete manufacture

An important problem during the manufacture of mass concrete is the workability of mixes with the maximum aggregates. Especially large capacity drum mixers with a nominal content of between six and nine cubic metres are used for this task. In some cases pan mixers and double shaft forced action mixers are used as well.

The achievement of satisfactory homogeneity in a freshly mixed mass concrete within an optimal mixing-time is the deciding criterion for the use of the latter machines in the mixing process.

3   Concrete transport

The following machines are the basic elements in the transport system mentioned above:

- Cable cranes
- Turret cranes
- Transport vehicles
- Belt conveyors
- Other conveyor systems

3.1   Cable cranes
The load-bearing cables are important for the functioning of both the low and high level cableways. They are also essential in the construction and design of the crane-towers.

The use of a crane is normally in an accelerating and decelerating phase. Due to this a better efficiency cannot be achieved by using higher trolley speeds. It is often more efficient to increase the load capacity of the crane instead of the speed.

3.2   Turret cranes
Turret cranes with a loading capacity of up to 9 cubic metres now represent a modern alternative in the transport and placing of mass concrete.

The use of a modern turret crane makes it possible to reach all building areas at the dam site. With this system the mass concrete can be placed according to the quality requirements. Modern control systems assure relatively high trolley speeds and a larger load capacity in addition to a high operating safety.

3.3   Transport vehicles
The simplest way to transport concrete is to use tipper trucks. The design of purpose-built trough-tippers as dumper or silo type vehicles is more complicated.

Silo vehicles with a capacity of six to nine cubic metres (or more) are often used on large building-sites. In such cases their use is generally of an intermediary nature, such as for the transfer of concrete from the mixing plant to other intermediary containers or supply pumps.

Dumper trucks with two containers can also be used with simultaneous charging from two crane skips to optimise the performance.

Alternatively, mobile mixers such as the well-known truck-mixers used in the ready-mix industry can be used. The speed of rotation and the mixing speed of the vehicles can be adjusted according to the type of concrete and its composition.

3.4 Belt conveyors

Special care has to be taken in the movement of stiff and plastic concrete. The speed and slope of the conveyor belt must be set up so that the coarse aggregate neither rolls of the transport, nor gets thrown too far off at the end of the supply route. The slope of the conveyor belt is restricted to between fifteen and thirty degrees. At the upper guide pulley a guiding sheet of metal and a scraper to remove cement grouts from the conveyor must be installed. Further, the width and speed of the conveyor belt are dependent on the size of the maximum aggregates in the batch.

When using conveyor belts in regions with extreme environmental conditions, it may be necessary to provide them with a shelter from weather.

3.5 Other conveyors

Depending on the amount of concrete, the transportation, the topographical conditions and the consistency of the concrete, other conveying methods such as the monorail or drag-bucket conveyors, or a combination of the aforementioned methods, can be economically used.

4 Decision making criteria

Concrete technology, efficiency and process technology criteria must be taken into consideration for the evaluation of transport systems.

4.1 Concrete technology

To prevent any poor quality of fresh or set concrete there are certain requirements for the composition of mass concrete.

In order to keep temperature increases to a minimum, it is possible among other things to use concretes with a lower heat development and to keep the cement content to a minimum. In addition, pre- and post-production cooling systems can be installed.

The restriction of the temperature of fresh concrete for use in the building of dams to within the range of six and seven degrees Celsius is now state of the art. Consistent homogeneity and workability of concrete is necessary to prevent concrete segregation.

Different grading curves (coarse aggregates from 60 mm to 200 mm) are chosen depending on the section of the dam-wall. The cement content varies between 40 and 200 kg/m^3 according to the intended use of the mix and the grading curve. Fly ash can also be combined in the mixture (Fig. 1).

The quality requirements of mass concrete mentioned above demonstrate the current standards of concrete technology. The following demands have to be fulfilled during the construction process (Figs. 2 and 3):

- Standardisation of the concrete mix design
- Undisturbed hydration process
- Lower temperature development

Design grade	Design age (day)	Aggregate type	Coarse aggregate gradation	Water (kg/m³)	Cement + fly ash (kg/m³)	Coarse aggregate (kg/m³)	Sand (kg/m³)	Additive	Slump (cm)
150	90	Artificial	4	115	136 + 34	1640	550	compound agent	4–6

Fig. 1 Basic mix proportions of conventional concrete for the dam

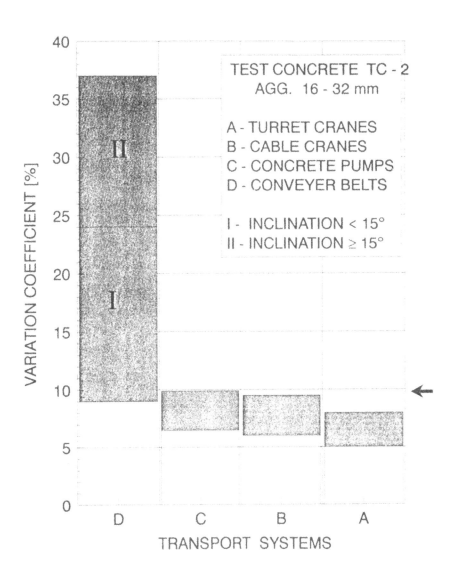

Fig. 3 Homogeneity

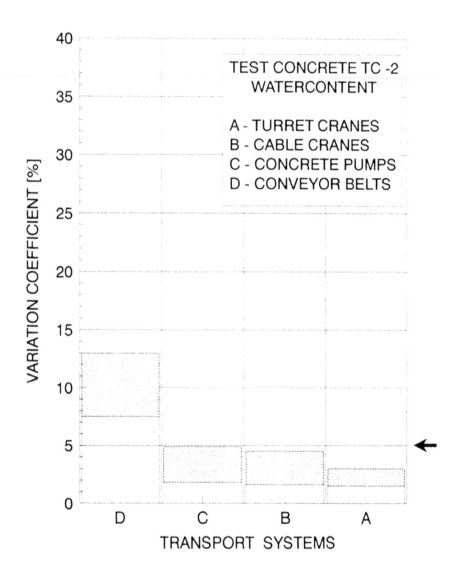

Fig. 2 Homogeneity

- Optimum homogeneity
- Constant water/cement ratio
- Freedom from segregation
- Freedom from agglomeration
- No loss of water or humidity
- Avoidance of brittleness
- Low placement temperatures
- Less shrinkage
- Good workability

4.2 Process engineering

For the evaluation of the transport system, in consideration of process technology aspects, it is important to adhere to the chosen system. This means that continuous concrete production should be linked to continuous conveyor movement, and intermittent concrete production should be linked to intermittent conveyor movement.

The following aspects have to be considered:

- Segregation-free flow of material
- Flow of the material free of disruption
- Flow of the material independent of weather conditions
- Adaptability to various forms of building geometry
- Slopes and falling distances independent of the grading curve
- Minimisation of the interfaces
- Abrasion reduction
- Minimisation of the damage susceptibility
- Easy maintenance
- Adaptability to extreme weather conditions
- User-friendly
- High operating reliability
- Minimisation of work disruption

4.3 Efficiency

The following criteria are essential to judge the efficiency of the transport system to be used:

- Short construction time
- Availability
- Quota of completion machines
- Short assembly time
- Short disassembly time
- Easy usage
- Lesser chance of damage
- High operating reliability
- Minimisation of susceptibility to damage

- High level of resistance to abrasion
- Large element durability
- Low investment costs
- High placing rate

It is of particular importance that the quality of the mass concrete is considered in assessing the efficiency.

5 Assessment of the transport system for the Three Gorges Dam

The second phase of construction for the project is taking place on the left bank of the dam. The final height of the dam will be 185 m. The concrete works for this phase of the project are subdivided into production for factory buildings, power station buildings, the leading wall of the factory building, the dam wall and the box dam.

Considering the topography, the best-possible construction time, the distances covered by the transport, the quality requirements and the decision-making criteria as mentioned in Section 4 it is possible to examine the particularities of the various transport systems.

5.1 Cable cranes
For the placement of concrete, the use of both the low and high level drag lines is possible. In both systems a silo vehicle with the same capacity as the hoist cable system has to be used for delivery.

Low level cable cranes

- High assembly costs at installation.
- Little disruption of other installations.
- Number of cycles for the drag-line limited.
- Shorter poles and cables compared to the high level conveyor system.
- Saving in construction time in comparison with the use of both bridge and turret cranes.
- No change-over of mixing towers.
- Problems with the capacity of the drag line during high-performance installation.
- Less availability.

High level cable cranes

- Higher assembly costs at installation.
- Less disruption of other installations.
- Number of cycles of multiple drag-lines are limited.
- Problems regarding the weight of the drag construction.
- Silo vehicles needed to bring the load to the drag-line.
- Able to supply all construction levels, in contrast to the lower drag-line.
- Saves construction time, in comparison with the use of a crane or turret crane.

- No transfer of mixers needed.
- Less availability.

With the available span and required placing performance of the above systems a larger carrying capacity, which is accompanied by the relatively large diameter of the cableway, is more likely to be required. The time gained by the use of higher speeds is partly lost due to the amount of care taken when transferring the loads on to the transport system at the loading point.

5.2 Large turret cranes - transport vehicles

- Turret cranes must be placed on bridges.
- The duration of the concrete placing is slightly extended by the extra assembly time needed.
- The quality of the concrete is guaranteed.
- The 28 tonne capacity of the large tower crane means that high placing output can be achieved.
- Flexibility of transfers according to the building progress.
- Flexibility in lifting work (bulldozers, formwork and other heavy loads of up to 80 tonnes).
- Due to the discontinuous nature of the use of the crane in high-performance construction it is preferable to include it in the production and haulage system.
- The nominal content of the transportation vehicles is adjusted to the capability of the large cranes.
- Supply backlogs to the transport vehicles are easily settled.
- The roads on construction sites, which are in constant use by the transport vehicles, must be of good quality.
- The capacity of the vats is fixed according to the capabilities of the cranes and the capacity of the mixers of the concrete preparation site.
- The advantage of the silo vehicle and the mobile concrete mixers is, among other things, that the concrete transport is protected to a greater degree from environmental conditions.
- Segregation-free transport and placing of concrete.

With regard to the assembly and disassembly, as well as to availability, this system shows distinct advantages over the other transport systems.

5.3 Turret cranes - conveyor belt
This system achieves the transport of mass concrete from the concrete placing by means of conveyor belts and the distributive conveyors supported by a rotary-tower crane.

- The flexibility of the heavy crane tower placement cannot be guaranteed to be safe in all areas.
- There are restrictions regarding the use of conveyor belts when the diameter of the particles on the belt exceeds 80 mm.

- No guarantee that the concrete will reach the site without segregation.
- Discontinuous concrete production, continuous transport system.
- Savings in construction time because the building of a bridge is unnecessary.
- Supply to the up- and downstream sides of the dam wall by the same crane is not possible.
- Many additional machines are required, especially at the foot of the dam, in order to complete the construction.
- On relatively long stretches of conveyor, storm-proof covering is required.
- Limited availability.
- Latent danger of supply backlogs.
- Transport of mass concrete with large aggregates (150 mm) is impossible on slopes exceeding thirty degrees.
- An additional reduction in the size of the particles means that the cement content will be increased. This means more danger of shrinkage and higher cement and aggregate costs.
- By placing the turret cranes with the conveyor belt in areas 7-14 (factory building) there is a saving on the 50 metre high bridges.
- This frees up more construction space.
- Due to the absence of assembly time for the bridges, the concrete construction time can be reduced or it can be started earlier.
- There are greater cooling costs with the use of mass concrete.

6   Conclusion

Many of the transport systems used in the installation of mass concrete have been described. The particularities of each transport system, with reference to the system-specific evaluation criteria and in respect of the Three Gorges Dam Project, have been explained.

7   References

1.   Kuhn and Beitzel, H. (1981) Itaipu - Bauwerk und Ausführung, VDI 25 Jahre, VDI Verlag.
2.   Beitzel, H. (June 1995) Geräteeinsatz Huites Staudamm, Mexico, BMT.

7 HIGH PERFORMANCE CONCRETE FOR EXPOSED CIVIL WORKS STRUCTURES – GUIDELINES FOR EXECUTION

F. MEYER
Carl Bro Civil & Transportation A/S, Odense, Denmark
J. FRANDSEN
4K-Beton A/S, Copenhagen, Denmark

Abstract
In an attempt to ensure durable structures, increasingly complex concrete mixes have been designed. Well established execution methods may as a consequence no longer be adequate, and contractors have experienced severe difficulties in achieving the results desired. The Danish Concrete Institute established in 1993 a task group to carry out a development programme with focus on the contractor's technology related to the use of high performance concrete.

The first phase of this programme, published in November 1993, identified the most important production properties which have given difficulties during execution. The production properties are: Workability, Stickiness, Stability of air entrainment, Vibration need, Crust formation, Setting time and Form pressure.

The second phase of the programme, published in August 1995, contains newly developed methods to determine the magnitude of the above mentioned production properties and gives guidelines for their expected magnitude based on the concrete composition. Guidelines for influences from each of the production properties on execution methods are also included.

This paper explains the general purpose of the programme and how it was accomplished. A part of the published guidelines concerning Vibration need is presented.
Keywords: Production properties, Workability, Vibration need, Guidelines, Execution, High Performance Concrete.

Production Methods and Workability of Concrete. Edited by P.J.M. Bartos, D.L. Marrs and D.J. Cleland. Published in 1996 by E & FN Spon, 2–6 Boundary Row, London SE1 8HN. ISBN 0 419 22070 4.

1 Introduction

Resistant concrete with a long service life is often ensured by requirements on low w/c-ratio and use of microsilica.

To obtain satisfactory workability the contractor must add a significant dosage of water reducing additives. This may lead to long setting time, and in some cases the concrete can loose its workability in a very short time.

The composition may also lead to stickiness, high viscosity and crust formation, all of which may cause problems during execution of the concrete works. Well established execution methods may as a consequence no longer be adequate when such high performance concretes are used.

The Danish Concrete Institute has carried out a two-phase development programme in order to systematically describe and clarify these production properties.

The first phase of this programme, published in November 1993 [1], identified the most important production properties which had given difficulties during the execution of 10 selected civil works structures, all cast with high performance concrete.

The production properties are:

- Workability
- Stickiness
- Stability of air entrainment
- Vibration need
- Crust formation
- Setting time
- Form pressure

The second phase of the programme, published in 1995 [2], [3], [4], [5], [6], [7], [8], [9], [10], [11], [12], [13] and [14] contains newly developed methods to determine the magnitude of the above mentioned production properties. The expected magnitude based on information of the concrete composition is also included.

The main results are published in a Guide for execution of civil works structures with high performance concrete [2].

2 Description of phase 2

Phase 2 of the programme consisted of 16 tasks as shown Fig. 1.

The general sequences for the tasks 2-8, which dealt with the production properties, were:

- Definition of practicable methods for the testing of the production property in question.
- Sensitivity testing of the methods by using concrete/mortar with extreme compositions. The chosen test methods were described in detail and investigated for reproducibility.

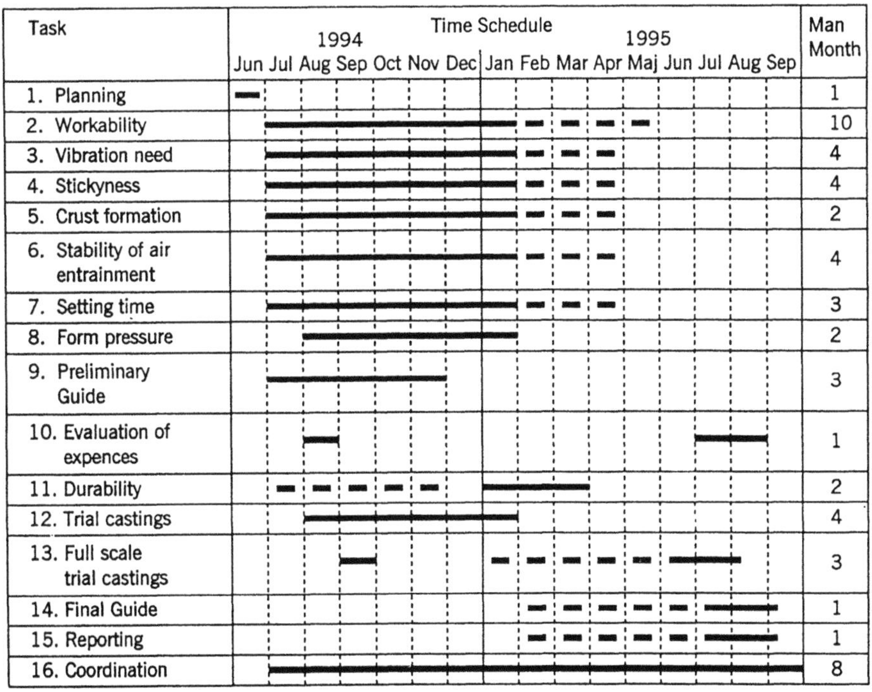

Task	Time Schedule 1994 / 1995	Man Month
	Jun Jul Aug Sep Oct Nov Dec | Jan Feb Mar Apr Maj Jun Jul Aug Sep	
1. Planning		1
2. Workability		10
3. Vibration need		4
4. Stickyness		4
5. Crust formation		2
6. Stability of air entrainment		4
7. Setting time		3
8. Form pressure		2
9. Preliminary Guide		3
10. Evaluation of expences		1
11. Durability		2
12. Trial castings		4
13. Full scale trial castings		3
14. Final Guide		1
15. Reporting		1
16. Coordination		8

Fig. 1. Time schedule for phase 2.

- By means of the selected method the magnitude of the production property in question was determined for concrete/mortar with different compositions. The tests were carried out in the laboratory, and the specific conditions which had the most important influence on the property was investigated.
- The methods were tested on 14 concrete mixes, each with different compositions, by small-scale trial castings (1-3 m³ each). The testing consisted of a quantitative determination of the production property by the developed method and by observations during execution of e.g. trowelling, floating and vibration. Relationships between the production properties and difficulties in execution were hereby established.
- Preparation of the guide for execution.
- Preparation of a detailed report.
- Pretesting of 3 selected concrete mixes which were used for full-scale testing of the guidelines for execution. The full-scale tests were carried out on 3 bridges, each of approx. 200 m³ of concrete.

3   Results

The aim of the development programme was a guide for execution with high performance concrete.

The guide [2] contains descriptions of equipments and test methods for measuring:

- Workability by Rheometer
- Stickiness by Unto Meter
- Crust formation
- Setting time

The influences of different execution methods are described for each of the production properties.

As an example some of the main results concerning vibrating with poker vibrators are presented in the following.

4   Guide for Poker vibration

4.1 The equipment
The construction of a poker vibrator is shown Fig. 2.

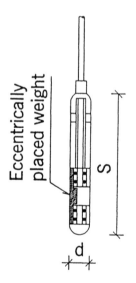

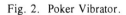

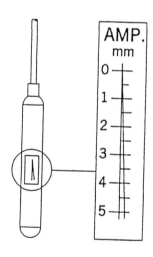

Fig. 2. Poker Vibrator.

Fig. 3. Label for testing the amplitude of a poker vibrator.

The vibrations are generated by a rotating eccentrically placed weight. The frequency indicates how fast the plump is rotating. The usual frequency for an electric poker vibrator is 12,000 rpm.

A freely suspended poker vibrator with a frequency of 12,000 rpm should not have an amplitude of less than 0.5 mm. The amplitude can be tested by a label shown Fig. 3.

The test is performed by observing the label, which is placed on the freely suspended poker vibrator. The two lines on the label will be seen as only one line when the distance between the lines is small. The maximum distance between the lines which gives the impression of only one line indicates the amplitude of the poker vibrator.

A freely suspended electric poker vibrator should not be allowed to run for more than 1 minute.

4.2 The vibration need

The vibration need is an expression for the energy, which is required to ensure that the fresh concrete has filled out the form completely. Furthermore the reinforcement, spacers and embedded items shall be completely surrounded.

The concrete shall be vibrated in such a way that most of the entrapped air is driven out leaving the concrete without cavities but with a satisfactory content of fine air bubbles.

The size of a poker is defined by the outer diameter (d) and the length (s). The extent of the effect of the vibrator is proportional with the diameter.

Vibrating is performed by immersion in a number of positions. The centre to centre distance between these positions is named the immersion distance (a). The immersion distance, a, is given as a multiple of the outer diameter $a = n \cdot d$.

After immersion of the poker, the concrete is vibrated a certain time, named the vibration time. Subsequently the poker vibrator is pulled out and moved to the next position. The vibration time is the time between start of immersion and start of pulling out.

The vibration need is not a well defined quantity, but may be expressed by combining the immersion distance and the vibration time.

The form complexity (including complexity of reinforcement, prestressing anchors, boxouts, couplers and cast-in items) determines the immersion distance.

The workability of the concrete determines the vibration time.

4.3 Usual vibration methods

The concrete shall be placed in such a way that large horizontal movements by poker vibration are unnecessary.

The usual vibration method is:

- The poker vibrator is submerged vertically in the concrete until only the top is visible.
- The poker vibrator is kept in this position for 5-20 sec. (vibration time).
- The poker vibrator is pulled out slowly in order to avoid traces without coarse aggregate.

4.3.1 Thickness of concrete layers

The concrete shall be placed in thin layers to ensure systematic vibration of all parts of the concrete.

The chosen thickness of the layers depends mainly on the length of the poker vibrator.

It is essential that the operator knows the exact vertical position. This is possible when the top is visible during vibration as shown Fig. 4.

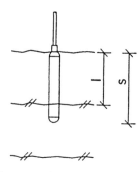

Fig. 4. Maximum thickness of concrete layers.

To ensure sufficient vibration between the layers it is recommended that the poker vibrator penetrates the preceding layer by approx. 20% of the poker length. The maximum thickness of the concrete layers should therefore not exceed 80% of the poker length.

Most poker vibrators have a length of 300 to 400 mm, corresponding to concrete layers of 240 to 320 mm.

If thicker layers are to be vibrated systematically it is necessary to perform the vibration with the poker placed in two different positions. In order to ensure correct placing in the lower position (1) a visible mark must be placed above the top of the poker vibrator as shown Fig. 5.

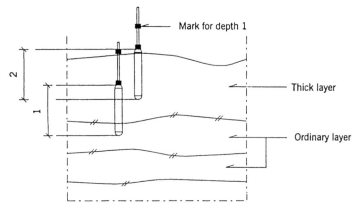

Fig. 5. Mark for thicker layers.

By vibration in two positions for each immersion the maximum thickness of the layers may be extended to 1.8 times the poker vibrator length.

The method, shown Fig. 5, is not recommended for ordinary use because the transportation distance for entrapped air is more than doubled and the risk of blow holes in the surface is consequently higher.

4.3.2 Vibration time

The vibration time depends mainly on the workability of the concrete, identified by the slump.

The workability decreases in time, and therefore the vibration time is not constant but varies by the age of the concrete as does the slump. If vibration times are constant it follows that a homogeneous vibration is dependent on small variations of the slump during placing and vibration.

The vibration time is also dependent on the stickiness and viscosity of the concrete.

The viscosity of the concrete is increased by use of microsilica, low w/c-ratio and high dosage of water reducing additives.

The necessary vibration time for ordinary and viscous concrete may be estimated on basis of Fig. 6. A more specific determination should be based on trial castings.

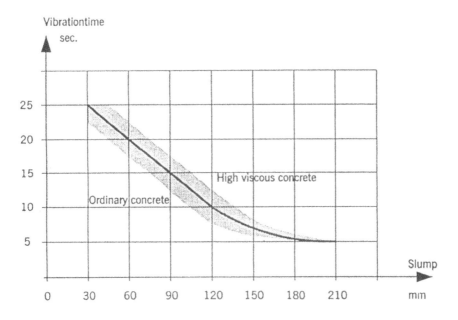

Fig. 6. Relation between vibration time and slump.

4.3.3 Immersion distance

The immersion distance a = n·d depends on the concentration of rebars, prestressing anchors, boxouts, couplers and cast-in items. The degree of form complexity may be estimated based on Table 7.

A more specific determination should be based on trial castings.

Table 7. Relation between the degree of form complexity and Immersion distance

Degree of form complexity	Example	Immersion distance a = n·d
1 Simple	Plain concrete and low steel percentage	10 d
2 Normal	Usual steel percentage	8 d
3 Difficult	High steel percentage and some congested areas	6 d
4 Very difficult	As 3 but many congested areas	4 d
5 Extremely difficult	Should not be used	2 d

4.4 Productivity

The effective volume, V, which a poker vibrator can vibrate for each immersion may be calculated by using the immersion distance, a = n·d and the thickness of the concrete layer l. The expression is:

$$v = l(nd)^2. \text{ See Fig. 9.}$$

The estimated vibration time for each immersion can be found in Fig. 6.

The time for pulling out and moving the poker vibrator to the next position is estimated to approx. 10 sec. The corresponding cycleperiods and number of immersions per hour are shown Table 8.

Table 8. Cycleperiods and number of immersions per hour.

Vibration time (sec.)	Cycleperiod (sec.)	Number of immersions per hour n_t
5	15	240
10	20	180
15	25	144
20	30	120

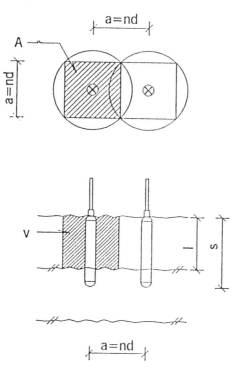

Fig. 9. Relation between Effective Volume, V, immersion distance, n·d, and thickness of concrete layer, l.

The productivity, Y, can be expressed as:

$$Y = \Phi \, n_t l (nd)^2 \qquad\qquad (1)$$

where

Y is the productivity [m³ per hour].
Φ is the factor of effectivity (waiting time).
n_t is the number of immersions per hour.
l is the thickness of the concrete layers [m]
d is the outer diameter of the poker vibrator [m]
nd is the immersion distance [m]

Fig. 9 shows the productivity for poker vibrators with diameter d between 20 and 80 mm. The factor of effectivity Φ is assumed to 1.0 and the thickness of the concrete layers 0.30 m.

The four diagrams are related to vibration times of 5, 10, 15 and 20 sec. On each diagram lines are shown for immersion distances 4d, 6d and 8d.

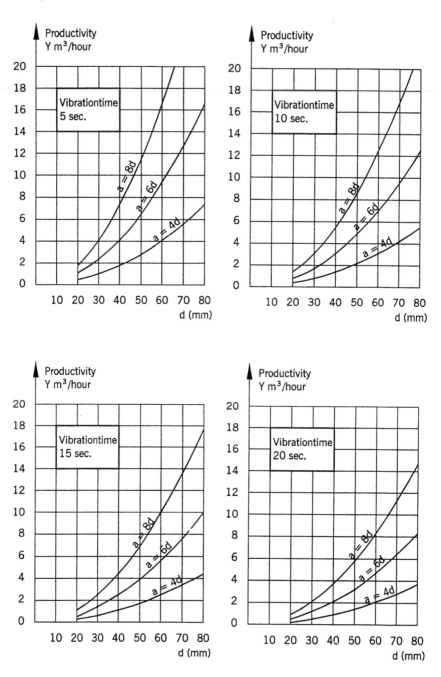

Fig. 10. Productivity, Y, for vibration times 5, 10, 15 and 20 sec.

Example on determination of productivity.

A slab is cast by use of a high viscous concrete. The slump is 90 mm. Fig. 6 indicates a necessary vibration time of approx. 15 sec.

The outer diameter of the poker vibrators is 40 mm and the length is 0.38 m. The maximum thickness of the concrete layers is 80% of 0.38 m equal to 0.30 m.

The degree of form complexity is "very difficult" corresponding to an immersion distance of 4 d = 4·40 = 160 mm.

Using the diagram Fig. 9 the productivity is found to 1.2 m³/hour per vibrator. If the planned casting rate is 12 m³/hour 10 vibrators should be used.

The form complexity of another part of the slab is "normal" corresponding to an immersion distance of 8d = 8·40 = 320 mm.

For this part of the slab the productivity is 4.2 m³/hour per vibrator and if the casting rate is 12 m³/hour only 3 vibrators are needed.

4.5 Selection of poker size

The effectivity per vibrator is increasing by the square of the poker size, d (ref. (1)).

It is therefore economically advantageous to use large poker vibrators.

However there are limitations for the selection of poker size.

The poker vibrator must not be placed closer than a distance of 3d from the shuttering. Closer placing may lead to formation of blow holes in the concrete surface.

The selection of poker size for casting of walls, columns and beams is therefore dependent on the minimum dimension as indicated Table 10.

Table 11. Relation between wall dimension and maximum poker size.

Wall dimension	Maximum poker size
120 mm	20 mm
180 mm	30 mm
240 mm	40 mm
300 mm	50 mm
360 mm	60 mm

5 Conclusion

The Guide for execution with high performance concrete contains methods to optimize the interdependence between the concrete composition, the production properties, the contractors execution methods and the economy.

6   References

1.   *High Performance Concrete for Exposed Civil Works Structures.* State-of-the-Art Report. Dansk Betoninstitut A/S (1993).
2.   Meyer, F. *Anvisning i brug af højkvalitetsbeton.* Dansk Betoninstitut A/S (1995).
3.   Frandsen, J. and Schultz, K.I. *Vejledning i vibrering med stavvibrator.* Dansk Betoninstitut A/S (1995).
4.   Meyer, F. *Højkvalitetsbeton til udsatte anlægskonstruktioner, Hovedrapport.* Dansk Betoninstitut A/S (1995).
5.   Goldermann, P., Pade C. and Thaulow, N. *Bearbejdelighed.* Dansk Betoninstitut A/S (1995).
6.   Frandsen, J. and Schultz, K.I. *Vibrering.* Dansk Betoninstitut A/S (1995).
7.   Kjær, U. *Klæbrighed.* Dansk Betoninstitut A/S (1995).
8.   Skorpedannelse. Dansk Betoninstitut A/S (1995).
9.   Geiker, M. and Rostam, S. *Luftindholdsstabilitet prøvestøbninger i Kalundborg og sammenfatning af litteratur.* Dansk Betoninstitut A/S (1995).
10.   Wegmann, T. *Afbindingstid.* Dansk Betoninstitut A/S (1995).
11.   Meyer, F. and Brinchmann, I. *Formtryk.* Dansk Betoninstitut A/S (1995).
12.   Geiker, M. *Styrke og holdbarhed.* Dansk Betoninstitut A/S (1995).
13.   Wegmann, T. *Prøvestøbning, Kalundborg.* Dansk Betoninstitut A/S (1995).
14.   Geiker, M. *Fuldskala-afprøvning.* Dansk Betoninstitut A/S (1995).
15.   *Standard Practice for Consolidation of Concrete.* ACI 309-72, revised 1982.

8 STEEL FIBRE CONCRETE FOR UNDERWATER CONCRETE SLABS

H. FALKNER and V. HENKE
iBMB, Technical University of Braunschweig, Braunschweig, Germany

Abstract
For the erection of a new multi-functional town quarter in the heart of Berlin, the area of the Potsdamer Platz, the construction of deep building pits with underwater concrete slabs becomes necessary. In order to contribute to the overall construction and operation safety of such a back-anchored underwater slab, it is planned to use steel fibre concrete for the first time for the construction of these slabs. This paper will give a short report about the large scale underwater and additional test necessary to solve the questions about the workability and material behaviour of the tested concrete mixes. Furthermore, these tests were necessary to satisfy German building regulations for the use of non-codified building materials.

1 General

In the centre of Berlin, Debis, a subsidiary company of Daimler Benz Ltd., is at present constructing a new town quarter on a large-scale building site extending over an area of 70,000 m². This new multi-functional town quarter, which will cover the area of the former Potsdamer Platz, the old heart of Berlin, will not only provide offices, housing, shops and shopping-centres but also hotels, theatres and cinemas.

The dimensions of this new town quarter, with a length of approximately 560 m and a width of up to 280 m are remarkable, as the whole construction has to be carried out in one single building pit. This building pit, with a depth of up to 20 m poses a special challenge to the planning engineers and the construction companies.

Production Methods and Workability of Concrete. Edited by P.J.M. Bartos, D.L. Marrs and D.J. Cleland. Published in 1996 by E & FN Spon, 2–6 Boundary Row, London SE1 8HN. ISBN 0 419 22070 4.

Those who know Berlin are aware of the fact that the ground-water level lies 2 -3 m below the surface. Furthermore, the Berlin authorities impose strict conditions on every construction measure interfering with the ground-water level. Only a few 100 m from the building site the large park area of the Berlin Zoo is situated, with valuable, old and rare trees. In order not to endanger this plant live, any lowering of the ground-water level is strictly forbidden. Additionally, Berlin - in the meantime again a city with over 4 million inhabitants - draws its drinking water from this ground-water reservoir. This is another reason for the stringent non-interfering conditions imposed on such a construction measure.

For depths of up to 10 - 12 m, it is possible to construct a "dry building pit" with an artificial sealing layer and deep sheet piling or slotted walls. For greater depths a different construction method has to be used; underwater excavation after the erection of the sheet piling or slotted walls with the subsequent placement of an back-anchored underwater concrete slab.

At present, this underwater excavation takes place with large floating dredgers in the construction section B of Debis and P, where the future central station Potsdamer Platz will be erected. This means that the building site B is dominated for several month to come by a large artificial lake covering an area of 15,000 m² and a depth between 14 and 17. This water is enclosed by 1.2 m thick slotted walls or sheet pile walls with a depth of 1.0 m. The second lake in section P, with an area of approx. 15,000 m², has a depth of up to 20 m.

Fig. 1: View of the building pit B with „Lake debis"

After the excavation approx. 2,000 tension piles for anchoring the underwater concrete slab, which later have to carry tension forces of up to 1,500 kN, will be driven from pontoons into the ground. The overall buoyancy force for this building pit amounts to 3 million kN. After completion of this pile driving, approx. February/March 1996, the 1.2 m thick underwater steel fibre reinforced concrete slab will be poured in sections. If this work finishes on schedule, the building pits will be pumped out at the beginning of April 1996.

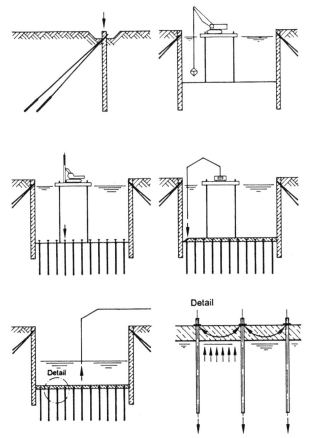

Fig. 2: Construction of the building pit „Potsdamer Platz"

This paper will report on preliminary tests concerning the concrete technology and procedural engineering for the construction of such an underwater, steel fibre reinforced concrete slab.

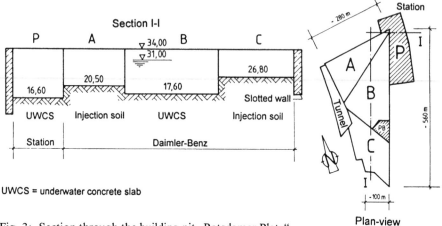

UWCS = underwater concrete slab

Fig. 3: Section through the building pit „Potsdamer Platz"

2 Approval for special cases

Under German building regulations, building materials or construction methods, which are not covered by the normal building regulations and/or codes have to be approved under the so called "special case approval regulations". This procedure makes sure, that these new building materials or methods conform with the regulations in existing building codes.

As there has been no previous experience about the handling of steel fibre concrete under such extreme circumstances, it had to be proven in a large scale test under building site conditions that steel fibre concrete can be produced and poured under water, satisfying all concrete technology and procedural engineering conditions.

This large scale test was carried out in the autumn of 1995. The following conditions for a special case approval had to be met:

- The large scale test had to prove, that the long pumping distances of up to 150 m and the relatively large water depth enable a proper concreting of the underwater slab, especially with regard to the concrete enclosure of the pile heads.

- Furthermore the properties of the different concretes have to be checked in such a way that a decision about the final concrete mix can be carried out with regard to the concrete material and workability properties.

- These tests had to be carried out for steel fibre concrete as well as for normal concrete.

- For this large scale test two test slabs per concrete mix, each test slab including two pile heads had to be concreted under water and recovered after the concrete hardening.

- The concrete mixes had to consider a possible high sulphate content of the ground water

These large scale tests were based on laboratory tests on point supported normal concrete and steel fibre reinforced concrete slabs, carried out at the Technical University of Braunschweig, in order to examine the load carrying and deformation behaviour of these slabs. It could be shown that for a steel fibre reinforced concrete slab the ultimate load carrying capacity, in comparison to normal concrete, could be doubled. Additionally the deformation values were 20 to 40 times larger compared to those of a normal concrete slab. This means that steel fibre reinforced concrete slabs have an extremely ductile deformation behaviour, which, as different ground movement in such a large building site can not be excluded, will add to the overall building element safety and reduce the construction risk of a sudden failure.

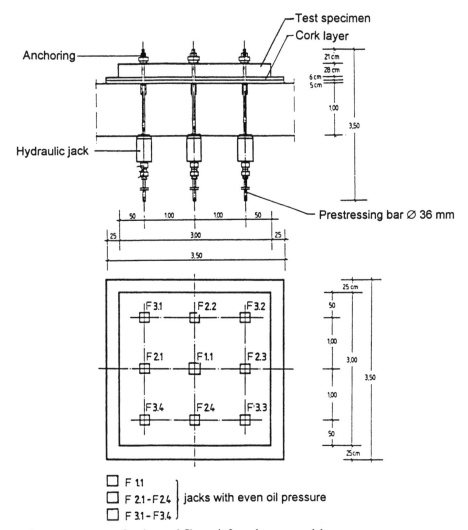

Fig. 4: Test set-up for the steel fibre reinforced concrete slab

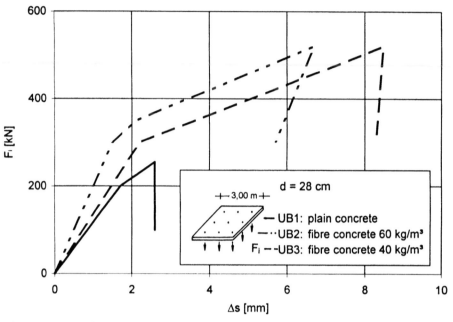

Fig. 5: Results of test loading - plate with plain and steelfibre concrete

Fig. 6: Failure with plain concrete

3 Concrete mix and production

For the preparation of these large scale tests, the basic suitability tests for different steel fibre concrete mixes were carried out at the central material testing laboratory of the Hochtief AG in Frankfurt. Basically, these mixes have a cement content of 200 kg/m^3 , a fly ash content of 150 kg/m^3, 210 l water per m^3 and an additive N 10 MC Bauchemie (fluidifier). Whereas the laboratory suitability tests proofed a perfect hardening behaviour of these concrete mixes, the first application under building site conditions showed considerable hardening delays for certain concrete mixes. Fig. 7 shows the steel container of one test specimen with the two anchors sinking in the building pit.

Fig. 7: Steel container for large scale test

From a concrete technological point of view there were strong reservations to name this test a success. The experts involved were of the opinion that after such a time a disintegration of the concrete structure would be unavoidable, especially with regard to the heavier steel fibres sinking to the bottom of the slab. Disregarding the hardening delay, the test result was nevertheless successful from the concrete technological and concrete engineering point of view. Even after a retarding time of 200 hours (one extreme case had 500 hours), the concrete structure did not show any signs of disintegration and the anticipated concrete strength was reached. Moreover, the fibre distribution was even over the cross section.

Fig. 8:   Cutting the test specimen with a diamond saw

With a slight modification of concrete mix, the main alteration being the concrete additive, it could be shown that steel fibre concrete with Portland Cement and sulphate resistant Portland Cement can be produced, pumped and poured successfully under the described building site conditions. Table 1 shows the concrete compositions of the tested concrete mixes.

4 Measurement programme

In order to record the concrete hydration process during the large scale test, the following temperature measurements were carried out:

- Partly adiabatic temperature measurements
- Temperature measurement in the underwater concrete
- Measurements on temperature regulated test specimens

These different measurement programmes will be discussed in the following.

Table 1: Concrete compositions

Concrete Mix	3106 a	3117 b
Concrete Strength	B 25 (Steel Fibre Concrete)	B 25 (Steel Fibre Concrete)
Cement Kind	CEM I 32.5 R	CEM I 32.5 RHS
Cement Content	280 kg/m^3	300 kg/m^3
Steel Fibres	Dramix ZC 50/0.60	Dramix ZC 50/0.60
Steel Fibres Content	40 kg/m^3	40 kg/m^3
Fly Ash-Content	220 kg/m^3	200 kg/m^3
Water Content	197 kg/m^3	207 kg/m^3
w/c	0.70	0.69
w/(c+f)	0.39	0.41
Gravel Size	0 - 16 mm	0 - 16 mm
Additive	VZ 0.4 % v.Z..	VZ 0.3 % v.Z.
	FM 1.5 % v.Z.	FM 1.1 % v.Z.
Consistancy	KR/F	KR/F
a_0	46 cm	50 cm
a	67 cm	75 cm

a_0 = before addition of fluidifier and fibres ,   a = after addition of fluidifier und fibres

4.1 Partly adiabatic temperature measurement

For all concrete mixes of the test specimens partly adiabatic temperature measurements were carried out. For these measurements 10 l of each concrete mix were filled into a specially insulated container and the time dependent temperature development was recorded. These partly adiabatic temperature measurements are used for a comparison of the temperature development of the different concrete mixes and for the conversion of mechanical concrete properties for different temperature conditions. With the aid of these calibrated insulated containers it is possible to calculate the adiabatic temperature curve, used for the later calculation of building member temperatures and the resulting stresses.

4.2 Underwater test specimen temperature measurement

For the determination of the temperature development in the underwater concrete of all test specimens, the time dependent temperature development and distribution over the cross section was recorded using thermo couples. These investigations are used for the determination of the temperature flux in the concrete and the calculation of the concrete strength development under water. Fig. 9 shows the increase of temperature during hydration process.

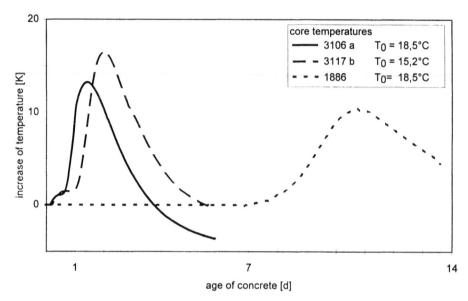

Fig. 9: Increase of temperature during hydration process

4.3 Temperature regulated test specimens

In context with the temperature measurements in the underwater test specimens, additional temperature regulated test specimens were produced. The temperature in these test specimens was regulated according to the temperature in the core of the underwater concrete. From these test specimens drill cores were taken at different times in order to determine the time and temperature dependent development of the mechanical concrete properties. These investigations supplied information for a material model, able to describe the concrete properties and hardening conditions of a concrete at early ages.

4.4 Mechanical properties

Parallel with the concreting of the underwater large scale test specimens, additional test specimens for the determination of the mechanical concrete properties were produced. The following cases have to be regarded:

- Standard test specimens, stored according to the conditions specified in DIN 1048 at 20 °C for the determination of the cube compression strength and flexural tensile strength for all concretes.

- Test on the drill cores taken from the temperature regulated test specimens for the determination of the temperature and age dependent cylinder compression strength, central tensile strength as well as the compression and tensile modulus of elasticity.

- Test on drill cores taken from the underwater test specimens for the determination of the cylinder compression strength as well as the compression and tensile modulus of elasticity in the actual test specimen after recovery from the building pit.

5 Workability proof

5.1 Test pumping

This test was designed to prove that it is feasible to pump steel fibre concrete in large quantities over a distance of 150 m.

In this test the concrete, after the addition of the steel fibres and a concrete fluidifier was pumped over a distance of 150 m through a straight pipe with a diameter of 125 mm. After this distance the pipe divided into two pipes with the same diameter and the concrete was pumped into a normal concrete pump, which finally pumped the concrete into the underwater large scale test specimens.

This test showed, that it is possible to pump steel fibre concrete over large distances without major difficulties. A concrete separation or other negative interferences resulting from the steel fibres did not occur.

5.2 Modified concrete mixes

As mentioned already in chapt. 3, some concrete mixes showed a strongly delayed hardening behaviour. In another large scale test a slightly modified concrete mix was used, i.e. the fluidifier FK 88 instead of N 10 was used. This concrete showed a perfect flowing property and hardening behaviour. The results of fresh concrete tests are given in table 2.

Table 2: Concrete properties of the modified mix

Concrete Mix	3106 a	3117 b
Date	24.10.1995	24.10.1995
a_0	46.0 cm	49.5 cm
a	67.0 cm	74.5 cm
T_0 (°C)	18.5 cm	15.2 cm

a_0 = before addition of fluidifier and fibres , a = after addition of fluidifier und fibres

6 Concrete hardening

Due to the delayed hardening of the original concrete mix for the underwater test specimens and the additional test specimens, the original time schedule for the determination of the time dependent concrete properties (after hardening) had to be abandoned. The additional test specimens and drill cores to be taken from the underwater test specimens could only be tested after the start of the hydration process, i.e. with a delay of approx. 14 days in the recovery of the underwater test specimens.

6.1 Drill core resutls from the underwater test specimens

In order to check the actual concrete properties of the underwater test specimens, drill cores with a diameter of 100 mm were taken from these blocks for subsequent testing. The test results are given in table 3.

Table 3: Drill core test results (underwater test specimens)

Concrete Mix	1885/95			1886/95		
	age (d)	fc (N/mm^2)	Mod. of E (N/mm^2)	age (d)	f_c (N/mm^2)	Mod of E (N/mm^2)
Cylinder Compression	15	17.3	23700	15	20.7	17100
Strength $f_{c,cyl}$	28	27.4	19800	28	36.3	19700
Central Tensile	15	2.0	-	15	1.5	-
Strength $f_{ct,ax}$	28	1.9	-	28	2.5	-

* Cylinder d/h = 10/20 cm   (Mean values of a series of 3 test specimens)

Concrete Mix	1910/95			2490/95		
	age (d)	fc (N/mm^2)	Mod. of E (N/mm^2)	age (d)	f_c (N/mm^2)	Mod of E (N/mm^2)
Cylinder Compression	14	11.9	20100			
Strength $f_{c,cyl}$	28	16.8	15700			
Central Tensile	-	-	-			
Strength $f_{ct,ax}$	-	-	-			

* Cylinder d/h = 10/20 cm   (Mean values of a series of 3 test specimens)

In comparison to the separately produced concrete cubes, stored under isothermal conditions, the drill cores taken from the underwater test specimens show a reduced concrete compressive strength. Nevertheless, the normal concrete and the fibre reinforced concrete fullfil the requirements for a concrete grade B 25. Compared to concrete stored under isothermal conditions, the strength development of fibre reinforced concrete is delayed.

The results of the strength investigation make it clear that the observed delay in the start of hydration process results in an other delay in the strength development. But the results show as well that the final concrete compression strength is not influenced by this delay. Apart from one concrete mix with a HS-cement (high sulphate resistant cement) all concretes tested showed the necessary cube strength for a concrete grade B 25 after 28 days. Furthermore, it can the assumed that due to the high fly ash content an additional hardening of these concretes will take place.

7 Evaluation of the underwater test specimens

This large scale test was designed in order to investigate the temperature development and the hardening behaviour of different concrete mixes intended to be used for the underwater fibre reinforced concrete slab in the building pit B debis, in order to prove the feasibility of underwater concreting with fibre reinforced concrete and to determine an optimal concrete mix for this purpose.

According to the results obtained so far, the concrete mixes No. 1886, 3106 a and 3117 b (see table 1 and 2) fullfil all requirements stated for underwater steel fibre reinforced concrete. The usefulness of these concrete mixes for underwater steel fibre reinforced concrete slabs was proven in the described large scale tests. Altogether, the material behaviour of a fibre reinforced concrete can be classified far above of a normal, plain concrete grade B 25.

The measurements show, that the underwater temperature has a certain influence on the hardening behaviour and strength development. For low water temperatures a delayed hydration process and strength development has to be considered. This can result in a later pumping out of the building pit, causing delays in the construction work.

The concreting of the underwater test specimens and the pumping tests showed, that only a comprehensive quality control mechanism ensures a perfect and faultless concreting, thus guaranteeing an even overall quality of the underwater concrete slab. Such a quality control system should not only comprise the concrete production and concreting, but also the supervision of the cement, fly ash and aggregate quality.

8 Summary

The positive test results obtained during this large scale test boosted the confidence of all parties involved (planning engineers, construction companies and building authorities) in the use of steel fibre concrete for underwater concrete slabs. Whereas for normal concrete underwater slabs sudden and unexpected failures have already occurred, the use of steel fibre concrete ensures that a possible failure will be announced by large deformations of the slab. This enables the use of countermeasures, in order to prevent additional damages and delays.

As countermeasures additional anchors or an injection of leaking slab parts are feasible. The main purpose in the use of steel fibre concrete has to be seen in reducing the risk involved in constructing such an underwater concrete slab to such a level that the need for possible countermeasures becomes unlikely. The first use of steel fibre concrete for such an underwater slab will contribute towards safer building pits and give an example for further construction measures of this kind.

If the construction of this building pit proceeds according to schedule, the pumping out of 350,000 m^3 of water is planned for April 1996. This opens the chance to report about this process at the symposium, even if this question is not directly connected with the material properties of the steel fibre concrete used for the construction of this building pit.

9 Literature

1   Falkner, H.; Teutsch,M.: Aus der Forschung in die Praxis. Beitrag zum Braunschweiger Bauseminar 1994, Heft 113 des iBMB der TU Braunschweig.

2   Falkner, H.; Huang, Z.; Teutsch, M.: Comparative Study of Plain and Steel Fibre Reinforced Concrete Ground Slabs. Concrete International, January 1995.

3   Falkner, H.: Bauwerke in Wechselwirkung mit dem Baugrund. Geotechnik 3/1994, Seite 121.

4   Winselmann, D.; Duddeck, H.: Gründungsmaßnahmen für die Verkehrsanlagen im zentralen bereich von Berlin. Beitrag zum Braunschweiger Bauseminar 1994, Heft 113 des iBMB der TU Braunschweig.

5   Fröhlich, H.; Mager, W.: Qualitätskontrolle durch baubegleitende meßtechnische Maßnahmen. Baumaschinentechnik 2/1995, Seite 26.

6   Karstedt, J.: Baugrund- und Gründungsgutachten Potsdamer Platz. Ingenieurbüro Elmiger & Karstedt, Berlin.

7   Falkner, H.: New Technology for the Potsdamer Platz. Beitrag zum Eurolab-Symposium 6/1996, Berlin.

8   Laube, M.; Onken, P.: Untersuchungsbericht Nr. 8900/95. Amtliche Materialprüfanstalt für das Bauwesen beim iBMB der TU Braunschweig (unveröffentlicht).

9   Falkner, H.; Baugrube Potsdamer Platz - Herstellung und Meßtechnik; 22. Lindauer Bauseminar 1996

10 Falkner, H.; Meßprogramm für die Baugrube Potsdamer Platz Berlin - Meßtechnische Überprüfung von Lastansätzen und Verformungen; VDI Berichte Nr. 1196; 1995

PART THREE
SPRAYED AND
VERY DRY
PRECASTING MIXES

9 SHOOTABILITY OF FRESH SHOTCRETE

D. BEAUPRÉ
Department of Civil Engineeering, Laval University, Ste-Foy (Quebec), Canada

Abstract

This paper presents a summary of the evolution of shotcrete technology. It describes the important properties related to both wet- and dry-mix shotcrete in the fresh state. It gives a definition of shootability that applies to both processes. Other testing procedures usually used on fresh shotcrete are described and discussed in an optic of future standardization. Rheological properties that can be used to give a numerical estimation of shootability are presented.

Keywords: Shotcrete, sprayed concrete, rheology, fresh properties, shootability, rebound, build-up thickness.

1 Shotcrete process

Shotcrete should not be considered as a special material, it should rather be regarded as a special process used to place and compact mortar or concrete. Over the years, different processes have been developed and all of them use compressed air to spray concrete or mortar onto a receiving surface. The two most popular methods are the wet and the dry processes. Hybrid processes have also been developed.

Production Methods and Workability of Concrete. Edited by P.J.M. Bartos, D.L. Marrs and D.J. Cleland. Published in 1996 by E & FN Spon, 2–6 Boundary Row, London SE1 8HN. ISBN 0 419 22070 4.

With the dry-mix shotcrete, compressed air is used to carry, at high velocity, a mixture of cementitious material and aggregates to a nozzle where some water is added. The amount of water is controlled by the nozzleman to obtain an appropriate consistency for the application. At the nozzle, in addition to water, it is possible to add other materials such as latex, air-entraining admixtures, mixed with the water.

With the wet-mix shotcrete, fresh concrete is pumped to the nozzle where compressed air is added to shoot the fresh mixture onto the receiving surface. Accelerators are sometimes added at the nozzle to increase the layer's thickness applied in a single pass and also to speed up the strength development.

With the development of new shotcreting equipment, other processes have emerged, though they are all more or less related to one or the other of the two original processes. All of these processes, including the "pure" dry-mix and wet-mix, produce the same final result: a stream of air, water, cementitious material and aggregates (which may also include fibers, additives and/or admixtures) projected at high velocity onto a surface

The difference between the wet and dry processes is the point where the air and the water are added to the mix to form the stream. Other processes can also be differentiated by the manner in which the components are introduced to form the spray. Most of these newer hybrid processes address some of the disadvantages of either the "pure" dry-mix or the "pure" wet-mix. They have originated from both the dry-mix and the wet-mix. Those originating from the dry-mix can be referred to as dry-to-wet-mix. Those originating from the wet-mix can be referred to as wet-to-dry-mix.

In the dry-to-wet-mix, the mixing time of the cement and the water is increased by adding the water at an earlier stage of the process. It is usually recommended that some water (between 3 to 5 % of the weight of the concrete mix) be added before the contact between the material and the compressed air, in order to reduce dust and to improve the homogeneity of the in-place material. This procedure is known as pre-wetting or pre-dampening. At present, the dry-mix without pre-wetting is not frequently used. Also, a special nozzle (with the water ring placed some distance from the nozzle end, and called "long nozzle") can be used to produce a more homogenous material, with less dust and less rebound, by increasing the mixing time of the cement and the water. This dry-to-wet-mix shotcrete maintains the advantage of having lighter hoses.

In the wet-to-dry-mix, the compressed air is added some distance from the end of the hose or directly at the pumping equipment. This process maintains all of the advantages of the wet-mix, plus the lighter weight of the hoses which is a characteristic of the dry-mix. For example, the "Top-Shot" method (Von Eckardstein, 1993) can be classified as a wet-to-dry-mix.

2 Definition of Shootability

All of these processes have the same final objective: the placement of well compacted cementitous material onto a surface. This material must be specially proportioned in order to stay in place right after shooting until setting takes place. It must also develop appropriate characteristics in its hardened state. The "stay in place" requirement is related to fresh properties and shootability.

Shootability is a property which incorporates parameters such as adhesion (the ability of plastic shotcrete to adhere to a surface), cohesion (the ability of plastic shotcrete to stick to itself and to be built-up in thick sections) and rebound (the material which ricochets off the impacted surface). The efficiency with which concrete can be applied is also dependent on the equipment used; this issue is not considered in this paper.

Shootability can be considered in terms of the efficiency with which a mix sticks to the receiving surface (adhesion) and to itself (cohesion). Thus, in this case, the build-up thickness is used to assess shootability: a mix that can be built-up to a great thickness in a single pass without sloughing will be referred to as possessing a good shootability. In other situations, where thickness is of less concern, the amount of rebound might be critical and use to assess the performance of the mix or its shootability: in this case, a mix with low rebound will also be referred to as possessing good shootability. Often, a mix with a good shootability requires low rebound as well as high build-up thickness.

3 Fresh shotcrete properties

From the above definition of shootability, one can feel the importance of build-up thickness and rebound characteristics of a mixture with respect to the efficiency of the shooting operations. Often, it is also important to consider other fresh properties or characteristics. For example, setting time can be a very important properties for application such as early ground support or repair within the tidal zone (to avoid washout of newly placed shotcrete). For placing purposes, important fresh properties include: build-up thickness, rebound, compaction, setting time, fiber content and rheological oriented properties. These properties, which are related to shootability, and their corresponding testing procedures will be described in the following sections.

At this point, it is important to note that most of these testing procedures are not widely standardized. For example, the present version of the "Guide to Shotcrete" of ACI 506 (Committee 506 of American Concrete Institute) does not possess any directive on how to measure the build-up thickness, the amount of rebound and the fiber content. In Norway, a document on underground application of shotcrete describes a washout test to measure the in-place fiber content of shotcrete and gives some guidelines for the determination of rebound. The European Committee for Standardization CEN/TC 104 on concrete has a sub-committee working on shotcrete (WG 10). It is the hope of the author that this sub-committee will come along with some standard procedures to measure the fresh shotcrete properties.

3.1 Build-up thickness

Vertical application of shotcrete is usually not a problem especially if the shotcrete is placed from the bottom in a self supporting manner. However, for overhead applications, this property is very important because it affects greatly the economical side of the application. Problems associated to low build-up thickness include: application in more than one layer, fall of fresh shotcrete or undetected debonding of fresh shotcrete creating voids behind the hardened shotcrete, especially when wire mesh is used.

Mix composition is the most important parameter controlling build-up thickness. This property is improved by the use of silica fume and set accelerator. Compaction process also increases the build-up thickness. For wet-mix, lower slump usually gives a higher build-up thickness. For dry-mix, water addition technique is also very important. The assessment of build-up thickness in numerical value is very dependent of the testing procedure used.

From the author knowledge, there is no standard test to measure this important characteristic. Depending on the type of application (vertical wall, overhead ceiling, presence of reinforcement, etc.), this thickness may vary for the same mixture. Thus, it is very difficult to provide a numerical (quantitative) build-up thickness value for a given mixture, especially when no standard testing procedure exists. However, several authors have tried to measure the build-up thickness.

Morgan (1991) defined a <u>thickness-to-sloughing</u> test measurement on wall (vertical build-up thickness). He observed two failure mechanisms which may cause the freshly applied concrete to behave in an unstable manner: adhesion failure and cohesion failure. As mentioned before, adhesion is the ability to adhere to another surface, while cohesion is the ability to adhere to itself. Adhesion failure occurs when, for a vertical application, the shotcrete starts sliding or sloughing under its own weight. Cohesive failure occurs when the fresh shotcrete ruptures within itself.

Figure 1 shows the set-up used by Morgan to measure the thickness-to-sloughing. Depending on the shape and the size of the base of the applied shotcrete, the measured thickness can be very variable. In case (a), because of a bigger base and a better shape, the shotcrete would probably exhibit adhesion failure compared to case (b) which is more likely to exhibit cohesion failure. In practice, most failures of fresh shotcrete are cohesion failures, especially if some sort of reinforcement is present. In this case, the shotcrete ruptures within itself at the level of the reinforcement or mesh.

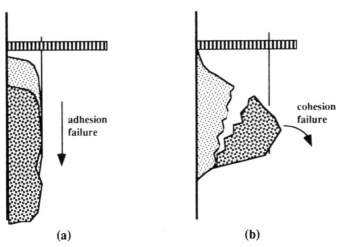

(a)   (b)

Figure 1: Thickness-to-sloughing test set-up

Beaupré and Mindess (1993) have used thea frame shown in Figure 2 to measure the vertical build-up thickness. This set-up consists in a fixed base of 200 mm x 230 mm mounted at 100 mm from the wall. Four steel angles are fixed to the base to prevent adhesion failure. During the test, the shotcrete is projected horizontally onto the base without moving the nozzle until the in-place shotcrete falls under its own weight. A video camera is used to record the test. This way, the build-up thickness can be determined precisely by reviewing the experiment.

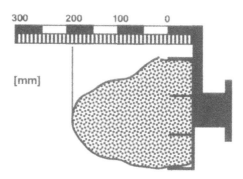

Figure 2: Set-up for vertical build-up test.

Overhead build-up tests can also be conducted but are more difficult to perform. The Industrial Chair on Shotcrete and Concrete Repairs of Laval University uses a frame as shown in figure 3. The complete test set-up and test procedure are described in the Appendix. That set-up also allows the evaluation of the build-up in term of "cohesion" of fresh shotcrete and not only in term of thickness. The "cohesion" of shotcrete represents, in some way, its fresh tensile strength.

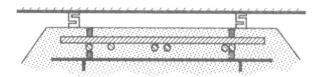

Figure 3: Set-up for overhead build-up test.

3.2 Rebound

During shooting, some particles called rebound (aggregates, cement grains, fibers ...) do not remain in place after the impact. The rebound is represented as a percentage of the total mass of shot material. Many parameters influence the amount of rebound. They have been studied by many researchers for the dry-mix shotcrete (Parker, 1977; Crom, 1981; Morgan and Pigeon, 1992; Banthia et al., 1992; Jardrijevic, 1993), and for the wet-mix shotcrete (Morgan, 1991; Beaupré et al., 1991; Morgan and Pigeon, 1992; Banthia et al., 1994).

The parameters influencing rebound can be separated into two categories: parameters related to the shooting technique, and those related to the mix composition. The most important parameters related to the shooting technique are: the process (wet-mix or dry-mix), equipment, air pressure, shooting position, angle, thickness and presence of reinforcement (bars or mesh). The most important parameters with respect to the mix composition are: workability of the mix, aggregate content and presence of silica fume and fibers.

From the author knowledge, no standard test exists to evaluate the amount of rebound. It is then very difficult to compare results obtained from different studies. To measure rebound, for vertical or overhead projection, most researchers have used closed chambers, large plastic sheets or, as in Bochum University (Germany), instrumented rebound collector.

When rebound is measured for payment purposes, the evaluation must be conducted in a way to represent the actual shooting configuration used during the real application. For example, the reinforcement, if any, should represent as close as possible the one of the in-place shotcrete. The test described in Appendix can also be used to measure rebound for repair applications. It can also be used to evaluate the reinforcement embedment characteristics of the mix-equipment-operator set-up. For mining application, rebound is often measured without any reinforcement. When rebound test is used to compare the influence of mix composition, it is very important to maintain the same shooting parameters for all mixtures.

3.3   Compaction

Compaction in not really a property of the fresh shotcrete. It is a reduction of the air content caused by the impact of the material upon the receiving surface or during pumping. This phenomenon is only significant for wet-mix because the determination of the fresh air content on dry-mix shotcrete is not possible before shooting.

The air content after shooting can be determine by using a modified ASTM C231 test (modified by shooting directly into the airmeter base). This procedure works well for workable wet-mix shotcrete but is not recommended for dry-mix shotcrete (Beaupré and Lamontagne 1995). One can then determine the amount of compaction that takes place while shooting wet-mix shotcrete. It is recognize that for non-air-entrained shotcrete, the final air content of the in-place shotcrete is always around 2-4 percent. When air-entraining agents are used, the final air content of the in-place shotcrete is around 3-6 percent, irrespective of the initial air content before pumping for the wet-mix.

Although the exact mechanism of air loss or compaction is not known, it is a fact that compaction increases the compressive strength and generally improves all mechanical properties. Physical properties, such as permeability, are also improved by compaction. It is probably the speed of the particles , which depends on the amount of compressed air used at the nozzle, and by their impact (the release of energy) on the receiving surface that produces the compaction by eliminating large air voids.

3.4   Setting time

At this point it is important to differentiate setting and hardening. When setting takes place, the concrete undergoes a rapid stiffening. After setting has taken place, the hardening or the gain in strength may be more or less rapid. The set accelerators used in shotcrete technology always affect the setting characteristics of the shotcrete, allowing application of greater thickness. Some of them also accelerate the strength development and can be considered as an hardener.

Different standard tests exist to measure setting time of paste or mortar. In shotcrete technology, the penetration needle test (ASTM C403) has been used the most. There should be some sort of relationship between the setting characteristics and the build-up thickness: is has been often observed that accelerated shotcrete can be build-up in very large thickness. It is also mentioned that accelerators help in reducing rebound, but this has not yet been demonstrated by strict measurements.

3.5 Fiber content

For fiber reinforced shotcrete, it is very important for job site quality control to measure the amount of fiber on the in-place shotcrete (Banthia et al, 1994). The Norwegian standard on shotcrete (Sprayed Concrete for Rock Support: NB #7) describes a washout test for steel fiber content determination. Beaupré and Lamontagne (1995) also used a similar test for steel fibers and a different wash out test for synthetic fibers.

3.6 Rheological oriented properties

The behavior of normal fresh cast in-place concrete can be represented by two rheological properties: the flow resistance and the viscosity (Tattersall and Banfill, 1983). Non accelerated fresh wet-mix shotcrete can also be represented with these parameters (Beaupré, 1994). Figure 4 shows a good relationship between g (flow resistance) and the maximum build-up thickness. In this Figure, the black squares refer to the flow resistance (g) obtained by considering a true Bingham behavior (linear relationship), while the white squares represent the real measured flow resistance (g'). Usually, g and g' are very close.

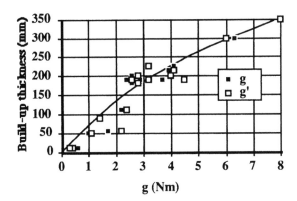

Figure 4: Relationship between the build-up thickness and the in-place flow resistance (g and g')

Dry-mix shotcrete is generally too stiff to be tested in a rheometer. Different penetration needle tests, similar to the setting time apparatus, have been used to estimate the mobility of in place stiff shotcrete. For example, Master Builders Technologies has developed a penetration needle test apparatus but this test is not yet standardized.

3.7 Pumpability

For wet-mix shotcrete, there is a permanent conflict between pumpability and shootability: when the pumpability increases (by increasing the slump, for example), the shootability decreases (smaller build-up thickness) and when the pumpability decreases, the shootability increases. It is always a challenge to find the best compromise between pumpability and shootability. Moreover, when this optimum is reached, one may avoid the use of accelerators or, for more difficult applications, it may at least allow a reduction in their addition rates.

There is not a single to measure pumpability because of the definition its definition which calls for many parameters: mobility and stability under pressure. Mobility can be evaluated by measurement of the rheological properties but the stability is more difficult to evaluate. Browne and Bamforth (1977) have developed a pressure bleed test in order to asses this issue but very few results have been published on this subject. Beaupré (1994) has try this test to asses stability of shotcrete mixture without much success.

4   Conclusion

It is obvious that the shotcrete industry needs more standard procedures to evaluate fresh shotcrete properties. Since research in shotcrete technology has increased a lot in the last five years there is hope that in the next few years, standard testing procedures will be adopted to evaluate shootability of shotcrete.

5   References

Banthia N., Trottier J.-F., Wood D. and Beaupré D, (1992), **"Steel Fiber Reinforced Dry-Mix Shotcrete: Influence of fiber Geometry"**, Concrete International, Vol. 14, No. 5, May, 1992, pp. 24-28.

Banthia N., Trottier J.-F., Beaupré D. and Wood D., (1994), **"Influence of Fiber Geometry in Steel Fiber Reinforced Wet-Mix Shotcrete"**, Concrete International, Vol. 16, No. 6, June, 1994, pp. 27-32.

Beaupré D., (1994), **"Rheology of High Performance Shotcrete"**, Ph. D. Thesis, University of British Columbia, Vancouver , 250 p.

Beaupré D., Pigeon M., Talbot C. and Gendreau M., (1991), **"Résistance à l'écaillage du béton soumis au gel en présence de sel déglaçants"**, Proceedings of the Second Canadian Symposium on Cement and Concrete, University of British Columbia, Vancouver , July 24-26, 1991, pp.182-196.

Beaupré, D., Lamontagne A. (1995) **"The Effect of Polypropylene Fibers and Aggregate Grading on the Properties of Air-Entrained Dry-Mix Shotcrete"**, Fiber Reinforced Concrete: Modern Developments, Second University-Industry Workshop on Fiber Reinforced Concrete, Toronto, Canada, 26-29 mars, pp 251-161.

Beaupré, D., Mindess, S. (1993) **"Compaction of Wet Shotcrete and its Effect on Rheological Properties"**, International Symposium on Sprayed Concrete, Fagernes, Norvège, 22-26 octobre, pp. 167-181.

Browne R.D. and Bamforth P.B., (1977), **"Test to Establish Concrete Pumpability"**, Journal of American Concrete Institute, May 1977, pp. 193-207.

Crom T.R., (1981), **"Dry-Mix Shotcrete Nozzling**", Concrete International, Vol. 3, No. 1, January, 1981, pp. 80-93.

Jardrijevic A., (1993), **"Decomposition of Shotcrete Mixes"**,Proceedings of the International Symposium on Sprayed Concrete, October 17-21, 1993, Fagernes, Norway, pp. 92-101.

Morgan D.R., (1991), **"Steel Fibre Reinforced Shotcrete for Support of Underground Opening in Canada"**, Concrete International, Vol. 13, No. 11, November, 1991, pp. 56-64.

Morgan D.R. and Pigeon M., (1992), **"Proceedings from the Half-day Presentation of the 4th Semiannual Meeting of the Network of Centers of Excellence on High-performance Concrete"**, Toronto, Ontario, Canada, October 6, 1992, pp. 31-56.

Parker H.W., (1977), **"A Practical New Approach to Rebound Losses"**, in ACSE and ACI SP-54, Shotcrete for Ground Support, Detroit, pp. 149-187.

Tattersall G. H. and Banfill P.F.G. (1983) **"The Rheology of Fresh Concrete"**, Pitman, London, 1983, 365 p.

Von Eckardstein K.E., (1993), **"Technology of the Top-Shot Wet-Mix Shotcrete System"**, Proceedings of the International Symposium on Sprayed Concrete, October 17-21, 1993, Fagernes, Norway, pp. 310-321.

APPENDIX

This appendix describes a test set-up used by the Industrial Chair on Shotcrete and Concrete Repairs (Laval university, Quebec, Canada) to evaluate shootability. The set-up allows the determination of rebound, build-up thickness and "cohesion".

Physical description

The test set-up consists in an instrumented mold mounted to simulate overhead shotcrete application into a rebound chamber (2,4 m x 2,4 m x 2,4 m). The steel mold is 600 x 600 mm and 3 mm thick. It has four sides inclined at 45° angle up to a height of 65 mm (see figure 5a).

The reinforcement is composed of ten 500 mm long reinforcement bars with a diameter of 15 mm and mesh (50 mm x 50 mm x 3 mm) as shown in figure 5b. The reinforcement is fixed to the mold by four bolts. The test can also be conducted without the reinforcement.

Three load cells with a capacity of 2,5 KN each are placed between the rebound chamber ceeling and the mold. A system of connectors allows easy removal of the mold with the help of a fork lift. The load cells are connected to an acquisition system an a video camera is also installed on the side of the rebound chamber to record the experiment.

Test procedure

The testing procedure allows the determination of both the amount of rebound and the maximum build-up thickness. This test is now used very intensively on many research projects.

To measure the amount of rebound, the weight of the empty mold is recorded with the load cells and the acquisition system. After starting the video camera, the shotcrete is sprayed into the mold until a even thickness of 90 mm is reached (several reference marks help to obtain an even thickness). The rebounds in the chamber are then collected. With the weight of rebound and the one of the shotcrete into the mold, it is possible to calculate the amount of rebound. Because of the position and the presence of reinforcement, high rebound rates are usually obtained.

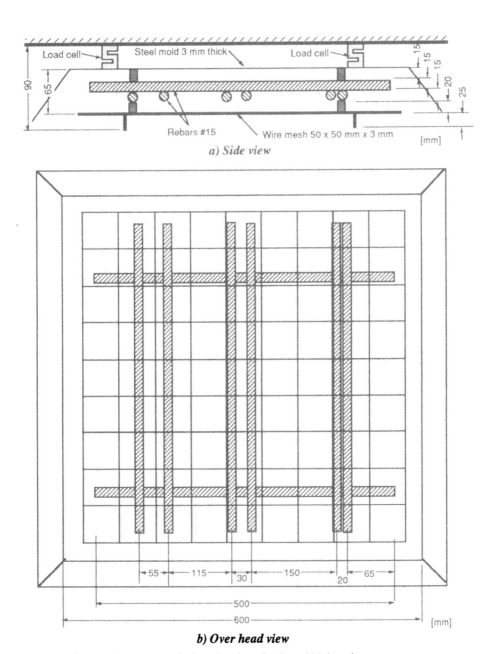

a) Side view

b) Over head view

Figure 5: Overhead test set-up designed and used at Laval University

Ten minutes after the rebound test has been completed, additional shotcrete is sprayed into the mold until the shotcrete falls under its own weight or until a thickness of 300 mm is reached. The maximum thickness is then assessed by the nozzleman. The exact thickness is also determined by reviewing the experiment on monitor. Because of the presence of the mesh, all failure are cohesion failure (not adhesion).

Because the acquisition system records the weight of the mold and the shotcrete, it is possible to evaluate the amount of shotcrete that fell from the mold. By evaluating the area of the ruptured fresh shotcrete, it is possible to calculate some sort of fresh shotcrete tensile strength or the "cohesion" of the fresh shotcrete.

10 PRODUCTION OF STEEL FIBRE REINFORCED SPRAYED CONCRETE AND THE INFLUENCE OF SETTING ACCELERATOR DOSAGE ON DURABILITY

A. MØRCH
Fundia Bygg AS, Oslo, Norway
J. LØKEN
Hamar-regionen energiverk, Hamar, Norway
T. FARSTAD and K. REKNES
Norwegian Building Research Institute, Oslo, Norway

Abstract

The control of the setting accelerator dosage is normally dependent on the operator's experience. This paper reports the results of an investigation of the influence of setting accelerator dosage on durability of wet mixed sprayed concrete used in tunnel linings. The experimental program includes tests on samples collected from tunnel linings in construction and from sample panels sprayed in the tunnel.

The setting accelerator used was a sodium silicate based accelerator. Some concretes were sprayed without setting accelerator, some with a normal dosage of setting accelerator and others with a very high dosage of setting accelerator. The concretes investigated had a compressive strength of approximately 60 MPa. The dosage of steel fibres in the sprayed concrete was 50 kg/m^3. All the concretes contained water reducing admixtures.

The following parameters were investigated: water intrusion, frost/salt scaling durability, pore structure and compressive strength.

The concrete strength decreased with increasing setting accelerator dosage.

An increasing dosage of setting accelerator caused a decreasing frost/salt durability and an increasing water intrusion. An increasing amount of cracks in the sprayed concrete were observed with increasing dosage of setting accelerator.

Keywords: Sprayed concrete, steel fibre, setting accelerator, durability (frost/salt scaling, water penetration, water porosity and compressive strength).

1 Introduction

Tunnel linings, like other large structures, are expected to be designed for maximum lifetime. However, little effort is often put into the documentation of the sprayed concrete and the actual durability of the structure is not known. This paper reports the results of an investigation of the influence of the setting accelerator dosage on the du-

Production Methods and Workability of Concrete. Edited by P.J.M. Bartos, D.L. Marrs and D.J. Cleland.
Published in 1996 by E & FN Spon, 2–6 Boundary Row, London SE1 8HN. ISBN 0 419 22070 4.

of wet mixed sprayed steel fibre reinforced concrete used in tunnel linings. The control of the amount of setting accelerator added to the concrete in the spraying nozzle is normally very poor, and adjustments of the dosage to the concrete flow is to a high degree dependent on the operator's experience. On this background the present investigation was carried out to evaluate the need of better dosage control equipment for controlling the amount of setting accelerator added during the spraying.

2 Experimental

2.1 Concrete mix design and mixing
The mix proportions are given in Table 1. The concrete mix was designed to give a compressive strength of 60 MPa without accelerator.

The concrete was mixed at a mixing plant, transported to the construction site and sprayed using the wet spraying technique. The steel fibres were added to the concrete during mixing at the mixing plant. The setting accelerator was added in the spraying nozzle during spraying. The amount added was adjusted by the operator.

2.2 Test procedures
The samples for concrete testing were cores drilled from the sprayed tunnel lining and from sprayed sample panels. The following were investigated:

- the influence of the setting accelerator dosage on the compressive strength
- the influence of the setting accelerator dosage on the steel fibre rebound
- the influence of the setting accelerator dosage on the water capillary absorption, air and total porosity of the sprayed concrete
- the influence of the setting accelerator dosage on the water intrusion depth of the sprayed concrete

Table 1 Mix proportions

Material	Dosage (kg/m$^{3)}$	
Cement (Norcem HS 65)	500	
Silika fume	30	
Water	234	w/(c+s) = 0.42
Aggregate 0-4 mm	1168	
Aggregate 4-8 mm	390	
Superplasticizer (melamin)	9	
Plasticizer (lignosulphonate)	4	
Steel fibres (EE 25)	50	
Setting accelerator (sodium silicate)	10 litres	
	20 litres	
	28 litres	

The testing was done using the test methods described in Table 2.

Table 2 Methods of testing

	Test method used
Compressive strength	Norwegian Standard, NS 3668
Frost/salt durability	Swedish Standard, SS 13 72 44 procedure A, with 3% NaCl
Porosity	[1]
Water intrusion depth	ISO/DIS 7031

The steel fibre content of the concrete was measured by washing the fibres out of the fresh concrete and by crushing the hardened concrete and collecting the fibres. The collected steel fibres were then weighed.

3 Results and discussion

The influence of the setting accelerator dosage on the compressive strength of the concrete is shown in Fig. 1.

A small dosage of setting accelerator, 10 l/m³, had no effect on the concrete strength. The concrete strength decreased with increasing dosage of setting accelerator, using more than 10 l/m³.

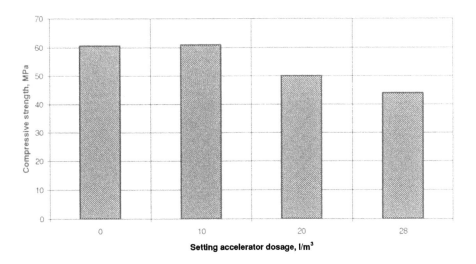

Fig. 1 Influence of setting accelerator dosage on the compressive strength of the concrete with a steel fibre dosage of 50 kg/m³. Samples drilled from sprayed sample panels

The influence of the setting accelerator dosage on the steel fibre rebound of the concrete is shown in Fig. 2.

An increase in setting accelerator dosage from 10 to 20 l/m^3 reduced the steel fibre rebound because the total concrete rebound was reduced.

The results of the frost/salt durability testing are shown in Fig. 3.

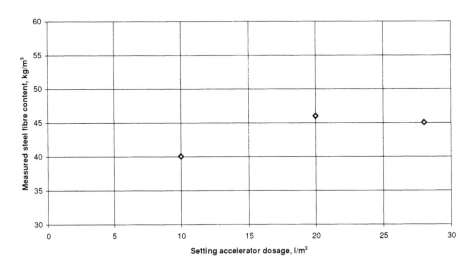

Fig. 2 Measured steel fibre rebound of the hardened sprayed concrete. Concrete sprayed on sample panels.

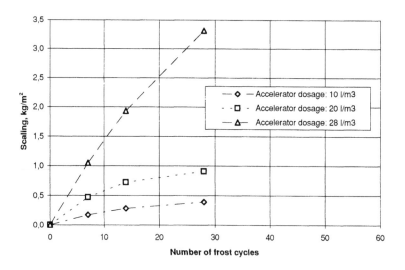

Fig. 3 The influence of the setting accelerator dosage on the frost scaling of the sprayed concrete. Samples drilled from the tunnel lining.

Only the concrete containing the lowest dosage of setting accelerator tended to an acceptable frost/salt durability. There was a pronounced increase in frost/salt scaling caused by an increase in setting accelerator dosage from 20 to 28 l/m^3, Fig. 3.

A normal dosage of setting accelerator is 20 l/m^3. A slight increase in dosage or an overdose of setting accelerator will cause a reduction in frost/salt durability of the concrete.

The results of the porosity testing are shown in Fig. 4.

The water capillary absorption (suction) porosity increased with increase dosage of setting accelerator, Fig. 4. The air porosity was not affected by the setting accelerator dosage, and the total porosity increased with increasing setting accelerator dosage because of the increased water capillary absorption porosity.

The results of the water intrusion tests are shown in Fig. 5.

In increase in setting accelerator dosage from 10 to 20 l/m^3 doubled the water intrusion depth, Fig. 5. Any further increase in setting accelerator dosage had no effect on the water intrusion depth.

On thin sections of the sprayed concrete an increasing amount of micro cracks in the concrete was observed with increasing dosage of setting accelerator. The thin-sections were prepared using the FLR-technique [2] which does not introduce cracks in the sample. This correlates with the increasing water capillary absorption porosity and the decreasing frost/salt durability of the concrete with increasing setting accelerator dosage. The increased water intrusion with the increased setting accelerator dosage is probably caused by the increased micro cracking observed.

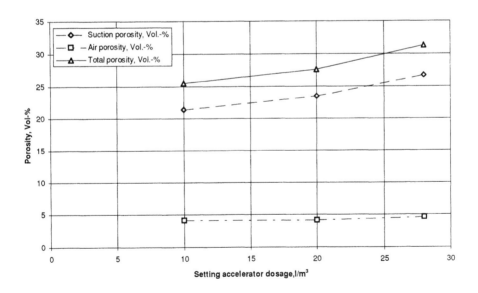

Fig. 4 The influence of setting accelerator dosage on the porosity of the sprayed concrete. Samples drilled from sample panels.

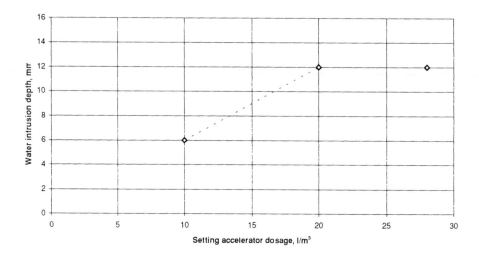

Fig. 5 The influence of setting accelerator dosage on the water intrusion depth
for the sprayed concrete. Samples drilled from sample panels.

4 Conclusions

The concrete strength decreased with an increasing dosage of setting accelerator dosage.

An increasing dosage of setting accelerator caused a decreasing frost/salt durability and an increasing water intrusion. An increasing amount of cracks in the sprayed concrete was observed with increasing dosage of setting accelerator.

To produce a durable and good quality concrete the addition of setting accelerator during spraying has to be controlled very precisely and accurately. The addition of setting accelerator has to be adjusted in proportion with the concrete flow.

5 References

1.   Sellevold, E.J. (1986) Herdet betong, bestemmelse av luft/makro og gel/kappilærporøsitet samt relativ bindemiddelinnhold, *Nordtestprosjekt 571-85, rapport 0 1731*, Norwegian Building Research Institute, Oslo.
2.   Gran, H.C. (1995) Fluorescent Liquid Replacement Technique. A Means of Crack Detection and Water:Binder Ratio Determination in High Strength Concretes. *Cement and Concrete Research*, Vol. 25, No. 5.

11 NEW LIQUID ALKALI-FREE FAST SETTING AGENT: A STEP FORWARD IN SPRAYED CONCRETE TECHNOLOGY

E. PRAT and L. FROUIN
Rhône-Poulenc Group, France
J. DUGAT
Bouygues Group, France

Abstract
Due to the dramatic increase in economic competition as well as safety and environmental concern, tunnelling and underground construction activities are currently confronted with major new technical challenges. In particular we now see very strong expectations for shotcrete technology to satisfy new requirements: better work conditions, increased spraying speed, improved bonding of shotcrete, early strength development, higher long term compressive strength. Traditional alkaline accelerators such as water glass and aluminates being unable to meet those requirements, demand for a better alternative is growing rapidly .
Rhone-Poulenc and Bouyges research teams have jointly developed a new alkali-free liquid accelerator, primarily designed for use in wet process. It is a fully stable, strictly inorganic, low viscosity, aqueous suspension which can be processed on any standard dosing equipment. Besides offering its users maximum operating simplicity thanks to its liquid form and security (as a product proven to be non-irritating and non-ecotoxic), this new accelerator also provides project engineers and contractors with significant additional benefits: improved quality of final concrete totally adapted for use in structural support, improved worker security allowing steady excavating progress thanks to the high early strength of the shotcrete ring. The paper will give a broad summary of the key experimental data characterizing the physico-chemical properties of this new liquid agent, its impact on the mechanical performances and the durability of shotcrete.
Keywords: accelerator, alkali- free, durability,.early strength, setting-agent, shotcrete.

Production Methods and Workability of Concrete. Edited by P.J.M. Bartos, D.L. Marrs and D.J. Cleland.
Published in 1996 by E & FN Spon, 2–6 Boundary Row, London SE1 8HN. ISBN 0 419 22070 4.

1 Introduction

Because of its advantages over traditionnal concrete regarding flexibility and productivity of casting, shotcrete is very popular in underground construction and has now been used for nearly one century. Both wet and dry spraying technologies have improved through the years, giving way to continuously enhanced economics of this concreting method.

Nowadays, if contractors are still very interested in further increase of productivity, they are becoming more and more concerned with better work conditions together with improved quality of the concrete.

Traditional alkaline accelerators such as water glass and aluminates do not meet those new requirements:

Although aluminates behave better than silicates regarding early strength development (security of working crew) and thickness of the shotcrete layer, it is well known that a reduction of the 28th days strength by more than 30% can not be prevented. Furthermore, because of its aggressive alkaline character, liquid aluminate is responsible for many serious injuries due to accidental contact with eyes or skin.

This is why demand is growing fast for a new generation accelerator that will be friendly to man and environment while:

- improving the bonding of shotcrete, to reduce rebound and improve thickness of the layer.
- enhancing the early strength development to insure maximum security at maximum spraying speed.
- keeping close to the 28 -day strength of non accelerated concrete in order to obtain high durability concrete.

Rhone-Poulenc and Bouyges research teams have recently developed a new liquid alkali-free accelerator, RP 750SC, that has been successfully tried on Bouyges working sites and in the Hagerbach cave.

This paper presents the specific characteristics of this new liquid alkali free accelerator together with lab tests that were designed to predict the behavior on the working sites. The complete comparative results of the shotcrete trials performed in Hagerbach with RP 750SC and a conventional liquid potassium aluminate are discussed.

2 RP 750SC main characteristics and functions

2.1 RP 750 SC : aspect, toxicity and eco toxicity

We have been particularly concerned with developping an accelerator friendly to man and nature: Trials performed with RP 750SC proved it perfectly respectful of the skin and eyes [1] and of environment in case of accidental spilling [2].

RP 750SC is a fully stable, strictly inorganic, low viscosity, aqueous blend that contains a precipitated silica with specific morphology and surface chemistry. Resulting from interaction of silica with other mineral compounds of the blend, RP 750SC behaves as a stable slurry that tends to gel at rest and becomes fluid again under mild stirring.

(viscosity = 0,10 - 0,2 Pa.s after 1 mn shear at 50 s-1).

2.2 Hagerbach trial: experimental results on site
Let us try to get a better view of the way RP 750SC interacts with shotcrete from the first minute following mixing until 90 days after the spraying.
The results discussed herebelow were obtained in the   Hagerbach cave   in the following conditions:

nature of concrete walls: hard rock cave
quantity of concrete sprayed: 25 m3
Shotcreting equipment: SCHWING - Topshot mobile, MEYCO - Spraying mobile
Ambiant temperature: 14 - 15°C
Fresh concrete composition

Cement: CEM I 52,5 ECLEPENS	425 kg	375 kg
Aggregates 0-8mm:	1748 kg	1844 kg
W/C:	0,5	0,54

2.2.1 From 1 mn to 20 mn after spraying: sticking and begining of setting.

The shotcrete obtained in the wet process, after mixing RP 750SC at the nozzle with fresh concrete is characterized by a high sticking power that allows to cast successfully with very few rebounds, even fluid concrete with slumps ranging from 18 to 25.
This particularity results from the combined action of silica with the other mineral compound of the blend (see paragraph 2.3).
Together with the sticking comes the begining of the setting due to the chemical reaction of RP 750SC with the C_3S and C_3A components of the cement paste.
This chemical reactivity can be monitored on site through temperature evolution, as displayed on *fig1*.
The combined properties of sticking and immediate setting allows the spraying of thick layers (up to 40 cm in vertical linings and 10-15 cm in cap) with low rebound (< 7%).

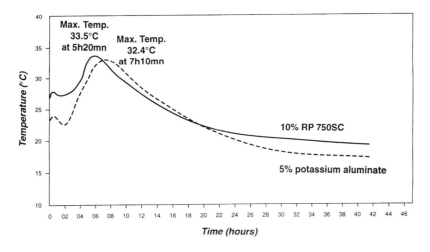

Fig. 1. Evolution of shotcrete temperature versus time. (Cement CEMI 52.5: 425 kg/m³.)

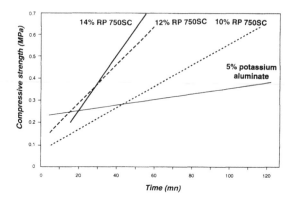

Fig. 2. Early age strength evolution. (Cement CEMI 52.5. 425 kg/m³. Method: 3 mm needle penetrometer.)

2.2.2 From 20 mn to 24 hours: early strength development.

The early strength development of the shotcrete was measured during the first two hours with a needle penetrometer (Dr Kusterle) of 3 mm diameter.
Fig 2 displays the comparative compressive strength obtained for shotcrete (425 kg cement/ m³) accelerated with various amounts of RP750SC and shotcrete accelerated with 5% liquid potassium aluminate:
From 30 mn on, one can notice that RP 750SC leads to compressive strength greater than those obtained with standard liquid aluminate.
Furthermore, the compressive strenth is all the more great in the early ages as the amount of RP750SC increases (from 10 to 12 and 14%), without impairing the final compressive strength of the shotcrete at 28 days .
Fig 3 shows the comparative early strength obtained with 12% RP 750SC by decreasing the cement content of the concrete from 425 kg/m³ to 375 kg/m³ .

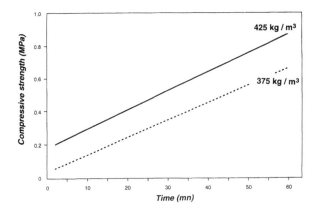

Fig. 3. Early age strength evolution. (12% RP 750SC. Method: 3 mm needle penetrometer.)

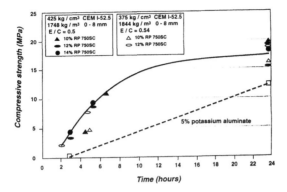

Fig. 4. Early age strength evolution. Incidence of cement and accelerator level.
(Method: Hilti gun.)

RP 750SC may allow a decrease of cement content without impairing early strength development, provided that the quality of cement be high enough.
From 2 hours to 24 hours, the compressive strength of the shotcrete is too high to be measured with a needle penetrometer, and we had to use a Hilti gun to measure the force needed to pull out nails.
Results obtained with varying amounts of RP 750SC for concrete with 425 and 375 kg cement per m³ concrete are compared to compressive strength development obtained with a 425 kg/m³ shotcrete accelerated with 5% standard liquid aluminate *(see fig 4).*
One can notice that compressive strength as high as 3 MPa at 3 hours, which insures security of the working crew, may be obtained with RP 750SC, while the aluminate activated concrete stays well below 1 MPa at 3 hours.

2.2.3 From 24 hrs to 90days: Almost no impairing of the final strength .

The final strength developments were followed by crushing core samples.
(f, h = 50mm). Comparative results displayed on *fig 5* show that RP 750SC does not significantly impair compressive strength of shotcrete at 28 days, even at dosage level as high as 14%.

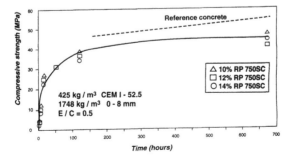

Fig. 5. Influence of RP 750SC level on final strength.

We obtain an average value of 85% of the compressive strength of unaccelerated concrete at 28 days, while aluminate accelerated concrete reaches 50% of the compressive strength of unaccelerated concrete (*cf fig 6*).

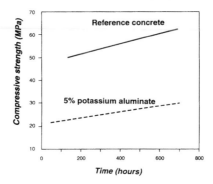

Fig. 6. Final strength of aluminate accelerated shotcrete. (Ref. concrete 425 kg/m³, CEMI 52.5 + 21 kg silica fume.)

2.2.4 Durability results:

Standard tests were performed in order to assess that the high values of compressive strength obtained (related to a densified microstructure) are indicative of a high durability concrete.

Permeability to water was measured according to DIN 1045 on core samples ($\varnothing$ = 200mm and h = 120mm): the depth of penetration of water ranged between 16 and 23 mm, which is indicative of a low permeability (see *micrograph 7*).

Resistance to accelerated carbonation was measured by exposing prisms (70/70/220mm) 28 days at a relative pressure of 70% CO_2: The depth of carbonation was 1mm to be compared to 5 to 10mm for a normal concrete.

Resistance to freeze thaw cycles was measured according to ANDRA 322 ET 09-03: After 25 cycles, the velocity of sound was not decreased by comparison with a non accelerated concrete sample

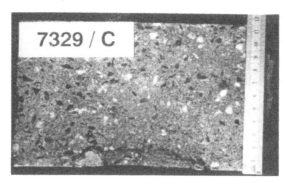

Fig. 7. Water permeability of shotcrete accelerated with 10% RP 750SC. Depth of penetration 16 mm.

2.3 How does RP750SC affect the rheological behavior, setting and final properties of shotcrete.

Immediately after spraying, shotcrete activated with RP750SC displays high cohesiveness and sticking properties, even before the setting takes place. This property results from the continuous mineral network of fine silica particles (*cf micrograph 8*), dispersed in the cement paste [3].

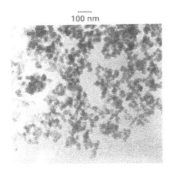

Fig. 8. Micrograph (TEM) of precipitated silica.

The setting starts immediately after spraying, due to chemical reaction of RP 750SC with the cement paste. One of our goal has been to measure accurately the stiffening of cement pastes in the first minutes following their mixing with RP750SC. This is of particular interest since immediate stiffening will guarantie sticking of the shotcrete to the wall.

This was achieved by using a Stevens texturometer, equipped with a one millimeter diameter needle. After thorough mixing of the cement grout with appropriate amounts of RP 750SC, the stiffening of the paste is measured by reading the resistance of the paste to penetration by the needle.

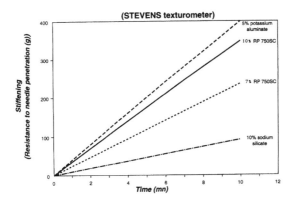

Fig. 9. Stiffening of grouts (w/c = 0.5) activated with RP 750SC and standard liquid alkali accelerators.

Fig 9 illustrates the behavior of cement pastes activated with varying amounts of RP 750SC, compared with that of a cement paste activated with classical liquid alkali potassium aluminate and sodium silicate.

It can be seen that stiffening may be modulated by adjusting the ratio of RP750SC to cement in order to match that obtained with. potassium aluminate in the first minutes..

On can notice that resistance to penetration by a needle of silicate activated paste stays well below RP 750SC and aluminate activated pastes.

In order to precise how RP 750SC interacts with the major components of cement, we used X ray spectroscopy to monitor the level of C_3S and C_3A in cement pastes activated with RP 750SC.

Comparative results with reference pastes are displayed in *fig 10*

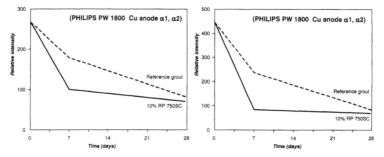

Fig. 10.a. Consumption rate of C_3S versus time in grouts (w/c = 0.5).

Fig. 10. b. Consumption rate of C_3A versus time in grouts (w/c = 0.5).

Although the amount of C_3S and C_3A consumption is equivalent at 28 days for both accelerated and unaccelerated concrete, RP 750SC increase their rate of consumption: 90% of C_3S consumption at 28 days for unaccelerated concrete is achieved at only 7 days for 10% RP 750SC activated cement pastes. These results confort the accelerating effect that could be observed in the early hours that could be evidenced by conductimetry measurements performed on grouts with a W/C = 4.

fig 11 shows comparative conductimetry results performed during the 5 first hours after beginning of hydration: The end of the sleeping period comes earlier by about 70 mn in the case of accelerated grouts.

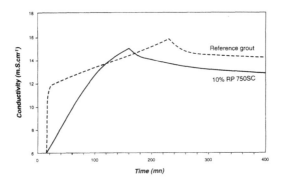

Fig. 11. Conductivity versus time of grouts (w/c = 5).

3 Conclusion

RP 750SC leads the way for new generation liquid alkali free accelerator that will improve security and handling operations of working crew, while increasing both productivity of casting and quality of final shotcrete:

RP 750SC presents a unique set of advantages:

Maximum security regarding health and environment, while easy to handle due to its liquid presentation.

The ability to satisfy the most demanding specifications regarding early ages or final strength development, while allowing variable cement and admixture dosage.

The insurance of optimum concrete efficiency thanks to high bonding to support.

The safety and durability of a superior quality concrete.

The authors wish to thank all the members of the research and application teams from Rhone-Poulenc and Bouyges who designed this product in partnership.

4 References

1. Directive 67/ 548/ EEC *Annex VI*
2. Journal of the European community *Method C2*, 29.12.1992
3. Iler, R. K. *The chemistry of Silica.* Wiley Interscience Publications.

12 VERY DRY PRECASTING CONCRETES

K.J. JUVAS
Partek Concrete Development, Pargas, Finland

Abstract
This paper deals with the special properties of dry concrete, particularly its workability. The exceptionally low workability of dry concrete might be a disadvantage, but it can as well create new opportunities for products, such as hollow-core slabs and concrete pipes. General requirements for testing methods are also presented. Four of them are discussed in more detail: Modified VB-test, Proctor test, Kango-Hammer test and Intensive Compaction test.
Keywords: Dry Concrete, Precasting, Workability, Testing Methods, Hollow-Core Slabs.

1 General

In some precasting processes it is normal to use a very low amount of water in concrete. A special method is needed to get such a concrete completely compacted. A typical water/cement ratio for such a concrete ranges between 0.20 and 0.40 depending on the amount and quality of fines and the admixtures used. The role of the water is mostly to take part in the hydration reaction. Very dry concrete has also been called by other names like no-slump or zero-slump concrete and earth moist or earth stiff concrete.

Very dry concrete is being used for many purposes in the precasting industry. Typical products are:
- extruded or slipformed hollow-core slabs
- several types of blocks, kerbs and tiles
- pipes
- slipformed thin-shell slabs

Production Methods and Workability of Concrete. Edited by P.J.M. Bartos, D.L. Marrs and D.J. Cleland. Published in 1996 by E & FN Spon, 2–6 Boundary Row, London SE1 8HN. ISBN 0 419 22070 4.

In other areas of the construction industry very dry concrete has been used in roads, pavements and massive dams. The concrete is usually named according to the compacting method i.e. Roller Compacted Concrete (RCC).

2 Character of workability

By using standardized tests or normal compaction methods the dry concrete can be considered to have a very poor workability. It cannot be compacted with a poker vibrator nor with a normal vibrating table. It is also difficult to pump and the aggregate has a tendency to segregate.

On the other hand, the cohesion of compacted concrete is mostly very good. The cohesion is good provided that the amount of cement paste (cement + water + fine aggregate) is right. The amount of cement paste can be minimized by developing very dense aggreagate packing and measuring the void content in compacted aggregate.Though in most cases the workability will be lower. If the amount of cement paste is insufficient, there will be difficulties during compacting and finishing of surfaces. A certain overfill of cement paste is required and after investigations /4/ the sufficient amount is 5...10 % in excess of the void content in compacted dry aggregate.

All production methods, from mixing of concrete to finishing of surfaces, must be designed specifically for the particular workability properties of dry concrete.

A comprehensive presentation on workability properties is given in the doctoral thesis of Christian Sørensen /12/.

Sørensen finds the following special characteristics for fresh dry concrete:

- resemblance to earth material. Like in soils the workability can be controlled by optimum moisture content and density in the uncompacted and compacted state.
- if the cement-paste is being increased beyond a certain limit, the concrete will become unstable.
- in the fresh state, interparticle forces resist the dispersion of fines and compaction of the granular material, but provide the desired stability. The surface forces are due to the air-water menisci and the polarity of fine particles.
- the amount of water required for good workability is very sensitive to the amount and quality of fines.
- there are three principal methods of machine compaction: vibration, dynamic pressure and plain static pressure. Of these, plain static pressure is the least effective method.

Sørensen has concluded in his investigations that higher air content increases the fresh strength (cohesion) of no-slump concrete. The main reason may be the effect of air/water-menisci. The same improving effect occurs also if part of the cement is replaced by an equal volume of inert filler which has coarser grading than cement. With silica fume a higher replacement is needed to achieve the same effect. Workability (IC-tester) and obtainable relative solid density are dependent on the yield strength rather than on the viscosity. Workability (IC-tester) and fresh strength indicate an inverse relationship.

Silica fume functions as a lubricant and decreases the friction in shear, provided a sufficient amount of superplasticizer is available for adequate dispersion.

Norden /13/, in his studies concerning workability of dry concrete used in pipe production, concluded that the workability can be improved by:

- adding silica fume (2.5%)
- pre-moistening of sand
- using cubic aggregate shape which decreases the required cement paste amount, if the aggregate is crushed
- using washed sand

3 Benefits of dry concrete

Why then has dry concrete been used despite the workability problems? Because the other properties are extraordinary. The cohesion and the fresh strength of newly compacted concrete is very good. Many production processes find this property necessary. It must be possible to remove pipes from their moulds right after compacting and to move them to the curing area in vertical position. Hollow-core slabs must keep their shape without any supporting mould around. It is even possible to walk on the fresh, compacted concrete directly behind the casting machine without causing any damage to the product.

The development of early strength is very fast, and the ultimate strength achieved is high compared with the amount of cement and other binding materials included. Of course the concrete must be properly compacted, which is even more important with dry concrete than with flowable concrete. The concrete always has extra strength potential to attain if its water/cement ratio can be lowered. The matter is visualized in Fig. 1 /2/. An example of typical strength development of an extruded hollow-core slab is presented in Fig. 2. /4/

Hardened concrete with a low water/cement ratio is very dense and contains no capillary pores. This results to good durability both against frost and chemical attack without any special admixtures. The concrete is also impervious to water, and the shrinkage is smaller, which will minimize the risk of cracking.

4 Testing of workability

4.1 General requirements
The term workability is very wide and can refer to different kinds of properties depending on the process, product and the phase of work. The more detailed and practice-based terms for different aspects of workability are:

- cohesion
- tendency to segregate and bleed
- compactibility
- finishability
- pumpability

For dry concrete the three first properties are the most important.

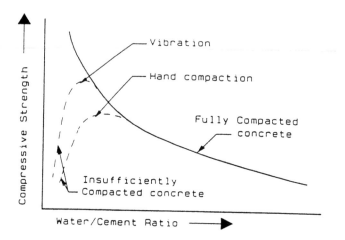

Fig.1. The effect of water/cement ratio and different compaction methods on the compressive strength of concrete /2/ .

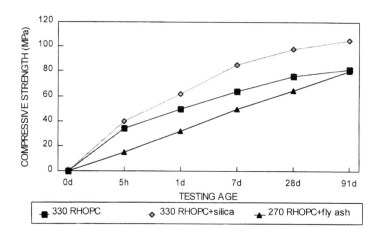

Fig. 2. The strength development of three different hollow-core slab concretes.
The water/cement ratio is 0.34 /4/.

There is probably no test method which could reliably characterize all the workability
properties. First it must be clarified which properties are important to know in the
particular case and choose the test method accordingly.

It is important that the test method simulates the real production process as good as possible. If the concrete is to be compacted by vibration, the compactibility must be tested by a vibrating type of test method. For the practical user of the test results it is not necessary that these are on rheological base. The most important is that they describe reliably the concrete and the differences between mixes in the particular process.

A test method is found suitable if:
- repeatability and reproducibility are good
- the test is fast and simple enough to be used in a process control
- the equipment is rather inexpensive
- the method simulates the real process and it is possible to be modified according to variations in the process.

The optimum compaction is an absolute condition for reaching good properties in dry concrete. It is very important to be able to test even small changes in workability which can be caused by changes in fine materials, water, temperature and shape of coarse aggregate.

Among conventional test methods there are very few suitable for dry concrete and even less are internationally standardized. Four test methods are presented in the following chapters:
- modified VB-test
- Proctor compaction test
- Kango-Hammer test (Vibrating hammer test)
- Intensive Compaction test (IC-test)

Some further tests are mentioned in the literature /3/, /7/, /13/17/.

4.2 Modified VB-test
Origin and principle
The VB-test was developed in Sweden by V. Bährner and the modification by CBI (Cement and Concrete Institute) and has been standardized in Sweden (SS 13 71 31). It differs from a normal VB-test by the use of two weights, 10 kg each, which make the compaction more effective and faster. The weights press the transparent plastic disc towards the concrete while it is subject to simultaneous vibration from below. /3/

Application
The test is applicable to dry concretes, particularly when the compaction in the real process takes place during simultaneous vibration and pressing. This type of process is used in the production of pipe and pavement blocks. The original VB test is not accurate enough when the concrete is dry, and the test lasts longer than 30 seconds to administrate.

Description
Normal VB-equipment with two weights totalling 20 kg and hanging over the plastic plate (Fig. 3 /4/) are needed. The sample size is about 6 litres.
The test procedure is similar to that of the normal VB-test. The time the plate needs to be entirely in contact with the concrete is measured.

Interpretation of the test results
The test result shows the compaction energy by vibration and pressing that the concrete sample requires.

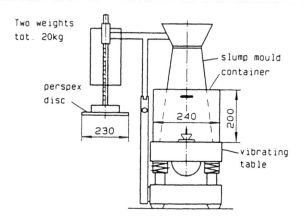

Fig. 3   Modified VB-test equipment /4/.
The relationship between slump, normal VB-test and modified VB-test is shown in Fig. 4 /4/, /5/. Also the relationship between IC-test and modified VB-test has been studied and is presented in Fig. 5 /4/.

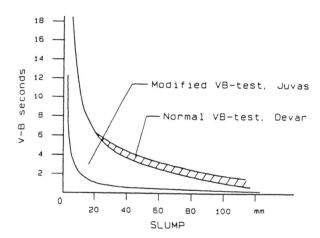

Fig. 4   Relationship between slump, normal VB-test and modified VB-test according to Devar /5 / and Juvas /4/.

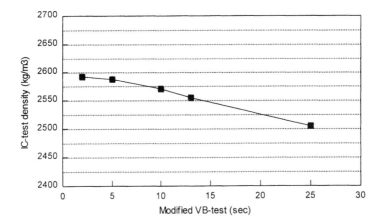

Fig. 5   Relationship between IC-density (80 compacting cycles with 4 bars pressure) and modified VB seconds. Concrete mix with 330 kg/m³ with water cement ratios of 0.25...0.45.

Precision
The timing is quite accurate, except for mixes which are too flowable (less than 1 sec.) and too dry (more than 20 sec.). The optimum result area is between 3 to 10 sec.Very small changes in the concrete mix are not observed in the results.

Advantages
The equipment is reliable and easy to use, even under practical production conditions. The procedure simulates the compaction method used in producing pipes and blocks. Both the pressure and the vibration are included in the test.

Disadvantages
The test result depends somewhat on the care and attention of the person performing the test. Very small changes in mix design will not be shown in the result and the accurate area of results is quite limited. The equipment is rather expensive.

4.3 Proctor-compaction test
Origin and principle
The test has been developed in the USA and was originally used for testing different soils. The target is to determine the water content providing the best compaction with constant compacting energy by tamping. The method is standardized in many countries such as the USA, Germany and Sweden. /3/.

Application
The test has mostly been used with lean dry concrete mixes for dams, roads, pavements and ground stabilizing concrete.

Description
In the modified test the sample is compacted in 5 layers, each with 25 strokes, using a 4.5 kg weight. The weight moves along a rod and the drop height is 45 cm. The principle of the standard and modified test is presented in Fig. 6 /3/. The density will be calculated on the basis of the weight, volume and water content of the sample. By changing the water content it is possible to determine the maximum density and the corresponding water content. Normally four to six tests are needed. The sample size for each test is about 3 kg. Both manual and mechanized equipment is available.

Interpretation of the test results
The test result correlates best with the properties of aggregate, such as differences in fine filler and in the shape of coarse aggregate particles. Very few comparisons between Proctor and other test methods are available. In their study, Törnqvist and Laaksonen /6/ have concluded that the Proctor compaction test and the IC-test (detailed description follows), show a relatively close resemblance. The relationship between the Proctor-test and the IC-test has been compared and presented in Fig.7 /4/.

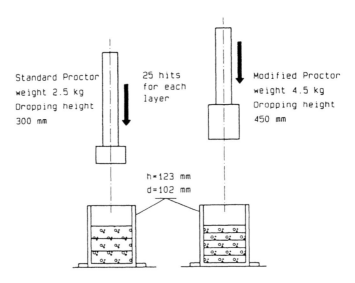

Fig. 6.   Principle of the Proctor compaction test /3/

Precision
The precision is accurate enough for production control. There is some sensitivity depending on who performs the test, i.e. how evenly he compacts the sample by strokes.

Advantages
The test is simple to perform both in the laboratory and the factory. The manual device is portable and inexpensive. The optimum water content is received as test result.

Disadvantages
This method does not consider the effect of vibration on the compaction. The test requires significant time and effort.

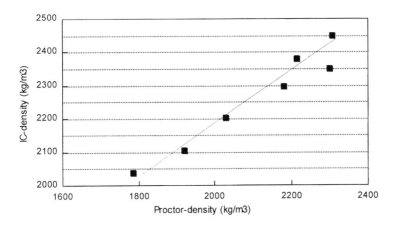

Fig 7. The relationship between the maximum density achieved with the IC-method and Proctor-method.

4.4 Kango-Hammer

Origin and principle
The method is based on the standards BS 1924:1975 "Methods of test for stabilized soils and BS 1377:1975 "Vibrating hammer method". The sample in a cubic or cylindrical steel mould is compacted by a constant pressing and vibrating force. The density of the compacted sample is then determined. The sample can be cured and compressed at a later age /3/, /7/.

Application
The method has usually been used for controlling the workability of dry road concretes.

Description
The sample size is 6 to 9 kg depending on mould type. The sample is compacted in 2 to 3 layers with a vibrating hammer type Kango 900, 950 (or the like). Pressure, vibrating time and frequency are constant. The calculation of the density is similar to that used in the Proctor compaction test. The equipment is illustrated in Fig. 8. /3/,/7/.

Interpretation of the test results
The test results have been found to correlate well with the compaction test results of samples taken from road pavement works where roller compacted concrete has been used /7/.

Precision
With this test, differences can be measured to a sufficient degree for practical use.

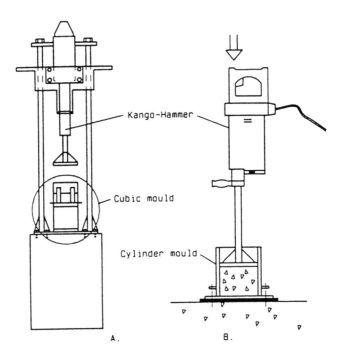

Fig. 8   Two types of Kango-hammer.
         A for cubic mould /3/, B for cylinder mould /7/.

Advantages
The test is easy to perform. The equipment is not very expensive. The test includes both the pressure and the vibration.

Disadvantages
The test is not accurate enough for detailed laboratory investigations.

5 Intensive Compaction test (IC-test)

Origin and principle
The method and equipment has been developed by I. Paakkinen 1984 in Finland and accepted as a Nordtest method (NT BUILDT 427). A small amount of concrete is compacted under constant compressive force and shearing motion.

The device measures the increasing density of the sample during the test. The fresh strength of the cylindrical specimen can be tested separately immediately after compacting, or the specimens can be cured and tested later. /8/ It is easy to produce several similar specimen with the constant and automatic compacting process. It is also possible to test aggregate gradings by compacting dry aggregate samples and to calculate the volume of voids left in the compacted sample.

Application
The test procedure simulates, for example, the production of extruded hollow-core slabs. This method is applicable to concrete of zero slump consistency or a consistency resulting in Vebe test of more than 20 seconds. During the test no vibration is used by the equipment.

Description
The tester has a control unit and a main unit, shown in Fig. 9.

Fig. 9   Intensive Compaction Tester /8/.

The different stages of the workability test are shown in Fig. 10.
The sample is placed in the work cylinder of φ 100 mm and its weight is recorded. The normal sample amount is 1900 to 2100 g. Test parameters are entered in the control unit. The compaction pressure and the number of compaction cycles can vary, normally they range from 0.2 to 0.4 MPa (2 to 4 bar, gauge reading) and 80 to 160 cycles, respectively. A sample is pressed between the top and the bottom plates in the work cylinder. The top and the bottom plates remain parallel, but the angle between the plates and the work cylinder changes constantly during the circular motion. The cyclic shearing motion together with the axial compressive force allows particles to realign into more favourable positions and air is forced from the sample. The tester continually measures the height of the concrete cylinder. During testing the control unit continues to calculate density on the basis of the height and the weight of the sample and as a function of the compaction cycles.

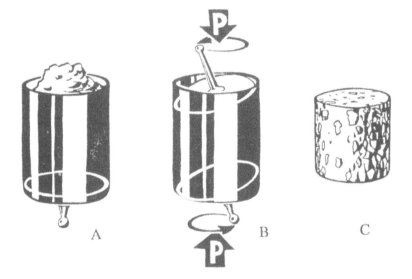

Fig. 10. Test stages /8/.
   A Weighed sample in work cylinder.
   B Sample under compaction.
   C Sample removed from the test cylinder.

After completed compaction the control unit prints out a test report. The sample's density is printed out after 10, 20, 40, 80, 160 etc. compaction cycles. The number of cycles can be adjusted for complete compaction of the sample. The sample is removed from the work cylinder and can be used for further testing, e.g. fresh strength testing, or cured and tested at later age. In a research model it is possible to measure the shear resistance of concrete samples. The maximum value of the shear resistance is the point of compaction, where cement paste starts to compress out from the specimen. The shear resistance decreases remarkably after that.

It is likewise possible to compact the samples to a constant target density. The test result shows how many compaction cycles are required in order to reach the target density /9/. Compaction curves of typical mixes are presented in Fig. 11 /10/.

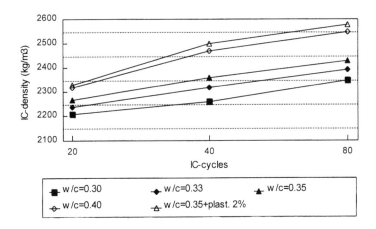

Fig. 11 The effect of water/cement ratio and superplasticizer amount on workability /10/.

Interpretation of the test results
The test results describe the compaction work which each concrete mix will need for complete compaction. Comparisons have been made between the Kango-hammer test /7/, Proctor test /6/ (Fig. 7), /7/, modified VB-test /4/(Fig. 5) and Tattersall's 2-point test with planetary motion impeller /4/. Correlation is good with all devices except the 2-point tester.

Precision
The device is able to identify a change in the compaction properties caused e.g. by an increase or decrease of three to five litres of water per cubic meter of concrete. It does not matter whether the water originates from the aggregate or whether it has been added in the concrete mixer. During the preparation of the Nordtest standard, a Round Robin test was arranged and the results of repeatability and reproducibility were at an acceptable level /14/.
Several researchers such as, Norden /13/ , Alasalmi /15/ and Johansen /16/ have found the apparatus to be reliable and having good correlation to properties in real production with dry concrete.

Advantages
The method is very accurate and simulates the extruded shear-compacting process and the roller compacting process. It can be widely used for quality control, mix design and research work. The compacting in the test can be adjusted after the real production process.

Disadvantages
The equipment is quite sophisticated and therefore rather expensive. The effect of vibration is not considered in the test.

6 Summary

Very dry concretes are particularly sensitive to a number of factors which affect to compaction and stability. Among these are overall grading, fines content, aggregate particle shape and both water and air contents. Four test methods for the design and control of this type of mixes are reviewed and are summarised as follows:

TEST METHOD	PRECISION	ADVANTAGES	DISADVANTAGES
MODIFIED VB	Reasonable	Easy to perform Pressure and vibration	Limited area of results Quite expensive
PROCTOR	Good	Easy to perform Optimum water content as result	Significant effort required No vibration
KANGO HAMMER	Reasonable	Easy to perform Pressure and vibration	Do not separate small differences
INTENSIVE COMPACTION	Very good	Adjustable after the casting process Gives remarkable lot of information about the workability	Expensive No vibration

References

1   Schwartz, S. (1984) Practical hollow-core floor slab production below 85 dB(A), Betonwerk und Fertigteil, 12, 807-813.

2   Neville, A. (1981) Properties of Concrete, London, Pitman 779 pp.

3   Andersson, R. (1987) Beläggningar av vältbetong (Concrete Pavements using Roller Compacted Concrete), CBI, Report number 87031, Stockholm, 59 pp.

4   Juvas, K. (1987) Jäykkien betonien työstettävyyskokeita (Workability tests with stiff concretes), Partek Concrete laboratory report (unpublished) Parainen, 10 pp.

5   Dewar, J. (1964) Relation between various workability control tests for ready mix concrete, C & CA Tech.Rep. 42, London 375 pp.

6   Törnqvist, J., & Laaksonen, R. (1991) ICT-koe maksimikuivatilavuuspainon mittauksessa (IC-test in measuring the maximum dry weight of volume), VTT Report TGL 1880, Espoo, 17 pp.

7   Magerøy, H. (1987) Kontrollmetoder för valsebetong. Praktiske forsök ved Tosentunneln (Control methods for roller concrete, practical tests at the Tosentunnel), Norcem Cement A/S, Oslo, 21 pp.

8   Paakkinen, I. (1986) Intensive compaction tester device for testing the compactability of no-slump concrete. Nordic Concrete Research Publication No 5, Oslo, 109-116.

9   Juvas, K. (1988) The effect of fine aggregate and silica fume on the workability and strength development of no-slump concrete, Master's Thesis, Helsinki University of Technology, 124 pp.

10   Juvas, K. (1990) Experiences to measure the workability of no-slump concrete, in Proceedings of Conference of British Society of Rheology and Rheology of Fresh Cement and Concrete, University of Liverpool, London, 259-269.

11   Juvas, K. (1990) Experiences in measuring rheological properties of concrete having workability from high-slump to no-slump, in Proceedings of RILEM Colloquium on Properties of Fresh Concrete, University of Hannover, London, 179-186.

12   Sørensen, Chr. (1992) Zero-slump concrete, Doctoral Thesis, University of Trondheim, 185 pp.

13   Norden, G. (1992) Komprimering av torrbetong (Compaction of dry concrete) Master's Thesis, University of Trondheim, 128 pp.

14   Nordtest, (1994) Nordtest Method (NT BUILD 427) for Fresh Concrete: Compactibility with IC-tester, Espoo, 4 pp.

15   Sarja, A., Alasalmi, M. (1988) Shear-Compactability Tests of Fresh Concrete for Hollow-core slab production, Research Report No. BET8964, VTT, Finland Espoo.

16   Johansen, Kj., Johansen, K. (1995) Tilpasning og bruk av maskinsand til betong (Suitability and use of crushed sand in concrete). Presentation at Norwegian concrete days 1995, 12 pp.

17   Bartos, P. (1992), Fresh Concrete. Properties and Tests, Elsevier, Amsterdam, 292 pp.

13 MECHANICAL BEHAVIOUR OF VERY DRY MIXED CONCRETE UNDER CONCRETE PLACEMENT

Y. MURAKAMI
Technical Research Institute, Hazama Corporation, Tokyo, Japan
T. TAUTSUMI and N. YASUDA
Tokyo Electric Power Incorporated (TEPCO), Tokyo, Japan
M. MATSUSHIMA
Tokyo Electric Power Services, Inc. (TEPSCO), Tokyo, Japan
M. OHTSU
Department of Civil Enginering, Kumamoto University, Kumamoto, Japan

Abstract

Very dry mixed concrete has a low unit quantity of water and unit quantity of cement, and the slump value, from which the consistency of concrete is generally determined, is 0 cm. As this concrete can be placed in large quantities, it is suitable for gravity dams and for pavements.

Unlike ordinary concrete, where compaction is achieved with compaction vibrators, compaction is carried out using vibrating rollers. The degree of compaction is evaluated from the density ratio, which is the relative density, measured using a radiation density gauge, to the theoretical density. The measurement of the density using the radiation density gauge is performed after completion of the compaction. As a result, the procedure does not provide a real-time evaluation.

The compaction of very dry mixed concrete has been studied in the past at the laboratory level, under conditions considerably different from those existing on site.

This research was carried out to investigate the compacted behaviour of very dry mixed concrete by performing on-site trials using a vibrating roller equipped with an electronic probe, accelerometer, and an acoustic emission (hereafter referred to as AE) sensor. The primary fundamental frequency, Fourier amplitude and distortion factor obtained from measurements of subsidence and acceleration, as well as measurements of the AE, were investigated to determine their relationships with the compaction of very dry mixed concrete.

Keywords: Very dry mixed concrete, concrete placement, subsidence, accelerometer, acoustic emission.

Production Methods and Workability of Concrete. Edited by P.J.M. Bartos, D.L. Marrs and D.J. Cleland. Published in 1996 by E & FN Spon, 2–6 Boundary Row, London SE1 8HN. ISBN 0 419 22070 4.

1 Introduction

Very dry mixed concrete has a low unit quantity of water and unit quantity of cement, and the slump value, from which the consistency of concrete is generally determined, is 0 cm. As this concrete can be placed in large quantities, it is suitable for gravity dams and for pavements.

Unlike ordinary concrete, where compaction is achieved with compaction vibrators, compaction is carried out using vibrating rollers. The degree of compaction is evaluated from the density ratio, the density measured using a radiation density gauge to the theoretical density. The measurement of the density using the radiation density gauge, is performed after completion of the compaction, so therefore is neither applicable to an evaluation in real-time nor at all locations.

It is known that the use of a vibrating roller for the compaction of very dry mixed concrete achieves compaction by increasing the density. It has been reported by Nakauchi et al[1], from results achieved by embedded accelerometers and pressure gauges in very dry mixed concrete, that when accelerations and pressures in excess of 0.2 G and 0.5 kgf/cm² respectively are induced by the vibration, adequate compaction is achieved. Kurita et al[2] reported that there is a linear relationship between vibrational energy and settlement, and confirmed a correlation between the quantity of vibrational energy and the efficiency of the compaction. Kagatani and Tokuta et al showed that there is a relation between placing time and compaction time for optimum compaction, and found an effective combination of vibration characteristics[3-5]. From the aforementioned research, some knowledge on the compaction mechanism of very dry mixed concrete has been obtained. However, most of this research has been performed at laboratory level where conditions are quite different from those on the construction site.

This research was carried out to investigate the compaction of very dry mixed concrete by performing on-site trials using a vibrating roller equipped with an electronic probe for subsidence measurements, accelerometer, and an AE sensor. The primary fundamental frequency, Fourier amplitude and distortion factor obtained from measurements of subsidence and acceleration, as well as measurements of the AE, were investigated to determine their relationship to the compaction of very dry mixed concrete.

2 Outline of the experiments

2.1 Mixing proportions of the very dry mixed concrete

Moderate heat Portland cement, fly-ash, slate type aggregate, and air-entrained water reducing agent were used in making the very dry mixed concrete. The maximum size of coarse aggregate used for the concrete, Gmax, was 120 mm, the values of unit quantity of cement were 110, 120 and 130 kg/m³, the unit quantity of water: 85, 90 and 95 kg/m³, the sand-aggregate ratios were 28% and 31%, and very dry mixed concrete consistencies of 10, 20 and 30 seconds determined by Vibration Compaction Test (VC value) were used. The VC value is the consistency indicator used for very dry mixed concrete. The VC test involves placing the concrete into a mould 24 cm internal diameter by 20 cm deep, placing a 20 kg weight on the surface and applying a vibration of frequency 3000 cpm with an amplitude of 1 mm, and then measuring the time for autogenous water rise to the surface.

2.2 Experimental equipment and measuring instruments
The vibrating roller used has a total weight of 10.2 tf and a vibrational force of 23 tf. As shown in Fig. 1, a steel sled mounting AE sensor is equipped, as well as an accelerometer and an electronic probe attached to the roller. The electric probe for subsidence measurements was a Nikon EPS-50A probe, the accelerometer was a strain gauge type accelerometer made by Kyowa Electronic Instruments (measuring range ± 20 G), and the AE measurements were made using a Nihon Physical Acoustics wide range type sensor. The steel sled was made from steel plate of 20 mm thickness, had a contact pressure of 29.1 g/cm^2, and was drawn between two rollers. The AE sensor was sensitive to the range 100~1000 kHz, and the threshold value was set to 80 dB (0.1 mV) to exclude the friction between the steel skid and the surface of the very dry mixed concrete.

An accelerometer was buried in the concrete to measure the vertical accelerations occurring when the vibrating roller was conducting the compaction.

2.3 The test site
The experimental block was 6 m wide and 14~28 m long. Placement of the very dry mixed concrete was carried out using a bulldozer in layers of 27 cm nominal thickness, and the concrete was placed in lifts of 75 cm, 100 cm, and 125 cm.

3 Experimental results
The acceleration waveform of the vibrating roller and the results of a frequency analysis using a fast Fourier transform (FFT) are shown in Fig. 2, and the acceleration waveform of the very dry mixed concrete during compaction and the frequency analysis are shown in Fig. 3. The shape of the acceleration waveform of the vibrating roller is relatively close to a sine wave of almost constant amplitude, and the spectral distribution is dominated by the primary fundamental frequency component. The amplitude of the acceleration within the very dry mixed concrete peaks when the vibrating roller is located right above, and then gradually decreases as the roller moves away.

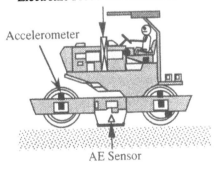

Fig. 1   Vibrating roller fitted with measurement instruments

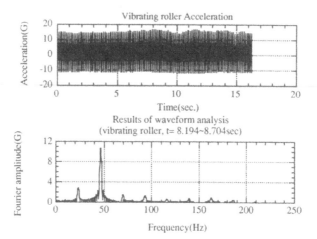

Fig. 2 Acceleration waveform and waveform analysis for vibrating roller

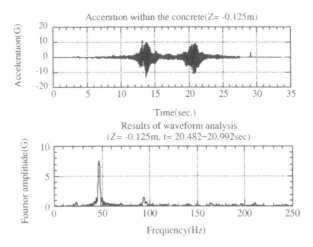

Fig. 3 Acceleration waveform and waveform analysis for very dry mixed concrete

The running spectrum of the concrete was determined, and the variation with time of the Fourier amplitude of the primary fundamental frequency component is shown in Fig. 4. The primary Fourier amplitude increases as the vibrating roller approaches, and reaches approximately 8 G. The primary Fourier amplitude also decreases with distance from the rolling surface indicating that it is being damped with distance from the point of vibration. It was also confirmed for each of the computation sections that the primary fundamental frequency closely matched the primary fundamental frequency of the vibrating roller.

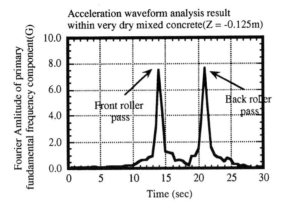

Fig. 4 Variation with time of the Fourier amplitude of primary fundamental frequency component

In Fig. 5, the AE hit rate, AE count, and AE energy are shown in relation to the number of compactions. It can be seen that there is a reduction in the AE hit rate, AE count, and AE energy with increasing number of compactions. The AE at 0 compactions is the value for the case of a roller pass without vibration. This AE value could result from the steel sled running over the undulations left by the bulldozer after placement of the concrete.

In Fig. 6, the relationship between subsidence and number of compactions is given. The subsidence increases with the number of compactions. This subsidence is the result of the air entrained within the very dry mixed concrete rising to the surface and being released under the action of the vibration, and the flow of the mortar around the aggregate. Furthermore, as a result of this subsidence, the density of the very dry mixed concrete, as well as its elastic modulus, increases.

4 Response analysis of the concrete during compaction

A single degree of freedom hypothetical model of a mortar particle in uniform very dry mixed concrete at a depth of Z (m) is shown in Fig. 7.
    Fig. 8 shows the equation of motion and transformation for the particle under the action of a forced vertical acceleration. If the values of f_o, m, ω, ω', and y are known, the damping factor h can be obtained from the final equation, Equation 6. The spring coefficient k can then be determined directly by substituting this value of h into Equation 5. In these equations:

f_o: vibrational force of the vibrating roller 23 (tG)
m: vibrating roller wheel loading 5 (t)
ω: radial vibration frequency of the vibrating roller 43.2 π (rad/s)
ω': primary fundamental radial frequency of the acceleration waveform measured within the very dry mixed concrete (rad/s)
y: Fourier amplitude of the ω' component (G)

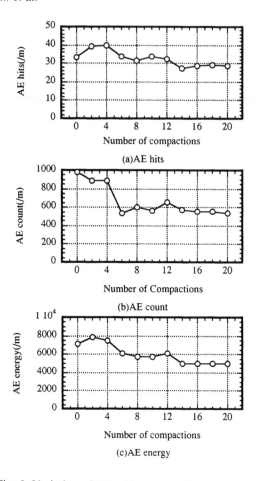

(a)AE hits

(b)AE count

(c)AE energy

Fig. 5 Variation of AE with number of compactions

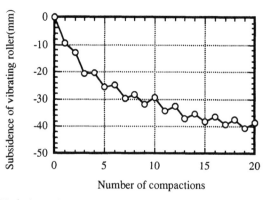

Number of compactions

Fig. 6 Variation of subsidence with number of compactions

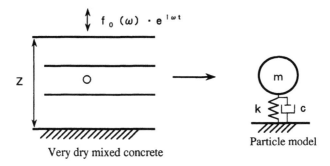

Fig. 7 Vibration model of very dry mixed concrete

Equation of motion:

$$m \cdot \ddot{y} + c \cdot \dot{y} + k \cdot y = f_0(\omega) \cdot e^{i\omega x} \qquad (1)$$

If the solution of the above equation is expressed in the form $y = A \cdot e^{i\omega x}$ then

$$A = \frac{1}{\sqrt{(1 - (\frac{\omega}{\omega_0})^2)^2 + 4 \cdot h^2 (\frac{\omega}{\omega_0})^2}} \cdot \frac{f_0}{m \cdot \omega_0^2} \qquad (2)$$

From Equation (2) the acceleration response of the particle can be expressed as

$$\ddot{y} = \frac{(\frac{\omega}{\omega_0})^2}{\sqrt{(1 - (\frac{\omega}{\omega_0})^2)^2 + 4 \cdot h^2 (\frac{\omega}{\omega_0})^2}} \cdot \frac{f_0}{m} \qquad (3)$$

Using $\quad \omega_0^2 = \dfrac{k}{m} \quad$, and substituting for m from equation (3)

$$k = \frac{f_0 \cdot \omega_0^2}{\ddot{y}} \cdot \frac{(\frac{\omega}{\omega_0})^2}{\sqrt{(1 - (\frac{\omega}{\omega_0})^2)^2 + 4 \cdot h^2 (\frac{\omega}{\omega_0})^2}} \qquad (4)$$

where k: spring coefficient of the system h: damping coefficient of the system, f_0: excitation force, w_0: natural undamped frequency of the system, w: excitation frequency, $\ddot{y}$: acceleration of the particle

If the natural frequency of the system with damping is expressed as

$$\omega' = \omega_0 \cdot \sqrt{1 - h} \qquad \text{then}$$

$$k = m \cdot \omega_0^2 = \frac{m \cdot \omega'}{1 - h^2} \qquad (5)$$

Using Equation (5) in Equation (4) and re-aarranging gives

$$\frac{\ddot{y}}{f_0} = \frac{(1 - h^2)(\frac{\omega}{\omega'})^2}{\sqrt{1 + 2(2h^2 - 1)(1 - h^2)(\frac{\omega}{\omega'^2})^2 + (1 - h^2)^2 (\frac{\omega}{\omega})^4}} \qquad (6)$$

Fig. 8 Equation of motion and transformation for particle with forced vertical acceleration

The Fourier amplitude and the primary fundamental radial frequency of the acceleration waveform measured by the accelerometer embedded within the very dry mixed concrete when the vibrating roller was located right above, were used for the values for ω' and y. The variations in damping factor h, and the spring coefficient k, calculated as described above, with number of compactions, are shown in Fig. 9. With the increase of compaction, there is a slight drop in the damping factor h, and a slight increase in the spring coefficient k, indicating that the stiffness of the concrete has increased.

5 Vibration analysis of the concrete using boundary element method (BEM)

As described in the previous section, it was confirmed that the elastic modulus of the very dry mixed concrete increased due to vibration compaction. It can be assumed that if the values of the physical properties of the concrete are varied, then the deformation characteristics also change. Therefore, an analytical model of the trial works

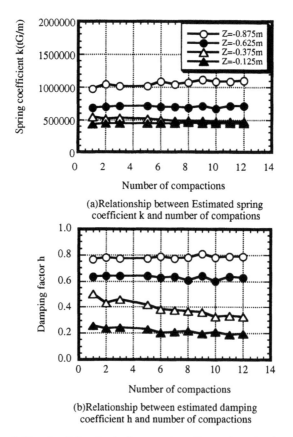

(a)Relationship between Estimated spring
coefficient k and number of compations

(b)Relationship between estimated damping
coefficient h and number of compactions

Fig. 9 Result of the physical properties of very dry mixed concrete

was made, the elastic modulus of the very dry mixed concrete varied and a vibration analysis was performed using BEM analysis.

The very dry mixed concrete was placed above the foundation concrete as shown in Fig. 10. The elastic modulus of the foundation concrete element was set at 100,000 kgf/cm^2 and the value of elastic modulus of the very dry mixed concrete was assumed as 10,000, 20,000, 30,000, 40,000 and 50,000 kgf/cm^2 respectively. The analysis was made considering the very dry mixed concrete as an elastic body and as a viscoelastic body.

Figs. 11 and 12 show the variation of central displacement of the analytical model with various elastic moduli. It can be seen that in the case of both elastic body and viscoelastic body, the central displacement decreases with increasing elastic moduli. The value of displacement became smaller for the case of the viscoelastic body model than for that of the elastic body model. In the case of the viscoelastic model, the value of displacement became high for elastic modulus of 30,000 kg/cm^2, due to the resonance vibration.

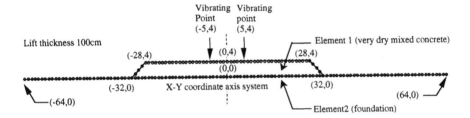

Fig. 10 BEM analysis model for vibrating analysis

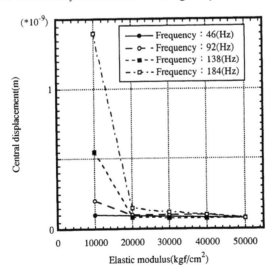

Fig. 11 Variation of concrete central displacement with elastic modulus (elastic analysis model)

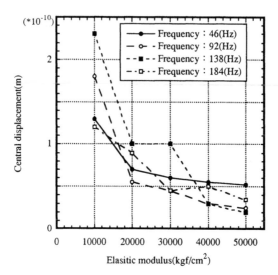

Fig. 12 Variation of concrete central displacement with elastic modulus (viscoelastic analysis model)

As described above, BEM analysis confirmed that displacements caused by vibration become smaller when the elastic modulus of the very dry mixed concrete increases. In addition, there is a reduction in the AE hit rate, AE count, and AE energy with increasing number of compactions. This is due to the increase in elastic modulus of the very dry mixed concrete caused by compaction with the vibrating roller, which results in smaller vibrational displacements, and therefore displacements of the steel sled decreases leading to a reduction in AE.

6 Conclusions

From the measurements of subsidence, accelerations and AE during the trial construction carried out for this research, the response analysis of the very dry mixed concrete during compaction, and concrete vibration analyses using the BEM, the following conclusions were reached:

1. The shape of the acceleration waveform of the vibrating roller is relatively close to a sine wave of almost constant amplitude, and the spectral distribution is dominated by the primary fundamental frequency component. The amplitude of the acceleration within the very dry mixed concrete peaks when the vibrating roller is located right above, and then generally decreases as the roller moves away.
2. According to measurements obtained by the accelerometer embedded within the very dry mixed concrete, the primary Fourier amplitude increases as the vibrating roller approaches, and reaches a value of approximately 8 G.
3. The action of the vibrating roller during compaction causes subsidence of the very dry mixed concrete and a subsequent drop in AE.

4. With progressing compaction, there is a slight drop in the damping factor h, and a slight increase in the spring coefficient k, indicating that the stiffness of the very dry mixed concrete has increased.
5. It was confirmed by BEM analysis that increasing elastic modulus of the very dry mixed concrete causes a reduction in the displacements caused by vibration.

Research on the changes in the physical properties of very dry mixed concrete due to vibration compaction, and of the related vibrational characteristics is in progress.

References

1. Kurita, Ono, Okumura, Shimada, and Hayashi (October 1983) Research into conditions during compaction of RCD concrete, *Technical Research Report of Shimizu Corporation,* No. 38.
2. Nakauchi, Nakagawa, and Shono (1981) Research into compaction of RCD concrete (Report 1), *Hazama Technical Report.*
3. Kagatani, Tokuta, and Kawakami (August 1987) Basic research into surface vibrators for compaction of very dry mixed concrete, *Proceeding of the Japan Society of Civil Engineers,* No. 384, Vol. 7, pp. 53-62.
4. Kagatani, Tokuta, Kawakami, and Tsujiko (August 1989) Basic research into the effects of methods of placing and compaction on very dry mixed concrete, *Proceeding of the Japan Society of Civil Engineers,* No. 408, Vol. 11, pp. 91-99.
5. Kagatani, Tokuta, Kawakami, and Tsujiko (January 1990) Determination of the degree of compaction of very dry mixed concrete from the pore water pressure system, *Concrete Research and Technology,* No. 1, Vol. 1, pp. 1-18.

PART FOUR
FIBRE REINFORCED CONCRETE

14 PROPORTIONING, MIXING AND PLACEMENT OF FIBRE-REINFORCED CEMENTS AND CONCRETES

C.D. JOHNSTON
Chairman, RILEM TC 145-WSM Subgroup on Fibre-reinforced
Mixtures, and University of Calgary, Canada

Abstract
Proportioning, mixing and placement procedures are identified and summarized for a wide range of fibre-reinforced cements and concretes based on glass, carbon, aramid, steel, polypropylene and acrylic fibres. They cover composites ranging from those made with relatively fragile fibres incorporated into an aggregate-free matrix utilizing special processes that minimize or eliminate mechanical mixing of fibres with the matrix to avoid damage to the fibres, to those made with robust fibres that can withstand the rigors of conventional mechanical mixing using coarse aggregate without damage to the fibres. Problems associated with achieving uniform fibre dispersion, dealing with the mixture-stiffening effect of fibres on workability, and developing techniques appropriate for measuring workability are identified and addressed. Typical mixture proportions and characteristics are given where appropriate to illustrate the uniqueness of the composite system. However, as most fibre-matrix combinations are special in some way, only the more widely used combinations are discussed. The information offered attempts to give guidance on the problems that may arise with the more common fibre-reinforced cementitious composites and the solutions that may effectively address them.

1 Introduction

Fibres in freshly mixed cement paste, mortar, or concrete act like very long slender needlelike particles of relatively high specific surface. Their extremely elongated shape and high surface area impart considerable stiffening and cohesion to the mixture. The mixture-stiffening or workability-reducing effect is a major factor

Production Methods and Workability of Concrete. Edited by P.J.M. Bartos, D.L. Marrs and D.J. Cleland. Published in 1996 by E & FN Spon, 2–6 Boundary Row, London SE1 8HN. ISBN 0 419 22070 4.

limiting the type, aspect ratio (ratio of length to equivalent diameter) and amount of fibres that can be uniformly distributed throughout a particular cementitious matrix, which in turn determine the degree of improvement in the mechanical properties of the composite in hardened state. Ideally, the amount and aspect ratio of the fibres should be as large as possible to maximize the improvement in mechanical properties, because the pullout resistance of the fibres from the matrix and their consequent reinforcing effectiveness improve with increase in their amount and aspect ratio. On the other hand, both the amount and aspect ratio should be as small as possible to minimize the mixture-stiffening effect of the fibres and associated difficulties in fabricating components from the fibre-reinforced mixture. What is possible in terms miscibility of fibres and matrix, placeability of the mixture, and property improvements in the hardened composite, is therefore a compromise that varies widely with the type, geometry and amount of fibres, the composition of the cementitious matrix, the manufacturing process and the nature of the intended application.

In any fibre-reinforced cementitious composite the physical compatibility of the fibres with the manufacturing process and their short and long term chemical compatibility with the moist alkaline environment of a cementitious matrix are of paramount importance. This report deals with physical considerations governing compatibility of fibres with the manufacturing process by attempting to synthesize information on proportioning, batching, mixing and fabrication processes for various kinds of fibre-reinforced cementitious composites.

1.1 Importance of fibre type

Most production processes used to combine fibers and matrix subject the fibres to varying degrees of bending, impact and abrasion which may damage the more fragile types of monofilaments causing breakage that reduces aspect ratio and consequently reinforcing effectiveness. Multifilament strands or rovings can suffer unintended separation with greatly increased fibre surface area and excessive mixture stiffening. Other multifilament strands or tapes that are intended to separate during mixing can also suffer damage by excessive mixing.

Most severe in terms of potential damage to fibres is blending fibres with the cementitious matrix in a conventional concrete mixer, especially when the matrix contains coarse aggregate. Fragile fibres like carbon monofilaments or multifilament glass strand or roving are generally unsuitable for such a process. Dry-process shotcreting is somewhat less severe in that the matrix is premixed, so the fibres escape the rigors of the mixing process, but still have to withstand the forces of bending, impact, and probably to a lesser extent abrasion, associated with spraying against a hard and often rough surface like rock. However, rebound and consequent loss of some fibres can be a problem. Various other premix processes based on mixing the constituents before adding the fibres are used to minimize the time during which fragile fibres like carbon are exposed to damage by the mixing process [1] (Fig. 1).

Least severe in terms of damage to the more fragile types of fibres is the spray process in which chopped fibres and a premixed cement-based slurry without coarse

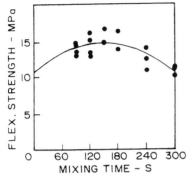

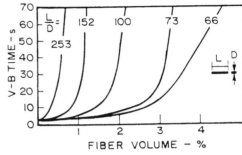

Fig. 1. Loss of strength with mixing
time using carbon fibres

Fig. 2. Effect of fibre amount and aspect
ratio on V-B time

aggregate are separately sprayed onto a stationary or moving formwork surface.

Another process especially suited to incorporating relatively large amounts of fibres without damaging them in the fabrication process is slurry infiltration. This involves preplacement of fibres in a mold followed by infiltration with a highly fluid slurry of paste or mortar poured into the mold following fibre placement.

1.2 Influence of fibre aspect ratio

The tendency of fibres to tangle in their packaging containers and fail to separate properly when dispensed into the matrix increases with aspect ratio. Their tendency to form clumps or balls in the mixture during the mixing process also increases with amount and aspect ratio. Both factors influence the uniformity of fibre distribution finally achieved in the hardened composite. Even when the fibres are initially dispensed uniformly into the matrix, the amount that can be incorporated with acceptable mixture workability is strictly limited by aspect ratio because of the more severe mixture-stiffening and workability-reducing effect of high aspect ratio fibres [2] (Fig. 2).

Innovations that combine reductions in fibre length or aspect ratio to minimize the mixture-stiffening effect with fibre features to improve resistance to fibre pullout from the matrix, and thus reinforcing effectiveness, are now quite common. They include surface texturing or deformation to roughen the fibre, crimping to a wavy rather than a straight profile, use of hooked, coned or flattened ends to improve fibre anchorage in the matrix, bundling of fibres with a water-soluble glue that allows them to separate during mixing, and partial splitting (fibrillation) to produce cross-linked multifilament strands that separate during mixing into small groups of branched monofilaments with enhanced mechanical anchorage. Consequently, straight smooth monofilaments of uniform cross-section are now quite rare as fibre reinforcement, except for glass and carbon fibres.

1.3 Nature of the cementitious matrix

The volume fraction of cement paste in a mixture determines the space available for fibres since they obviously cannot locate within the aggregate. The greater the paste volume fraction in the mortar the larger is the amount of fibres that can be accommodated for a given level of mixture workability [3] (Fig.3 left). Therefore, the maximum amount of fibres that can be uniformly distributed without excessive loss of mixture workability is larger for cement pastes and mortars than for concretes (Fig. 4). This has led to two quite distinct types of composites commonly termed fibre-reinforced cements and fibre-reinforced concretes. The former typically contain at least 3% by volume of fibres and up to 13% in the case of slurry-infiltrated products. The latter rarely contain more than 1% by volume of fibres and sometimes as little as 0.1%. In either case, the more fluid the paste phase the greater the fluidity or workability of the fibre-reinforced mixture [3] (Fig.3 right).

In addition, the nature of the matrix influences the effectiveness of fibres of any given aspect ratio and amount because experience suggests that fibres shorter than the maximum aggregate particle size have little reinforcing effectiveness. Consequently, fibre lengths used with concrete matrices are longer (25-60 mm) than fibre lengths used with cement matrices (5-30 mm).

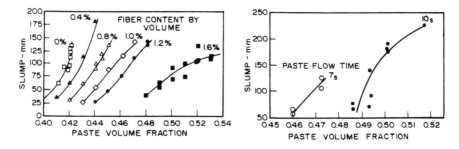

Fig. 3. Effect of paste volume fraction (left) and paste fluidity (right) on mixture workability.

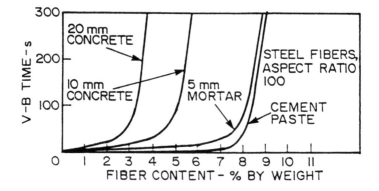

Fig. 4. Effect of aggregate size and fibre amount on V-B time.

2 Fibre-reinforced cements and mortars

Despite being commonly identified as fibre-reinforced cements, the matrix in these composites, while primarily cement-based, normally contains significant amounts of fine sand and occasionally filler materials such as fly ash, slag or silica fume. The paste volume fraction is high compared with concrete matrices, and consequently so is the maximum fibre amount possible without undue mixture stiffening. Fibre cements are often fabricated into components for thin-section applications where the section thickness is less than the fibre length and the fibres are consequently aligned to some extent in the plane of the section. They were developed primarily as alternatives to the historically successful but nowadays unacceptable asbestos cement.

2.1 Glass fibre-reinforced cement

Known as glass fibre-reinforced concrete (GFRC) in N. America and as glass fibre-reinforced cement (GRC) in Europe, this material is proportioned and fabricated in two different forms, both of which employ zirconia-based (minimum 16% zirconia) alkali-resistant (AR) multifilament glass strand as fibre reinforcement. Each strand comprises 102, 204, or 408 individual filaments bound together by a non-dispersable coating material called a sizing that is intended to prevent separation of the filaments and protect them from abrasion [4,5]. The sizing may incorporate a chemical inhibitor to reduce or delay the migration of calcium hydroxide into the strand which eventually bonds all of the filaments together and embrittles the composite [5]. A roving is a group of strands gathered together and wound into a package. The roving, typically with 20-40 strands [5,6], normally separates into strands when chopped to length. The effectiveness of the sizing in preventing separation of the strand into filaments (filamentising) and the extent to which the roving separates into strands during processing largely determines the resulting surface area of the fiber reinforcement and the associated mixture-stiffening effect. Dispersing the chopped roving with additives such as polyethylene oxide or carboxyl methyl cellulose [5] that lubricate the exposed fibre surface and increase the viscosity of the mixing water, thus improving resistance to segregation, have been found to facilitate mixing, particularly with soft-coated rovings that tend to filamentise easily [7]. Harder coatings/sizings tend to maintain the integrity of the strands. Naturally, minimizing or eliminating mechanical mixing in the manufacturing process also helps to reduce filamentising and abrasive damage to the fibre reinforcement. This has resulted in two distinct manufacturing processes, one which minimizes mechanical mixing (premix) and the other which eliminates it entirely (spray-up).

2.1.1 Spray-up GFRC

In the spray method [5] or spray-up process [4,6] the roving is fed through a chopper which cuts it to the required length and injects the chopped strands into the matrix slurry emitted through a spray gun. The sprayed slurry, atomized by compressed air, mixes with the chopped strand as it passes from the gun to the formwork surface against which successive layers of fiber-reinforced slurry are deposited. Each layer is sprayed to a thickness of 3-6 mm and then compacted by rolling to remove entrapped air and achieve maximum density. Successive layers are built up until the required component thickness, normally 13 mm minimum, is reached. This fibre-reinforced backing mixture may be preceded by an architectural face mixture without fibre reinforcement.

The minimum fibre content for spray-up GFRC is 4% by weight [4,6] and is typically 5% by weight of 25-50 mm long strands. Fibre contents more than 7% or strands longer than 50 mm lead to laydown and compaction problems, while smaller amounts of shorter fibres offer less reinforcing effectiveness although spraying and compaction are of course easier. Naturally, the consistency of the slurry reaching the spray gun is important in determining mixture workability at the formwork surface, and specifically its roller-induced compactability.

The consistency of the slurry for spray-up mixtures is established by filling a 80 mm long, 57 mm diameter, plexiglass tube placed at the centre of a plexiglass base with slurry, gently rodding the slurry to remove entrapped air, lifting the tube vertically to allow the slurry to slump and spread across a series of eight concentric rings of diameter 65 to 225 mm marked on the plate, and recording the spread in terms of rings covered on a scale of 0 to 8 [4]. The consistency may vary from relatively stiff, zero-ring spread, to relatively fluid, 8-ring spread, depending on water-cement ratio, sand-cement ratio, the presence of normal chemical admixtures (water-reducing, accelerating, retarding etc.) or a thixotropic additive such as methyl cellulose used to reduce mixture slump when spraying vertical surfaces. It also depends on the presence or absence of an acrylic thermoplastic copolymer dispersion often added to eliminate the need for the moist curing otherwise required. The acrylic copolymers are aqueous dispersions with a polymer solids content of 45-55% by weight [6], so the water they contain becomes part of the mixing water and should be allowed for in calculating water-cement ratio. Typical mixture characteristics [4,6] are given in Table 1.

Table 1. Typical portland cement-based GFRC mixtures

Constituent	Spray-Up	Premix
Chopped strand-%[a]	5[a]	3-3.5[b]
Strand length-mm	25-50	12-25
Sand[e]/cement	1.0	0.75-1.0
Water/cement	0.30-0.35	0.30-0.35
Acrylic copolymer-%[c]	4-15	0-10
Defoamer-%[d]	-	0.2
Superplasticizer-%[d]	0-0.5	0-0.5

a Minimum 4% by weight (approximately 2.8% by volume) of mixture. Fibre specific gravity 2.7.
b Maximum 4% because of mixing and compaction considerations
c Percent of polymer solids by weight of dry cement
d Percent by weight of mixture. May be incorporated with the acrylic copolymer.
e Close to Fuller maximum density gradation with maximum particle size 1 mm

2.1.2 Premix GFRC

In this process the matrix ingredients are mixed together in the first stage of the mixing process and chopped AR glass strand is added gradually towards the end of the mixing process [8]. Since the fibre reinforcement is exposed to potential damage during mixing, only high integrity chopped strand specially made with sizing appropriate for premix should be used. Strands from spray-up roving have a softer sizing and are less resistant to separation of the strand into filaments. The key to preventing damage to the fibre reinforcement is mixing the strands for the shortest time needed to achieve uniform distribution and thorough wetting by the slurry, usually less than 2 minutes. Any evidence of tangling or filamentising indicates a mixing regime that is too long, or too severe, or both. A two-speed high shear mixer with the matrix premixed at the high speed and the strand added under slow speed operation is effective [9]. Another effective mixer in terms of minimizing damage to the strand is the flexible-base omni-mixer [10] in which mixing occurs by an undulating action of the base that causes the mixture to turn over on itself rather than being sheared by blades.

Since the fibre reinforcement is mixed in one way or another with the matrix in the premix process, the mixture-stiffening and workability-reducing effect limits the amount and aspect ratio or length of strand that can be incorporated even with the best mixers. Consequently, fibre content is less and strand length is shorter than in the spray-up process (Table 1), and reinforcing effectiveness is consequently also less. Nevertheless, premix is useful for small highly sculptured architectural units and flat or thick-section mass-produced standard products where strength is not the prime requirement. It is usually cast into molds, similar to precast concrete products, and should be vibrated to help expel entrapped air and achieve proper compaction. A defoaming additive may assist in this regard.

Premix can also be sprayed with appropriate equipment, as in the spray-up process, but the end-product is usually inferior to that produced in normal spray-up where the fibre content is normally higher and the strand lengths are longer and more favorably oriented two-dimensionally in the plane of the section. Accordingly, premix is not considered an alternative to spray-up for thin-section applications where strength is important.

2.1.3 New GFRC matrices

To avoid the problem of long-term chemical incompatibility of AR glass fibres with Portland cement-based matrices associated with calcium hydroxide formation during hydration, two new types of matrix have been developed [6].

One system uses a calcium sulphoaluminate rapid hardening hydraulic cement with a retarding additive to control setting and a pozzolan to consume the calcium hydroxide produced during cement hydration [6,11]. The system cures rapidly under moist conditions to attain about 80% of its 28-day strength in 24 hours. Set time depends strongly on slurry temperature, optimum about 15°C, and on the amount of retarding additive. At a mixture temperature of 15°C and ambient temperature of 21°C the set time is about 30 minutes if no retarder is included [11]. Using the maximum amount of retardation extends set time to 2-3 hours depending on ambient temperature. Setting is exothermic with temperature rising to over 38°C, usually within less than 30 minutes. The composite must be water-cooled until it reaches ambient temperature, after which curing occurs without the need for additional water. Typical mixture characteristics for both spray-up and premix processes are given in Table 2 [11,12]. Water-cement ratios are higher than for Portland cement-based equivalents in Table 1, and acrylic copolymer curing additives are not permissible.

Table 2. Typical rapid hardening hydraulic cement-based GFRC mixtures (zircrete)

Constituent	Spray-up	Premix
Chopped strand-%[a]	5%	2-4
Strand length-mm	25-50	13-25
Sand/cement	1.0-1.14	1.0-1.5
Water/cement	0.43-0.50[d]	0.40-0.45
Retarding additive-%[b]	0-8	0-8
Methyl cellulose[c]	x	
Pumping aid		x

a   Percent by weight of mixture
b   Percent by weight of cement of proprietary combination of pozzolan, plasticizer, and set-retarder [12]
c   Optional thixotropic aid to stiffen mixture for spraying vertical surfaces
d   Value adjusted to produce a slurry with a 6-ring spread

Another system is based on portland cement enriched with 20-25% metakaolinite to consume as much as possible of the calcium hydroxide liberated during cement hydration, which is of course detrimental to the long-term performance of glass fibres in cement [6,13]. Unlike other pozzolans which react quite slowly with calcium hydroxide, metakaolinite is highly reactive and does not slow strength development at early ages. The high surface area of the metakaolinite stiffens the mixture and increases water demand or superplasticizer dosage needed for satisfactory workability [13]. Typical mixture characteristics [14] are given in Table 3. Acrylic copolymer curing additives are optional, and help to keep water-cement ratios lower while achieving adequate workability.

Table 3. Typical metakaolin-modified GFRC mixtures

Constituent	Spray-up	Premix
Chopped strand-%[a]	5	3
Sand/cement	1.0	1.0
Metakaoline-%[b]	25	25
Water/cement	0.43-0.34[c]	0.46-0.37[c]
Superplasticizer-%[b]	0.5-1.0[b]	0.5-1.0[b]
Acrylic copolymer-%[b]	0-15[c]	0-15[c]

a   Percent by weight of mixture
b   Percent by weight of cement
c   With polymer curing aid

Many other matrices modified by polymers or by replacing Portland cement with low lime or lime-free cements such as high alumina cement and supersulphated cement made from blastfurnace slag and calcium sulphate have been investigated for use with glass fibre-reinforcement [5].

2.2 Carbon fibre-reinforced cement

Nowadays carbon fibre-reinforced cements (CFRC) are usually made with the cheaper pitch-based carbon fibres as opposed to the more expensive PAN-type produced from polyacrylonitrile. Like glass fibres, these fibres are relatively fragile and prone to damage during mixing (Fig. 1), but, unlike glass, they are not affected by lime or alkalis, and are compatible with normal cement-based matrices.

The process of blending the fibres into the matrix is analogous to the premix process for GFRC previously explained, and may utilize a special mixer such as an omni-mixer [15,16] to minimize damage to the fibres or a conventional mortar mixer [17,18]. Matrices can be cement-only pastes or cement-filler pastes with fly ash, fine silica sand less than 0.5 mm [17], silica powder less than 0.15 mm [16], or silica fume [15,16,18]. Fibres are generally of high specific surface and aspect ratio and tend to ball and be difficult to disperse uniformly at fiber contents greater than about 1% by volume unless dispersing additives like carboxyl methyl cellulose and fine filler material are present. Water-reducing or superplasticizing admixtures are also normally needed to ensure adequate workability with reasonable water-cement ratios. Typical mixture characteristics are given in Table 4. Lightweight matrices with fly ash, silica powder and a foaming agent to reduce density have been used with and without a vinyl acetate polymer emulsion which is reported to decrease porosity and improve the interfacial bond between fibre and matrix [16].

Table 4. Typical mixture characteristics for carbon fibre-reinforced cements prepared using an omni-mixer [15,16] or a normal mortar mixer [17,18]

Constituent	Ref.[15]	Ref.[16]	Ref.[17]	Ref.[18]
Chopped fibre-%[a]	2-4.5	0.5-2.0	1-4	1-3
Length-mm	10	3-12	6-18	3-10
Diameter-μm	14.5	14.5	15-17	18
Aspect ratio	690	210-830	400-1060	170-560
Silica sand/cement				0.50
Silica powder/cement	0-0.5	0.4-0.5[c]		
Silica fume/cement		0.4[c]		0.2
Water/cement	0.30-0.53	0.30-0.53	0.45	0.35
Methyl cellulose-%[b]	1.0	0.5	0.25-1.0[d]	
Superplasticizer-%[b]	0-2	1-6	2.4	0.8-2.4
Defoaming agent-%[b]			0.1	

a   Percent by volume of mixture. Fibre specific gravity 1.63
b   Percent by weight of cement.
c   Either but not both
d   Depending on mixing regime.

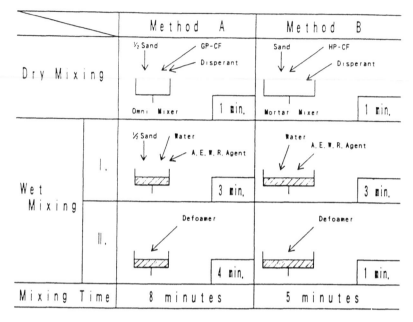

Fig. 5. Alternative mixing regimes for CFRC.

The mixing regime has been found to have an important influence on the amount of dispersing additive (carboxyl methyl cellulose) needed to achieve good fiber dispersion [17]. In the two regimes investigated [17] (Fig. 5) carbon fibre, cement, some of the sand, and dispersant were first dry-mixed before adding water, water-reducing and air-entraining admixtures, and defoaming agent. In Method A only half the sand was added at the dry-mix stage with the remainder added early in the wet-mix stage, while in Method B all of the sand was added at the dry-mix stage. The effect of changing from Method A to Method B was to reduce the amount of dispersant needed by 75%. The decreased amount of dispersant and resulting lower mixture viscosity increased mixture flow, improved trowel finishability, and produced much superior flexural strength in the hardened composite [17], so the mixing regime is clearly important.

For a given mixing regime the mixture-stiffening effect of carbon fibres measured in terms of flow determined using a standard mortar flow table depends on fibre volume fraction, fibre specific surface per unit volume of mixture which reflects the number of fibres per unit volume and their diameter [17] (Fig. 6), and to some extent on fibre length [18] (Fig. 7). Clearly, the combination of fibre aspect ratio and fibre size, which reflects number per unit volume percentage of fibre content, determines the mixture-stiffening effect of the fibres. In addition, the surface texture may influence the ease with which the fibres slide over each other and the associated fluidity of the mixture, as in the case of 6mm carbon versus 6 mm long polypropylene where the latter with a presumably smoother surface offers comparable mixture fluidity despite much higher aspect ratio (1500 vs. 170) and specific surface (Fig. 7).

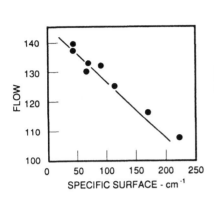

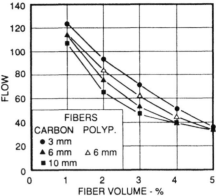

Fig. 6. Effect of fibre specific surface   Fig. 7. Effect of fibre volume fraction and
on workability of CFRC   aspect ratio on workability of CFRC

2.3 Aramid-fibre reinforced cement

Aramid fiber-reinforced cement (AFRC) uses fibres based on aromatic polyamides developed for use in tires, ropes, sails etc., for example Kevlar 49. Like carbon fibres, they are difficult to disperse in a cementitious matrix, but unlike carbon, they are not easily damaged during mixing.

Mixing regime has again been shown important in achieving proper fibre dispersion [19]. Using a conventional mortar mixer, the approach was to use silica fume to disperse the fibres, as a superplasticized silica fume paste is apparently effective for breaking apart fibre balls and coating the fibres. The optimal regime involved low speed mixing of silica fume with 2/3 of the water and superplasticizer for about one minute, gradual addition of fibres with continued mixing, addition of remaining water and superplasticizer followed by cement and mixing for a further minute, stopping the mixer for 30s, and final mixing at medium speed for another minute. The mixture-stiffening effect of 12 μm diameter aramid fibres in a cement-silica fume paste matrix determined in terms of flow shows the expected decrease in flowability with increase in fibre volume fraction and length or aspect ratio [19] (Fig. 8).

2.4 Other fibre-cement composites

Other fibres used in cementitious matrices, some of which are too fragile to withstand conventional mechanical mixing, have been incorporated into the matrix using the well-known Hatschek process developed originally for producing asbestos cement. Fibres used as reinforcement include natural cellulose-based fibres [20] and a variety of synthetic fibres such as acrylic (polyacrylonitrile), polyvinyl alcohol, rayon, polyester, nylon, polypropylene, polyethylene, aramid and carbon [21]. In this process the matrix is much more dilute and fluid than in any process described previously, with a fibre plus cement and other solids content of only 5-10% when initially mixed. To retain cement and other fines in suspension in such dilute

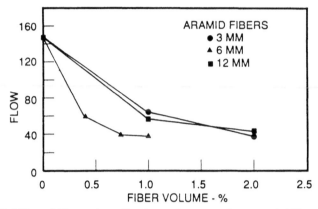

Fig. 8. Effect of fibre volume fraction and aspect ratio on workability of AFRC.

mixtures requires addition of filter fibres such as wood cellulose, acrylic pulp or polyethylene pulp. The highly dilute mixture is picked up on a moving belt and subsequently concentrated by pressing and vacuum dewatering. Mixture workability is less important in this process than in the manufacturing processes previously described in view of subsequent pressing and dewatering, and dispersion of the fibres is relatively easy in view of the initially very fluid nature of the mixture. Many combinations of fibre reinforcement, filter fibres, matrix fillers, dispersants and flocculants have been investigated [21], and are too numerous to discuss further in detail.

3 Fibre-reinforced concretes

Unlike fibre-reinforced cements, the matrix in these composites contains a significant volume of coarse aggregate larger than 5 mm and a much lower paste volume fraction available to accommodate the fibres, so the maximum fibre amount possible without excessive mixture stiffening and loss of workability is correspondingly lower, usually less than 1% by volume of mixture. Only the more robust types of fibre can withstand the forces of bending, impact and abrasion inherent in the conventional mechanical concrete mixing process that blends fibres with coarse and fine aggregate, cement, and sometimes fly ash, slag or silica fume to increase the volume fraction of the paste phase and facilitate accommodation of the fibres.   Steel fibres are the most commonly used, followed by fibrillated polypropylene and only rarely by other types.

3.1 Steel fibre-reinforced concrete
The fibres for steel fibre-reinforced concrete (SFRC) are manufactured in many mostly monofilament forms, often with modifications to increase the resistance of the fibre to pullout from the matrix, and its associated reinforcing effectiveness in the hardened composite, without increasing the fibre length and aspect ratio to the point of causing excessive mixture stiffening and loss of workability.   These

modifications include surface deformation, crimping to a wavy rather than a straight profile, and end anchorage improvement using, spaded, coned or hooked shapes. Bundling of fibres with a water-soluble glue that allows the fibres to separate during mixing is also used to reduce the aspect ratio and clumping tendency of the fibre units during packaging and batching.

3.1.1 Mechanically mixed SFRC

The concrete matrix for conventionally mixed SFRC needs to be proportioned in a way that maximizes its workability with a minimum of water while keeping the maximum coarse aggregate size no larger than the fibre length. Accordingly, from the viewpoint of workability smooth rounded sand with a low fineness modulus and a slight excess of particles passing the 600 μm and 300 μm sieves is preferable to coarser angular manufactured sand [22]. Coarse aggregate maximum size should generally not exceed 25 mm, but mixtures with aggregate up to 38 mm have been evaluated. Reducing the coarse aggregate volume fraction by up to 10%, as in concrete proportioned for pumping, allows the fine aggregate and the mortar volume fraction to increase. Alternatively, the volume of the paste phase can be increased by use of fly ash, slag or silica fume. Either or both measures increase the volume fraction of the mortar and facilitate accommodation of fibres (Fig. 3).

Unless the cement paste is supplemented by a pozzolan like fly ash, the cement and fine aggregate amounts should exceed the minimums needed to create a mortar volume fraction sufficient to accommodate the required amount and type of fibres without excessive loss of workability. However, since fibres stiffen the concrete mixture, particularly with respect to slump, adding water to restore workability should be limited to the amount needed to reach the maximum water-cement ratio for the applicable exposure condition, typically 0.45 to 0.55. Use of water-reducing or superplasticizing admixtures is a better way of meeting both workability and water-cement ratio requirements. ACI Committee 544 guidelines for water-cement ratio, cement content and fine to total aggregate are given in Table 5[23].

Table 5. ACI 544 recommended mixture proportions for SFRC

Mixture Characteristic	Coarse aggregate max. size		
	10 mm	20 mm	38 mm
Water/cement	0.35-0.45	0.35-0.50	0.35-0.55
Cement-kg/m^3	360-600	300-540	280-420
Fine/total agg.-%	45-60	45-55	40-55
Entrained air-%	4-8	4-6	4-5
Smooth fibre-%[a]	0.8-2.0	0.6-1.6	0.4-1.4
Deformed fibre-%[a]	0.4-1.0	0.3-0.8	0.2-0.7

a   Percent by volume of concrete. 1% $\simeq$ 80 kg/m^3

Before discussing more precise proportioning of fibres by amount and type, it is important to recognize their real effect on mixture workability and the consequences for fibre selection. The mixture-stiffening effect of fibres is particularly noticeable in terms of slump because the slump test assesses primarily the stability or cohesion of the mixture under static conditions, rather than its mobility and compactability under dynamic vibratory conditions [24]. Since fibres greatly increase mixture stability, the slump test does not reflect the placeability of SFRC using the vibratory consolidation equipment that is normal in practice, and consequently mixtures with fibres exhibit what appears to be unacceptably low slump when compared to mixtures without fibres that have the same placeability under vibration [25] (Fig. 9 right). Tests appropriate for measuring placeability of SFRC under vibration include the test for time of flow through an inverted slump cone (under the influence of an internal vibrator), ASTM C995 (Fig.9 left), and the V-B test standardized in Europe, both of which measure primarily mobility and

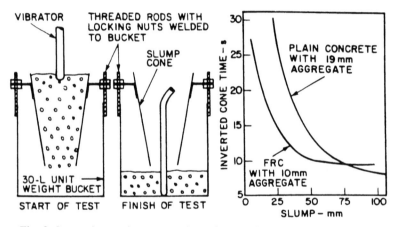

Fig. 9. Inverted cone time versus slump for SFRC and plain concrete.

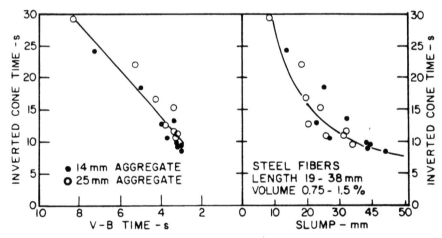

Fig. 10. Relationships between inverted cone time, V-B time and slump for SFRC.

therefore correlate strongly and linearly with each other (Fig. 10)[24,25]. When used for relatively stiff SFRC mixtures of workability appropriate for vibration, (V-B time 3-9s, inverted cone time 8-30s), the precision of the slump test is much poorer than precision for either of the other tests [24].

The mixture-stiffening and workability-reducing effect of fibres in mechanically mixed concretes reflect the primary influence of fibre amount and aspect ratio [25](Fig. 11), as was the case for mixed fibre cements (Fig. 6-8), namely that for

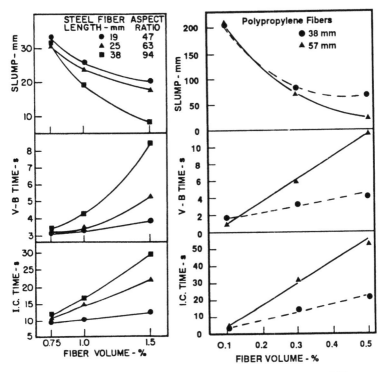

Fig. 11. Effect of fibre amount and length or aspect ratio on workability
of SFRC (left) and PFRC (right).

any particular matrix the effect increases with increase in both fibre amount and aspect ratio. Changing the matrix by reducing the coarse aggregate volume fraction and correspondingly increasing the fine aggregate, while keeping the volume fraction and water-cement ratio of the paste constant, also improves the workability for any particular amount and aspect ratio of fibres because the mortar volume is increased (Fig. 12)[26]. This corresponds to the practice of reducing the coarse aggregate volume fraction by up to 10% for improving pumpability of ordinary concretes. As shown previously (Fig. 4), maximum coarse aggregate size is also important because the number of straight rigid fibres like steel that can be accommodated within a unit volume reduces with increase in the maximum aggregate size, as shown schematically [2](Fig. 13).

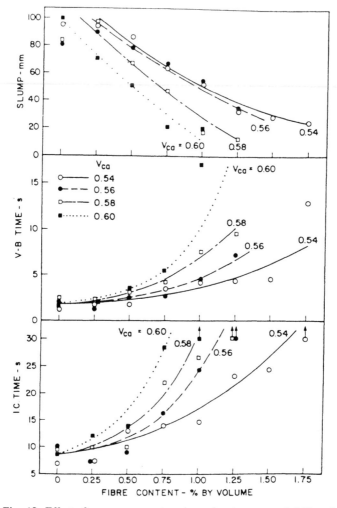

Fig. 12. Effect of coarse aggregate volume fraction on workability of SFRC.

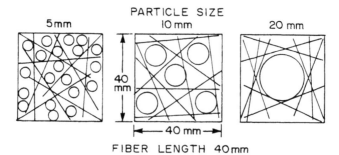

Fig. 13. Schematic of the effect of aggregate size on fibre
arrangement in the concrete matrix.

The recommendations of one steel fibre manufacturer that relate the maximum fibre content possible without encountering workability or fibre balling problems to fibre aspect ratio, maximum aggregate size, and normal or pumped delivery (Table 6), are consistent with the relationships in Fig. 4, 11, 12 and 13. For other fibre types the pattern should be similar with probably only minor modifications to reflect differences in profile, surface texture, and nature of end anchorage.

Table 6. Maximum recommended[a] steel fibre content[b]-kg/m^3

Max. coarse aggregate size-mm	Steel fibre aspect ratio					
	60		75		100	
	Normal	Pumped	Normal	Pumped	Normal	Pumped
4	160	120	125	95	95	70
8	125	95	100	75	75	55
16	85	65	70	55	55	40
32	50	40	40	30	30	25

a N.V. Bekaert S.A.
b 1% by volume of concrete $\simeq$ 80 kg/m^3

Prior to the mixing process fibres must be effectively free of any balling or tangling that results from densification in packaging or simply from inherent fibre characteristics such as high aspect ratio, crimping, surface roughness or end anchorage entanglement [23]. Monofilament types are more prone to this problem, sometimes called dry balling. Passing the fibres through a 50-100 mm mesh screen, depending on fibre length, is effective, particularly if the screen is connected to a source that induces vibration. One manufacturer (Eurosteel) provides a vibro-pneumatic system consisting of a vibrating screen on top of a hopper, a blower, and a discharge pipe. The vibrating mesh separates fibre balls, and the air blower conveys the separated fibres through the discharge pipe to the mixer or aggregate conveyor belt. Another approach is to reduce the potential for dry balling or tangling prior to mixing by bundling the fibres with water-soluble glue which effectively reduces the aspect ratio of the bundled units to the point where they do not ball or tangle. However, the mixing time must be sufficient to ensure complete separation of the bundles into individual monofilaments without introducing the possibility of overmixing. Overmixing to the point where the individual monofilaments start to ball in the mixture, sometimes termed wet balling [23], may happen no matter whether the fibres are initially batched in bundled or monofilament form when the recommended maximum amount-aspect ratio limits are exceeded (Table 6). Other causes of wet balling are adding fibres too fast to a matrix of initially too low workability, a matrix with too much coarse aggregate, or using a mixer with worn blades or a design inappropriate for use with fibres.

When possible, the preferred method of introducing fibres is addition to the aggregate conveyor followed by mixing of fibres with coarse and fine aggregate prior to addition of cement and water. Simultaneous feeding of aggregate, fibres, cement and water to the mixer is also effective [22]. When preblending of fibres is not possible they can be effectively introduced by addition to the freshly mixed concrete, provided it has appropriate initial workability, which according to various fiber manufacturers means a slump of at least 120 mm and as much as 200 mm for high amount-aspect ratio fibres. This should be achieved by use of water-reducing or superplasticizing admixtures rather than excess water. It is recommended [23] that slump before addition of fibres be 50-75 mm greater than the final slump desired, but in view of the comments made previously about the validity of the slump test for low-slump SFRC mixtures this probably holds only for final slumps greater than 50 mm where V-B and inverted cone times are too small to be determined accurately.

Some manufacturers provide convenient and manoeuvrable conveyors for transporting fibres packaged in small (up to 40 kg) bags to the aggregate belt or mixer. The unit developed by N.V. Bekaert has a self-powered conveyor and can be towed behind a small truck. The Skako system [27] developed for large volume production in a plant handles packages of up to 1000 kg of fibres, provides hopper storage up to 2000 kg, automatically controlled weighing, and conveyance of fibres to the aggregate belt or mixer.

Another way of delivering fibres to the mixer that avoids the need for fibre packaging, handling of the packages, and separation of the fibres after opening the packages, is to eliminate use of precut (manufactured) fibres by cutting of the fibres from coiled wire as they are fed into the mixer. An automated high speed cutter, fed from four coils of wire that processes approximately 9000 metres of wire per minute, has been developed to feed 30-40 kg of steel fibres per minute directly into concrete in a stationary central or truck mixer [28].

The type of mixer and the mixing regime are important for achieving complete and uniform fibre dispersion throughout the concrete matrix. Mixers unsuitable for low-slump concrete in general are not likely to work well for concrete with fibres. Constant flow drum-type or screw-type mixers where the materials move axially through the mixer as they are mixed are not suitable. Pan mixers where the blades rotate in the same direction as the pan are less suitable than mixers where the blades and pan rotate in opposite directions. Addition of fibres in a manner that fails to distribute them throughout the load, may lead to poor fibre distribution. For example, adding fibres to the rear end of a truck load rather than using a conveyor to distribute them along the length of the load, tends to prolong the mixing time needed. Regardless of the mixer type, the mixing regime must ensure uniform fibre distribution without overmixing and the consequent possibility of wet balling of the fibres.

3.1.2 Shotcrete SFRC

The principles applicable to proportioning, batching and mixing conventionally mixed SFRC apply in large part to shotcreting by the wet process. Prevention of blockages due to dry or wet balling by placing a screen over the pump hopper is particularly important [29]. Likewise, for the dry process a screen should be placed over the receiving hopper to intercept any balls formed in the dry mixture prepared by first mixing fibres with fine aggregate and following with cement and other ingredients. Screw-type mixers have been found satisfactory for mixing the dry ingredients prior to discharge into the shotcrete hopper [29]. It has also been found that a good electrical ground to the gun and nozzle reduces fibre clumping and plugging at the nozzle. For both processes the hose diameter should be at least 1.5 times the fibre length and not less than 50 mm.

Generally, steel fibres suitable for shotcreting have a shorter length and lower aspect ratio than those used for mechanically mixed SFRC to facilitate production of ball-free material and easier conveyance of it through hoses. Fibre contents are in the range 40-80 kg/m^3, and silica fume is often used in combination with fibres to reduce rebound, particularly in the dry process where fibre loss due to rebound tends to be much greater (20-50%) than for the wet process where it is typically 5-10% [30]. In the dry process there is evidence of significant preferential two-dimensional fibre alignment in the plane of the sprayed surface [31] which may be advantageous, but hardly offsets the higher fibre loss due to rebound. Typical compositions for dry and wet process mixtures are given in Table 7 [30].

Table 7. Typical steel and polypropylene fibre shotcrete mixture characteristics [30, 33]

Constituent material	Steel fibers[a]		Polypropylene fibers[b]	
	Dry-mix	Wet-mix	Wet-mix	
Cement-kg/m^3	400	420	402	157
Fly ash-kg/m^3			-	236
Silica fume-kg/m^3	50	40	-	-
10 mm. aggregate-kg/m^3	500	480	430	403
Concrete sand-kg/m^3	1170	1120	1285	1230
Fibres-kg/m^3	60	60	4	6
Water reducer-ℓ/m^3	-	2	2	-
Superplasticizer-ℓ/m^3	-	6	-	7
Water-kg/m^3	170	180	190	162
Air entrainer	No	Yes	Yes	Yes

[a] Aspect ratio 50-70
[b] 38 mm fibrillated strand

A variation of the wet process that permits fibre aspect ratios much higher than the typical 50-70 values (Table 7) to be utilized effectively involves cutting of long fibres from two coils of wire fed to the spray gun and injecting them into the pumped premixed concrete matrix [32]. It eliminates the need for cutting, packaging and separation of fibres prior to mixing, and of course improves the reinforcing effectiveness of the fibres because of their high aspect ratio.

3.2 Polypropylene fibre-reinforced concrete

Fibres for polypropylene fibre-reinforced concrete (PFRC) are manufactured mostly in multifilament fibrillated strands with the main longitudinal filaments crosslinked to each other by usually smaller transverse filaments. These strands are quite flexible and are intended to separate into mini strands comprising a few main longitudinal filaments crosslinked by the transverse filaments to form an open lattice that is penetrated by the matrix mortar, thus creating an improved mechanical bond. This mechanical bond is desirable since polypropylene is hydrophobic and therefore not easily wetted by cement paste to naturally develop an adhesive bond. However, some manufacturers use proprietary surface treatments to try to improve the adhesive bond.

3.2.1 Mechanically mixed PFRC

The concrete matrix needs to be proportioned with a minimum mortar volume fraction to accommodate fibres, so this means reducing the coarse aggregate and increasing the fine aggregate or adding pozzolan, as discussed previously for steel fibres. The mixture-stiffening effect of polypropylene fibres is very marked due to their very high surface area, and the consequent loss of workability depends on fibre amount and strand/fibre length (Fig. 11 right) just as for steel fibres (Fig. 11 left). Fibrillated polypropylene is used at volume fractions as low as 0.1%, mainly to control plastic shrinkage, but can be used at volume fractions up to about 0.7% where improvement in hardened concrete properties can be significant.

To achieve proper strand separation with adequate length for bond to develop, along with uniform fibre distribution in the mixture, two manufacturers recommended strand lengths related to coarse aggregate maximum size. However, the recommendations differ because of differences in the material and method of forming the strands. In one case the fibre length can be approximately 1.5-2.0 times the aggregate size, while in the other it is 3.0 times the aggregate size (Table 8). In the latter case, the manufacturer's reported experience shows that using long fibres with small aggregate, for example 63 mm fibres in a mortar matrix, leads to mixing and distribution problems. This is probably due to the small aggregate particles being unable to effectively separate the long filaments.

Table 8. Recommended polypropylene strand lengths

Coarse agg.[a] size-mm	Strand[a] length-mm	Coarse agg.[b] size-mm	Strand[b] length-mm
5	6-13	6	19
10-16	19	13	38
16-25	38	19	57
19-38	51	25+	63

a   Fibermesh flat strand
b   Forta twisted strand

The strands or bundles can be added with other materials as they enter the mixer, or may be added to previously mixed concrete taking care to distribute them uniformly throughout the batch. Matrix workability should be high before addition of the fibres, at least 100 mm slump for 0.1% volume of fibres and progressively more for higher fibre amounts, preferably by means of admixtures rather than excess water. However, the manufacturers recommend that when a superplasticizer is to be used the fibres should be blended with the matrix before the superplasticizer is added. Workability after the fibres are added may appear low in terms of slump and can be evaluated more realistically by V-B time or inverted cone time for the reasons discussed previously. However, the inverted cone test can sometimes be rendered invalid by long strands wrapping around the vibrator or a very cohesive mixture failing to exit the cone after unlimited vibration which simply creates a hole in the sample under test.

3.2.2 Shotcrete PFRC

Although polypropylene fibres, both monofilaments and fibrillated strands, have been tried at low volume fractions (0.1% or 1 kg/m^3) to a limited extent in dry and wet process shotcrete, most recent work has involved wet process application with relatively high volume fractions of 0.4-0.7% (approximately 4-6 kg/m^3). Increasing the paste volume fraction by using fly ash replacement of cement up to 60% has been shown effective in raising the maximum possible fibre content from 4 to 6 kg/m^3 without changing the total cementitious material [33]. This total should be at least 400 kg/m^3 [33]. Mixtures used to evaluate wet process PFRC typically contain 10 mm aggregate, sand, cement, water-reducing or superplasticizing and air-entraining admixtures, and 4-6 kg/m of 38 mm polypropylene strands (Table 7)[33,34].

3.3 Polyacrylonitrile fibre-reinforced mortar and concrete

High tenacity polyacrylonitrile (acrylic) fibres are produced in Europe [35] (tradename Dolanit) in a range of sizes for both fibre-reinforced cements and concretes. While the smaller (13-18 μm diameter) versions of these fibres are used in fibre cements prepared by the Hatschek process, larger versions (52-104 μm diameter) can be incorporated into conventionally mixed concrete. These larger fibres can sometimes be added without any special means of separation (vibrating screen etc.) if the concrete mixer effectively induces separation during the mixing process. However, like other monofilament types, these fibres tend to clump together in their packaging containers. Mixers that move the matrix on the counterflow principle or mixers with a high speed, high shearing, action are effective in separating the fibres, especially the longer ones (12-25 mm). Less efficient slow mixers may require a fibre separation machine. One such machine employs a high speed rotating brush which loosens bundled fibres and injects them into the matrix in the mixer [35]. Fibre dispersion in the matrix must occur quite quickly within short mixing times, as long mixing times may cause fibre balling. As for other fibre types, larger low aspect ratio fibres with moderate surface area are easiest to disperse. Typical fibres for concrete are 6 or 12 mm long and 0.1 mm diameter. Water-reducing or superplasticizing admixtures are recommended to combat the mixture stiffening effect of the fibres, otherwise the water-cement ratio becomes excessive.

Various mixing regimes have been investigated for acrylic fibre-reinforced concrete (ACFRC) [35] as follows:

(a) mixing the concrete first and then adding fibres

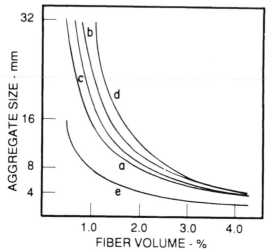

Fig. 14. Effect of mixing regime and aggregate size on maximum
fibre volume percentage for ACFRC.

(b)   mixing fine aggregate, cement and water, adding fibres, and finally adding
      coarse aggregate
(c)   as for (a) using a fibre separator
(d)   as for (b) using a fibre separator
(e)   dry-mixing of cement with fibres prior to the normal wet-mixing procedure.

The results (Fig. 14) indicate limiting fibre volume fractions for 6 x 0.1 mm
fibres according to aggregate maximum size and mixing regime (a) to (e). The
largest fibre contents are achieved with regimes (b) and (d) which involve mixing
the mortar fraction, adding fibres and then adding coarse aggregate last. Using
these regimes in concrete, for example with 16 mm aggregate, limits the fibre
amount to about 1.8% by volume or about 20 kg/m^3, a limit not untypical of similar
concrete made with steel fibres of the same aspect ratio.

4 Conclusion

During its preparation, it became clear that this report could not hope to cover all
conceivable types of special mixtures used to prepare fibre-reinforced cementitious
composites, so the information offered can only attempt to guide those unfamiliar
with such composites on the nature of the problems that may in general arise and
the solutions that in general may effectively address them.

5 Acknowledgements

Since the writer lacks close familiarity with many of the composites covered in this
broadbrush compilation, it would have been impossible to prepare without the
assistance of many individuals with expertise on specific fibre-matrix combinations.

The information provided by Hiram Ball, Xavier Destree, Mel Galinat, Bud Molloy, Dirk Nemegeer, Peter Tatnall, Åke Skärendahl, J.D. Wörner and Bob Zellers is gratefully acknowledged.

6 References

1. Sakai, H. et al (1994) Flexural behavior of carbon fibre cement composite, Fiber Reinforced Concrete Developments and Innovations, American Concrete Institute Special Publication SP-124, pp. 121-140.
2. Hannant, D.J. (1978) Fibre Cements and Fibre Concretes, John Wiley & Sons Ltd. U.K., 219 pp.
3. Peiffer, G. and Soukatchoff (1994) Mix design approach for fibre reinforced mortars based on workability parameters, Special Concretes Workability and Mixing (ed. Bartos), RILEM Proceedings 24, E&FN Spon, pp. 89-97.
4. GFRC Quality Control Manual Subcommittee (1991) Manual for Quality Control for Plants and Production of Glass Fiber Reinforced Concrete Products, MNL-130, Precast/Prestressed Concrete Institute, Chicago, USA.
5. Majumdar, A.J. and Laws, V. (1991) Glass Fibre Reinforced Cement, BPS Profession Books Division of Blackwell Scientific Publications, Oxford, U.K.
6. PCI Committee on Glass Fiber Reinforced Concrete Panels (1993) Recommended Practice for Glass Fiber Reinforced Concrete Panels, MNL-128, Precast/Prestressed Concrete Institute, Chicago, USA.
7. Ryder, J.F. (1975) Applications of fibre cement, Fibre Reinforced Cement and Concrete, RILEM Symposium 1975, Construction Press Ltd., UK, pp. 23-35.
8. Molloy and Associates (1994) A guide to premix, AR Glass Fiber Tech Topics, Issue #1, Molloy and Associates Inc., Hutchins, Texas, USA.
9. Peter, I.D. (1994) Mixing of glass fibre reinforced cement, Special Concretes Workability and Mixing (ed. Bartos), RILEM Proceedings 24, E&FN Spon, London, UK, pp. 73-79.
10. Editorial. (1986) Carbon Fiber Reinforced Curtain Walls, Concrete Construction, January, pp. 49.
11. Molloy, H.J. and Jones, J. (1993) Application and production using rapid hardening hydraulic cement composites, Proceedings of 9th Biennial Congress of the GRCA, Glass Fibre Reinforced Cement Association, Wigan, UK.
12. Molloy, H.J., Jones, J. and Harmon, T.G. (1994) Glass fibre reinforced concrete with improved ductility and long term properties, Thin Reinforced Concrete Products and Systems, American Concrete Institute Special Publication SP-146, pp. 70-90.
13. van der Plas, C., Yue, B. and Bijen, J. (1992) Effect of pozzolans on the durability of polymer modified glass fibre reinforced cement, Fly Ash, Silica Fume, Slag and Natural Pozzolans in Concrete, Proceedings of Fourth CANMET/ACI International Conference, Supplementary Papers, pp. 175-188.
14. Cem-FIL Star GRC-Material Properties, Cem-FIL International Ltd., Newton-le-Willows, Merseyside, U.K.

15. Akihama, S., Suenaga, T. and Banno, T. (1986) Mechanical properties of fibre reinforced cement composites, Int. Journal of Cement Composites and Lightweight Concrete, Vol. 8, No. 1, pp. 21-33.

16. Park, S.B. and Lee, B.I. (1991) Fabrication of carbon fiber reinforced cement composites, Fiber-Reinforced Cementitious Materials, Materials Research Society, Vol. 211, pp. 247-254.

17. Ando, T. et.al. (1990) Fabrication and properties for a new carbon fibre reinforced cement product, Thin-Section Fiber-Reinforced Concrete and Ferrocement, American Concrete Institute Special Publication SP-124, pp. 39-60.

18. Banthia, N., Moncef, A. and Sheng, J. (1994) Uniaxial tensile response of cement composites reinforced with high volume fractions of carbon, steel and polypropylene micro-fibers, Thin Reinforced Concrete Products and Systems, American Concrete Institute Special Publication SP-146, pp. 43-68.

19. Soroushian, P., Bayasi, Z. and Khan, A. (1990) Development of aramid fiber reinforced composites, Thin-Section Fiber Reinforced Concrete and Ferrocement, American Concrete Institute Special Publication SP-146, pp. 11-24.

20. Sorousian, P., Shah, Z. and Marikunte, S. (1994) Reinforcement of thin cement products with waste paper fibers, Thin Reinforced Concrete Products and Systems, American Concrete Institute Special Publication SP-146, pp. 25-42.

21. Gale, D.M. (1994) Synthetic fibers in thin-section cement products: a review of the state of the art, Thin Reinforced Concrete Products and Systems, American Concrete Institute Special Publication SP-146, pp. 11-24.

22. Schraeder, E.K. (1988) Fiber reinforced concrete, International Committee on Large Dams, Bulletin No. 40.

23. ACI Committee 544 (1993) Guide for specifying, proportioning, mixing, placing and finishing steel fiber reinforced concrete, American Concrete Institute Report ACI 544.3 R.

24. Johnston, C.D. (1984) Measures of the workability of steel fiber-reinforced concrete and their precision, ASTM Cement, Concrete, and Aggregates, Vol. 6, No. 2, pp. 74-83.

25. Johnston, C.D. (1994) Fiber-reinforced concrete, Significance of Tests and Properties of Concrete, ASTM Special Technical Publication STP 169-C, Chapter 51, pp. 547-561.

26. Johnston, C.D. (1994) Comparative measures of workability of fibre-reinforced concrete using slump, V-B and inverted cone tests, Special Concretes Workability and Mixing (ed. Bartos), RILEM Proceedings 24, E&FN Spon, pp. 107-118.

27. Nielsen, N.H. (1994) Proportioning fibres and mixing of fibre concrete, Special Concretes Workability and Mixing (ed. Bartos), RILEM Proceedings 24, E&FN spon, pp. 69-72.

28. Skarendahl, A. (1994) Rational steel fibre concrete production, Proceedings, International Symposium on Brittle Matrix Composites (ed. Brandt, et al), IKE and Woodhead Publishing, Warsaw, pp. 44-50.

29. ACI Committee 506. (1984) State-of-the-art report on fiber reinforced shotcrete, Concrete International, Vol. 6, No. 12, pp. 15-27.
30. Morgan, D.R. (1991) Steel fiber reinforced shotcrete for support of underground openings in Canada, ACI Concrete International, Vol. 13, No. 11, pp. 56-64.
31. Armelin, H.S. and Helene, P. (1995) Physical and mechanical properties of steel fibre dry mix shotcrete, ACI Materials Journal, Vol. 92, No. 3, pp. 258-267.
32. Skarendahl, A. (1992) Improving the performance of steel fibre shotcrete: the Swedish experience, High Performance Fiber Reinforced Cement Composites, RILEM Proceedings 15, E&FN Spon, pp. 156-163.
33. Morgan, D.R., McAskill, N., Carrette, G.G. and Malhotra, V.M. (1992) Evaluation of polypropylene fiber reinforced high-volume fly ash shotcrete, ACI Materials Journal, Vol. 89, No. 2, pp. 169-177.
34. Morgan, D.R. (1989) A comparative evaluation of plain, polypropylene fiber, steel fiber and wire mesh reinforced shotcretes, Transportation Research Record No. 1226, National Research Council U.S., pp. 78-87.
35. Wörner, J.D. and Techen, H. (1994) Mixing procedure of fibre concrete, Special Concretes Workability and Mixing (ed. Bartos), RILEM Proceedings 24, E&FN Spon, pp. 81-87.

15 PROPERTIES OF HIGH STRENGTH PLAIN AND FIBRE-REINFORCED CONCRETES

M.R. TAYLOR, F.D. LYDON and B.I.G. BARR
Division of Civil Engineering, Cardiff School of Engineering, University of Wales Cardiff, Cardiff, Wales, UK

Abstract

The apparent increased brittleness of high strength concretes can be significantly reduced by the use of fibre reinforcement. However, considering their high performance characteristics, comparatively little work has been carried out on high strength fibre-reinforced concretes. This paper reports on several aspects of the fresh and hardened properties of a range of such materials.

Mixes of strengths 40, 60, 80, 100 and 120 N/mm^2 at 28 days were made using crushed limestone and gravel coarse aggregates; for strengths above 80 N/mm^2, 10% silica fume and a superplasticiser were used. Mix proportions were chosen to give high workability and rheology suitable for the inclusion of three steel and three polypropylene fibre concentrations per strength. Eighty mixes were made for which workability, compressive and tensile strengths and toughness results were recorded and the optimum and maximum fibre concentration for each strength and fibre type ascertained.

Keywords: Fibre reinforcement, high performance concrete, high strength, toughness, workability.

1   Introduction

Currently there is a large amount of research being carried out on concretes which have enhanced properties in both the fresh and hardened states in comparison to the normal strength concretes (NSC) used in most practical applications at present. Of these high performance materials, the two main types are high strength concrete (HSC) and fibre-reinforced concrete (FRC).

HSCs are typically considered as those materials with compressive cube strengths

Production Methods and Workability of Concrete. Edited by P.J.M. Bartos, D.L. Marrs and D.J. Cleland. Published in 1996 by E & FN Spon, 2–6 Boundary Row, London SE1 8HN. ISBN 0 419 22070 4.

greater than 60 N/mm² at 28 days. The ability to produce such concretes easily and consistently is due to the relatively recent availability of chemical admixtures which allow high workability at low water contents and mineral admixtures which produce a relatively more homogeneous concrete. Though more expensive to produce than NSC, the advantages of HSCs, such as their high strength, high workability, increased early-age strength and greater long-term strength, has lead to their increased practical use. Such materials are inherently more durable and allow more economic construction through their ability to allow, for example, quicker stripping of formwork and more slender structural members. However, weighed against such advantages is the apparent increasing brittleness of concretes with increasing strength.

FRCs have been progressively developed since the 1960s [1]. Currently, a large number of types of fibre reinforcement and fibre composites are available. In parallel with the commercial development of FRCs, an intensive amount of research has been carried out to attempt to quantify their enhanced properties and to allow comparisons to be made between different fibre types. Though the workability of concrete tends to be increasingly reduced with the inclusion of greater concentrations of fibres, the improvements in impact resistance and enhanced toughness of concretes reinforced with low volumes of fibres and the increased strength and toughness of higher fibre volume materials, have received considerable attention; a range of impact and toughness tests has been proposed and evaluated.

There is presently a significant amount of research being carried out to investigate the properties of high strength fibre-reinforced concretes, in particular any increase in ductility which may be found. This paper reports on the practical experience gained from the study of a range of normal and high strength plain and fibre-reinforced concretes.

2   High performance concretes

2.1   High strength concrete
The ratio of water to cementitious binder is the most significant parameter affecting concrete strength with given materials. Consequently, low water/binder ratios are typical for HSCs. The advent of superplasticisers (SP) has permitted reasonable water and cement contents and low water/cement ratios to be used whilst allowing high workability if required. Furthermore, cement contents can be reduced to more practical amounts by modifying the Portland cement binder by the inclusion of mineral admixtures such as silica fume (SF), ground granulated blast-furnace slag (ggbs) and pulverised fuel ash (pfa). Usually, replacement of 10-12% of the Portland cement content with SF gives an economical concrete that has good characteristics.

Water demand, workability and strength are affected by the coarse aggregate maximum particle size. Though an increase in the volume concentration of the coarse aggregate tends to increase the strength of the concrete, there may be little opportunity to vary this and still meet the requirements of the fresh concrete. Coarse aggregate type can also significantly affect the strength of concrete. The stronger paste and denser aggregate/paste interface [2] afforded by the use of SP, SF and a reduced water/binder ratio produce a predominance of fractured particles at failure, regardless of whether the coarse aggregate is crushed or uncrushed. As a result, the strength of the concrete will show greater dependence upon more fundamental aspects of the coarse aggregate, such as its mineralogy and surface activity [3]. These will tend to limit the practical maximum strength of the concrete with given materials.

The fine aggregate affects water demand and stability of fresh concrete. Though various attempts have been made to optimise the fine aggregate content of HSCs [4,5,6], the proportions have invariably been adjusted empirically. This is especially so when further materials, such as fibre reinforcement, are included in the concrete mix.

2.2 Fibre-reinforced concrete

The enhanced properties of FRCs are well-known. They include improvements in impact strength and toughness, the control of cracking and failure mode by means of post-crack ductility and the reduction in workability compared to a similarly proportioned plain concrete. These properties are influenced by the strength, size, shape and orientation of the reinforcement.

The two most commonly used fibres are steel and polypropylene. To improve the bond between fibre and matrix, steel fibres are usually crimped along their length and/or have bent or hooked ends. Polypropylene fibres are normally used in a fibrillated form which allows the matrix to infiltrate the reinforcement.

The workability and stability of fresh concrete can show marked changes when fibres are included in the matrix due to increased internal friction between fibres and fibre/aggregate interaction. As a result, of the three most common methods of measuring the apparent workability (slump, compacting factor and vebe) only the latter is of practical value as it imparts sufficient energy to the concrete. Increasing size and aspect ratio of fibres and greater maximum coarse aggregate size have been shown to significantly increase the vebe time for a given concrete mix [7]. More fibres also increase the stability of FRC due to the tendency for water to be held around fibres which in turn reduces the amount available for bleeding [8].

Most research has shown that compressive and flexural strengths are reduced by polypropylene fibres [9,10] and increased by steel fibres [11] in comparison to similarly proportioned plain concretes. However, such differences are numerically small and of much greater significance are the enhancements in impact resistance and toughness gained from the inclusion of both fibre types.

These parameters are not easily quantified and a range of tests has been evaluated in recent years, of which greater attention has been paid to the measurement of toughness. This has generally been considered as being the energy absorption capacity as characterised by the load/deflection curve obtained from flexural tests. There is a good deal of variation in the methods used to interpret such curves and probably the most widely reported is the ASTM C1018 standard defined in Figure 1 [12]. Several problems are inherent within this and most of the other toughness measures available at present; currently deliberations are taking place to devise a better method using, for example, notched and compact test specimens.

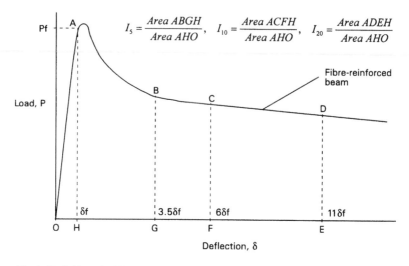

Fig. 1. Definition of ASTM C1018 toughness indices

3   Experimental programme

The main concretes used in this study needed to be:

1. Very workable.
2. Of 10 mm maximum coarse aggregate size.
3. Of sufficient stability to accept a reasonable amount of different types of fibre.
4. Of a good range of strengths; ideally 40, 60, 80, 100 and 120 N/mm² at 28 days, using only conventional materials, manufacture and curing.

Though it was realised that the use of pfa and ggbs as well as SF would produce more economic concretes with greater chloride resistance and relatively lower heats of hydration, it was decided that the main aim of the study, the fracture of HSC and fibre-reinforced HSC, could be carried out using solely SF and the results would probably not be affected greatly by changes in the binder. Consequently, for the 80, 100 and 120 grade concretes, 10% of the Portland cement binder was replaced by SF; these concretes also contained SP to achieve the required workability.

3.1 Materials used
A single batch of class 42.5 N Portland cement was used for the test programme. The cement properties, as supplied by the manufacturer, are shown in Table 1 and can be compared to published values for UK Portland cements [13].

The SF used was Elkem Emsac 500S, a 50:50 water:powder slurry. The SP was Conplast SP430, a polynaphthalene sulphonate-based admixture containing 40% active solids in solution. The fine aggregate consisted of a local sea-dredged sand of

Table 1. Portland cement properties

Constituent	%	Compound	%
SiO_2	19.8	C_3S	62.7
Al_2O_3	4.3	C_2S	10.2
Fe_2O_3	2.0	C_3A	8.0
CaO	63.9	C_4AF	6.2
MgO	2.4		
SO_3	3.0		
Na_2O	0.1		
K_2O	0.7		
Others	0.6		
Free lime	1.0		
LOI	2.2		

BS882 "F" grading and the coarse aggregates, both of 10 mm maximum size, were crushed limestone from South Wales and a pea gravel from the North Midlands of England. Some properties of the aggregates are given in Table 2. The fibres used were Bekaert Dramix ZL30/.50 made of low-carbon steel with length 30 mm, width 0.5 mm and hooked ends, and 19 mm long Fibermesh 6130 fibrillated polypropylene fibres.

Experience with the sand had shown that, when used in an air-dry condition, its absorption was extremely low and its effect on the free-water content of the concrete was negligible. Similarly, the effect of using air-dry coarse aggregates was negligible and mitigated by the use of a standardised mixing procedure.

3.2 Mix proportions
From the literature on HSC there is a wide range of mix compositions reported especially at higher strength levels. Typically, concretes with 28 day strengths greater than 90 N/mm^2 have cement contents in the range 450-500 kg/m^3 (though amounts significantly above and below these have been used) and water/binder ratios usually between 0.25 and 0.35, though lower values are not uncommon. The proportions of SF and SP vary depending upon specific types and, in the case of the latter, the ambiguity that exists in defining the amount used. Normally, the sand contents

Table 2. Aggregate properties

Aggregate	Relative density (oven dry)	Shape	Surface texture
Sand	2.60	Rounded/irregular	Smooth
Crushed limestone	2.65	Angular	Rough
Gravel	2.60	Rounded/irregular	Smooth

account for 35-45% of the total aggregate but again significantly less has been used. The actual amounts are dependent upon, amongst other things, sand grading.

From a range of trial mixes, plain concretes were chosen which matched the required strengths and the four criteria cited in the "Experimental programme". Two plain concretes and FRCs containing three concentrations of each fibre were to be made for each strength and coarse aggregate type. The exact amounts of fibre would depend upon the concrete grade and coarse aggregate and fibre type, though they would include, for each strength of concrete, the maximum fibre content. (This can be defined as that quantity of fibres which, without "balling", causes so great a reduction in workability that complete compaction of the concrete in a practical application would be difficult.) Using three different fibre concentrations would also allow an optimum amount to be ascertained (that is the quantity of fibres giving the maximum reduction in brittleness whilst retaining sufficient workability).

Because gravel coarse aggregate was used to allow comparison with the crushed limestone concretes, extra trial mixes were made to check for any effects on cube strength this might have. It was anticipated that, as strengths increased, higher strengths would be achieved by crushed limestone concretes relative to gravel ones of similar mix proportions due to better aggregate/paste bonding. Two further trial mixes were made with increased cement contents and reduced water/binder ratios and these confirmed that a practical 28 day ceiling strength of about 110 N/mm² appeared to exist for gravel concretes using the available materials. The chosen mix proportions and fibre contents are shown in Table 3.

3.3 Mixing details
The standardised mixing procedure consisted of the cement and sand being mixed first before the water, SF (where necessary) and coarse aggregate. For the grade 80 concretes and above, SP was added until high workability was achieved. The complete mixing time was about 5 minutes. For the FRC mixes, three plain concrete control cubes were taken out of the mix prior to the addition of the fibres. These cubes were tested at 28 days to confirm the strength and repeatability of the mixes. Fibres were added by sprinkling one handful at a time into the concrete whilst mixing continued. As experience of handling the FRCs increased, it was found that in some cases more fibres would be incorporated into the mix without balling. Consequently, a greater volume of steel reinforcement was included in the grade 60, 100 and 120 gravel concretes compared to similar strength limestone mixes.

3.4 Fresh concrete
The experience gained from the trial mixes was used to ensure that the concretes chosen had a satisfactory behaviour in the fresh state, namely, high workability and sufficient stability to avoid segregation and handling problems. It was not part of the research objectives to characterise the complex rheology of the HSCs. As noted earlier, the vebe test was thought to be the only relevant measure of workability of FRCs and tests were carried out about 5 minutes after the end of mixing (experience from trial mixes had shown that there was negligible change in workability in this time). No attempt was made to measure such values as yield stress and plastic viscosity, analogous to Bingham body behaviour, which might have been used to

Table 3. Mix proportions

Aggregate type	Nominal grade	Mix proportions						Fibre content (% by volume)	
		Cement (kg/m³)	Fine aggregate (kg/m³)	Coarse aggregate (kg/m³)	Water (kg/m³)	Silica fume (kg/m³)	Superplasticiser (ml/kg of cement)	Polypropylene	Steel
Limestone	40	400	800	1000	224	-	-	0 / 0.52 / 0.78 / 1.04	0 / 0.52 / 0.78 / 1.04
	60	400	724	1124	200	-	-	0 / 0.26 / 0.52 / 0.78	0 / 0.26 / 0.39 / 0.52
	80	340	721	1190	170	37	13.5	0 / 0.26 / 0.52 / 0.78	0 / 0.26 / 0.52 / 0.78
	100	400	708	1188	140	44	23.0	0 / 0.26 / 0.52 / 0.78	0 / 0.26 / 0.39 / 0.52
	120	510	653	1086	122	56	35.9	0 / 0.26 / 0.39 / 0.52	0 / 0.26 / 0.39 / 0.52
Gravel	40	400	800	1000	224	-	-	0 / 0.52 / 0.78 / 1.04	0 / 0.52 / 0.78 / 1.04
	60	400	724	1124	200	-	-	0 / 0.26 / 0.52 / 0.78	0 / 0.26 / 0.52 / 0.78
	80	370	714	1188	159	41	21.5	0 / 0.26 / 0.52 / 0.78	0 / 0.26 / 0.52 / 0.78
	100	455	696	1151	132	50	26.5	0 / 0.26 / 0.52 / 0.78	0 / 0.26 / 0.52 / 0.78
	120	510	653	1086	122	56	35.9	0 / 0.26 / 0.39 / 0.52	0 / 0.26 / 0.52 / 0.78

distinguish between the effects of water content changes and SF content changes [14], or the effects these values might have on such parameters as compactibility and finishability.

A consistent level of observed behaviour of the fresh concrete (that is satisfactory handling, compaction and finishing) was achieved using experience gained from the trial mixes and this appeared to confirm the methodology used. Immediately after measuring the vebe time, test specimens were made. These consisted of three 100 mm cubes, three 100 x 200 mm cylinders, nine 100 x 100 x 500 mm beams and three 150 mm cubes, which were compacted on a vibrating table. The operation took no longer than 30 minutes, over which time no significant loss in workability was noted. All test specimens were kept in their moulds and covered to prevent evaporation until the following day when they were demoulded and put in water at 20°C until required for testing.

4   Test procedures

All samples were tested at 28 days. 100 mm cube compressive strength (f_{c28}), modulus of rupture (MOR) (f_{b28}) and split cylinder (f_{s28}) indirect tensile strengths were determined. A further indirect tensile test, torsion of a beam, was also carried out. This has been developed at Cardiff and uses an elastic analysis of the shear induced during torsion to derive the tensile failing stress (f_{t28}) [15]. The testing arrangement is shown in Figure 2.

Toughness measurements were carried out using two fracture tests. The first was identical to the G_f test [16] and is illustrated in Figure 3. The beam is notched at mid-span (in this case the notch depth was 50 mm but other depths are valid) and pinned at both ends with one support free to move horizontally. Most importantly, a yoke was placed on the beam which allowed the mid-span deflection to be measured relative to the beam itself. According to Banthia et al [17], it is only by using such an arrangement that the real deflection of the beam can be determined (eliminating the extraneous deflections inherent in measuring cross-head deflections such as localised crushing at the supports).

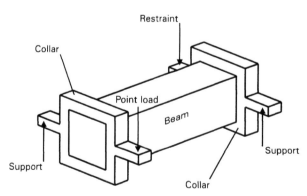

Fig .2. Torsional testing arrangement

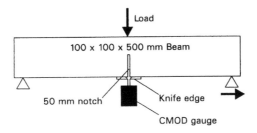

Fig .3. Notched beam toughness testing arrangement

Knife edges were attached on either side of the notch and used to mount a crack-opening displacement gauge. The test was carried out in a Schenck machine which allowed closed-loop control, that is the rate of crack-mouth opening displacement (CMOD), as measured by the gauge, was kept constant through continuous monitoring and variation of the load. The beams were loaded centrally and load/CMOD and load/displacement graphs were plotted autographically.

The second fracture test was based on earlier work by Barr et al [18] and consisted of a 150 mm cube with 50 mm deep notches in the middle of opposite faces loaded eccentrically above one of the notches as shown in Figure 4. The CMOD and crack-tip opening displacement (CTOD) of the opposing notch were measured using knife edges and two displacement gauges. Again, closed-loop testing controlled by the CMOD was used and load/CMOD and load/CTOD graphs were plotted autographically.

5   Results and discussion

The results are considered in three parts: the fresh concrete, strength test results and toughness test results.

5.1 Fresh concrete
The workability results are given in Table 4. It can be seen that all of the plain concretes have high to very high workabilities in the range 1-4 seconds. In all cases,

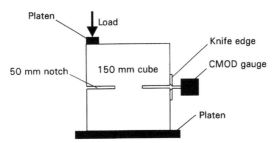

Fig. 4. Notched cube toughness testing arrangement

Table 4. Workability results

Nominal grade	Polypropylene fibre				Steel fibre			
	Limestone		Gravel		Limestone		Gravel	
	Fibre content (% by volume)	Vebe (sec)	Fibre content (% by volume)	Vebe (sec)	Fibre content (% by volume)	Vebe (sec)	Fibre content (% by volume)	Vebe (sec)
40	0	1	0	1	0	1	0	1
	0.52	3	0.52	2	0.52	1	0.52	1
	0.78	3	0.78	3	0.78	2	0.78	2
	1.04	8	1.04	8	1.04	2	1.04	3
60	0	3	0	3	0	4	0	2
	0.26	6	0.26	6	0.26	6	0.26	3
	0.52	7	0.52	8	0.39	7	0.52	5
	0.78	9	0.78	11	0.52	9	0.78	6
80	0	1	0	1	0	1	0	1
	0.26	5	0.26	5	0.26	3	0.26	2
	0.52	9	0.52	7	0.52	3	0.52	3
	0.78	16	0.78	11	0.78	8	0.78	4
100	0	2	0	1	0	4	0	1
	0.26	3	0.26	2	0.26	5	0.26	3
	0.52	5	0.52	7	0.39	6	0.52	3
	0.78	10	0.78	12	0.52	10	0.78	5
120	0	2	0	2	0	3	0	2
	0.26	7	0.26	5	0.26	4	0.26	4
	0.39	8	0.39	7	0.39	5	0.52	5
	0.52	12	0.52	10	0.52	7	0.78	6

though increasing the content of both types of fibre caused a reduction in the measured workability, the numerical values did not adequately describe the observed rheological behaviour. The inclusion of fibres made the concrete noticeably harder to work with and in all cases the concretes containing the maximum fibre content were difficult to handle; taking scoops from the mixer, for instance, was very hard.

The amount of SP used is dependent upon both the water content (normally the dominant factor affecting workability) and the binder, since it has particles which need deflocculating. It can be argued therefore that a good correlation between the amount of active solids in the SP per m³ of concrete and the water/binder ratio exists if the differences in the size and shape of the cement and SF particles are ignored. (These are likely to affect both the number of flocs formed and the beneficial effects the spherical SF particles may have on workability. However, as the SF was in slurry form, which would tend to minimise the number of flocs, and the amount used was only 16% of the volume of cement used, such differences should be minimised.) Figure 5 shows a plot of SP active solids against water/binder ratio which could be said to be linear between w/b ratios of 0.45 and 0.26, below which it appears to be non-linear.

Assuming the relative densities of cement and SF are 3.15 and 2.2 respectively, then the data from the mixes allow the volume of paste (cement + SF + water) per m³ of plain concrete to be calculated. All mixes had mean paste volumes of 293 l/m³ (± 2%) bar that for the highest strength concrete with 309 l/m³.

To obtain this high strength a low water/cement ratio was essential. The high cement content of just over 500 kg/m³ meant a significant increase in SP was needed. However, a linear correlation between SP active solids content and binder content per m³ of concrete is seen to exist (Figure 6). Therefore, it may be that the need for deflocculating the powder particles is the major requirement for workability and evidently much experimental work is necessary to check this.

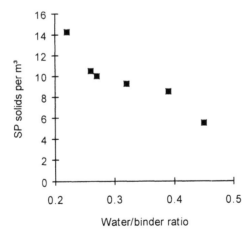

Fig. 5. SP solids per m³ against water/binder ratio

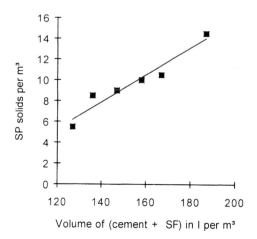

Fig. 6. SP solids per m³ against volume of binder

5.2 Strength test results
The results are given in Tables 5 and 6.

5.2.1   Cube strength
The coefficients of variation of the results were very low with most being within 3% and all less than 5%. Though the strengths of the plain gravel concretes were less than those of plain limestone concretes with similar mix proportions (grades 40, 60 and 120), such differences were smaller than expected and are not considered significant. For both aggregate types, a greater concentration of polypropylene fibres caused a decrease in the 28 day cube strength, whereas increasing volumes of steel fibre gave increased cube strength. For both fibre types, where comparison was possible (same mix proportions and fibre contents), the limestone-based FRCs were again stronger than similar gravel concretes.
 It should be possible from the given data to:

1. Estimate the binder content required for a given 28 day mean cube strength.
2. Estimate the volume of binder (cement and where necessary SF) per unit volume of concrete.
3. Estimate the amount of SP active solids per unit volume of concrete and hence the volume of SP solution required.
4. Estimate the water/binder ratio and the water content per unit volume of concrete.

The proportion of fine aggregate can only be estimated from the available materials but, with the common concreting sands used in normal weight concrete, a value of 35 to 40% of the total aggregate content is reasonable under practical circumstances.

Table 5. Strength and toughness results for polypropylene fibres

Nominal grade	Limestone						Gravel					
	Fibre content (% by volume)	f_{c28} (N/mm²)	f_{b28} (N/mm²)	f_{t28} (N/mm²)	f_{s28} (N/mm²)	I_5	Fibre content (% by volume)	f_{c28} (N/mm²)	f_{b28} (N/mm²)	f_{t28} (N/mm²)	f_{s28} (N/mm²)	I_5
40	0	41.0	5.5	-	3.1	3.4	0	41.0	5.8	3.6	2.4	3.7
	0.52	37.6	4.6	4.9	3.0	3.0	0.52	33.7	4.4	4.0	2.6	3.9
	0.78	35.6	4.4	4.6	2.8	3.0	0.78	32.3	4.0	4.0	3.3	3.4
	1.04	33.2	4.4	4.1	3.0	4.5	1.04	30.7	4.0	4.0	2.8	4.2
60	0	58.9	5.9	5.7	3.5	3.1	0	58.4	6.2	6.6	3.9	3.8
	0.26	54.2	6.5	6.7	4.0	2.8	0.26	51.1	6.0	6.1	3.4	3.6
	0.52	47.9	6.0	5.8	3.9	3.3	0.52	45.8	5.7	6.0	3.3	3.9
	0.78	54.7	6.8	6.7	4.0	2.9	0.78	46.7	5.9	5.8	3.2	3.9
80	0	80.4	7.7	8.1	4.0	2.6	0	81.3	7.7	7.3	3.8	2.9
	0.26	67.8	6.9	6.9	4.3	3.2	0.26	71.3	6.2	6.0	3.8	3.3
	0.52	68.5	7.6	6.8	4.7	3.3	0.52	65.2	7.2	6.4	4.4	3.7
	0.78	62.2	6.7	6.4	4.5	3.2	0.78	69.3	7.9	7.4	4.5	3.8
100	0	100.0	10.4	10.8	5.1	2.7	0	98.7	11.0	11.1	5.7	2.7
	0.26	96.0	11.0	11.0	5.4	2.4	0.26	92.9	11.2	11.0	5.4	2.3
	0.52	96.0	11.0	10.8	5.7	2.7	0.52	93.8	11.3	11.3	6.2	3.1
	0.78	90.0	10.4	10.2	6.5	3.2	0.78	92.9	10.7	11.0	5.9	3.4
120	0	115.2	11.7	10.5	5.1	2.5	0	113.2	10.6	12.0	4.7	2.9
	0.26	107.0	11.1	11.1	6.0	3.0	0.26	102.0	10.1	10.4	5.6	3.2
	0.39	110.5	12.3	11.5	6.2	2.7	0.39	92.3	9.5	9.7	5.1	3.0
	0.52	114.3	12.0	10.3	6.3	2.5	0.52	95.7	9.2	8.7	5.6	3.4

Table 6. Strength and toughness results for steel fibres

Nominal grade	Limestone Fibre content (% by volume)	f_{c28} (N/mm²)	f_{b28} (N/mm²)	f_{t28} (N/mm²)	f_{s28} (N/mm²)	I_s	Gravel Fibre content (% by volume)	f_{c28} (N/mm²)	f_{b28} (N/mm²)	f_{t28} (N/mm²)	f_{s28} (N/mm²)	I_s
40	0	43.8	6.6	6.1	3.4	2.6	0	41.5	5.1	5.0	2.8	2.5
	0.52	42.0	6.9	6.7	4.1	3.2	0.52	41.2	6.5	6.1	3.8	3.3
	0.78	44.3	9.1	6.8	5.2	3.7	0.78	42.3	8.7	6.1	3.9	4.5
	1.04	47.0	10.9	7.4	5.8	3.9	1.04	42.7	9.8	6.3	3.4	3.2
60	0	67.1	8.2	8.7	5.2	2.5	0	59.3	5.7	6.7	4.3	2.5
	0.26	67.7	7.8	8.6	5.2	3.3	0.26	58.7	7.1	7.1	4.6	3.1
	0.39	69.6	8.0	8.2	5.7	3.2	0.52	64.2	8.4	7.2	5.7	3.5
	0.52	71.0	9.5	8.8	7.3	3.4	0.78	64.7	9.4	7.3	6.8	3.8
80	0	85.3	9.0	9.4	4.5	2.2	0	83.3	8.3	9.3	5.4	2.5
	0.26	84.7	9.9	9.5	6.8	3.0	0.26	82.3	8.6	9.5	6.1	3.1
	0.52	86.0	10.2	9.7	7.0	3.4	0.52	85.7	11.4	10.7	7.1	4.0
	0.78	88.6	11.5	10.7	9.7	4.4	0.78	86.2	13.5	10.0	7.4	4.1
100	0	104.2	10.8	12.0	6.9	2.5	0	101.3	11.5	12.3	5.6	2.7
	0.26	104.4	11.1	11.8	7.8	3.1	0.26	98.5	10.6	11.9	7.6	3.6
	0.39	104.3	11.8	11.5	8.7	3.6	0.52	102.5	12.1	12.2	8.6	3.7
	0.52	107.7	10.9	11.4	8.7	3.3	0.78	105.2	12.8	12.8	9.7	3.5
120	0	110.0	13.2	13.2	6.4	1.7	0	110.8	12.5	13.1	6.0	2.3
	0.26	109.3	13.5	13.1	8.1	2.7	0.26	110.3	12.4	13.1	6.4	2.8
	0.39	113.3	13.6	13.9	8.7	2.9	0.52	112.0	12.2	13.6	8.7	2.8
	0.52	122.2	12.6	12.4	10.4	2.8	0.78	118.5	14.1	13.6	9.0	3.9

5.2.2 Indirect tensile strengths

The coefficients of variation for all tests were very good. For both aggregate types, increasing concentrations of polypropylene fibre gave reduced indirect tensile strengths. However, any changes were not significant, especially compared to corresponding changes in cube strength. A similar trend, but with increasing strengths as fibre content was increased, was noted for both limestone and gravel steel fibre-reinforced concretes.

The indirect tensile strengths evaluated from torsion tests gave encouraging results with values similar to the MOR and it would appear that the test is worthy of further investigation.

5.2.3 Toughness test results

Due to lack of space available, only the toughness results from the beam tests are presented here (Tables 5 and 6). However, though the testing arrangements look quite different, both the beam and cube have a ligament length of 50 mm and CMOD is measured at 50 mm from the root of the notch (and start of the crack). Consequently, the results from both tests are similar.

Failure of the 120 grade polypropylene FRCs occurred by rupture of the fibres and consequently failure was quite sudden. The remaining grades, as a result of a weaker fibre/paste bonding, failed in a more ductile manner by pull-out of the fibres. Due to the excellent bond afforded by their hooked ends, steel fibre-reinforced concretes of all grades failed by snapping of the reinforcement.

All FRC beams showed CMODs of over 3 mm without actual failure occurring (the fibre reinforcement spanning the crack and holding the beam together). In comparison, CMOD at first crack and CMOD at failure for the plain concretes were of the order of 0.05 mm and 1 mm respectively. Typical load/CMOD curves for increasing fibre content are shown in Figure 7.

Both I_5 and I_{10} toughness indices have been calculated of which the I_5 values only are presented here. It can be seen that the I_5 index does not extend far enough to capture the significant increase in toughness available from the use of fibres, nor does

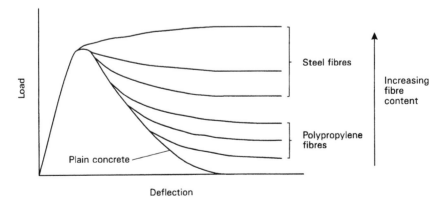

Fig. 7. Typical load/CMOD curves using a closed-loop testing arrangement

it allow different types of fibre to be differentiated. Unfortunately, the same is true of I_{10} and other higher indices, and it is clear that an improved definition of toughness requires development to describe better the enhanced toughness of the FRCs.

The optimum fibre content in all cases was taken as that volume below the maximum amount included; these concretes were significantly tougher than plain concrete whilst remaining easy to handle and compact.

6   Conclusions

Plain HSCs with high workability and good stability can be easily produced in the laboratory using good quality aggregates, SF and SP. The rheology of these concretes is such that they can be reinforced by sufficient volumes of polypropylene and steel fibre to increase significantly their toughness, while their strengths in compression and tension remain relatively constant. Work is required to develop a method which fully quantifies the enhanced toughness of such materials.

7   Acknowledgements

The work reported in this paper forms part of EPSRC grant GR/J79379 "Fracture characteristics of ordinary and high performance concretes by means of relaxation tests". The authors wish to express their sincere thanks to EPSRC for its support of this work.

8   References

1. Romualdi, J.P. and Batson, G.P. (1963) Mechanics of crack arrest in concrete. *Journal of Engineering Mechanics*, ASCE, Vol. 89. pp. 147-168.
2. Goldman, A. and Bentur, A. (1989) Bond effects in high-strength silica-fume concretes. *ACI Materials Journal*, Vol. 86, No. 5. pp. 440-447.
3. Aïtcin, P.-C. and Mehta, P.K. (1990) Effect of coarse-aggregate characteristics on mechanical properties of high-strength concrete. *ACI Materials Journal*, Vol. 87, No. 2. pp. 103-107.
4. de Larrard, F. (1993) A survey of recent researches performed in the French "LPC" network on high-performance concretes, in *Proceedings of the Symposium on High Strength Concrete, 20-24 June, Lillehammer, Norway*, (eds. I. Holand and E.Sellevold), Norwegian Concrete Association, Oslo, Vol. 1, p. 57.
5. Soutsos, M.N. and Domone, P.L.J. (1993) Design of high-strength concrete mixes with normal weight aggregates. *ibid*. Vol. 2, p. 937.
6. Powers, T.C. (1968) *Properties of Fresh Concrete*, John Wiley & Sons, New York, p. 32.
7. Edgington, J., Hannant, D.J. and Williams, R.I.T. (1974) Steel-fibre reinforced concrete. *Building Research Establishment Current Paper CP 69/74*.

8. Ritchie, A.G.B. and Rahman, T.A. (1974) The effect of fiber reinforcements on the rheological properties of concrete mixes. *An International Symposium: Fiber Reinforced Concrete*, SP-44, American Concrete Institute, Detroit.
9. Bayasi, Z. and Zeng, J. (1993) Properties of polypropylene fiber reinforced concrete. *ACI Materials Journal*, Vol. 90, No. 6. pp. 605-610.
10. Ramakrishnan, V., Gollapudi, S. and Zellers, R. (1987) Performance characteristics and fatigue strength of polypropylene fiber reinforced concrete. *Fiber Reinforced Concrete Properties and Applications*, (eds. S.P. Shah and G.B. Batson), SP-105, American Concrete Institute, Detroit, pp. 141-158.
11. Wafa, F.F. and Ashour, S.A. (1992) Mechanical properties of high-strength fiber reinforced concrete. *ACI Materials Journal*, Vol. 89, No. 5. pp. 449-455.
12. ASTM (1992) Standard test method for flexural toughness and first-crack strength of fiber-reinforced concrete (using beam with third-point loading). *ASTM C1018-92, ASTM Annual Book of Standards*, Vol. 04.02, ASTM, Philadelphia, USA. pp. 277-295.
13. Corish, A. (1994) Portland cement properties - updated. *Concrete*, Vol. 28, No. 1. pp. 25-28.
14. Helland, S. (1984) Slipforming of concrete with low water content. *Concrete*, Vol. 18, No. 12. pp. 19-21.
15. Timoshenko, S. and Goodier, J.N. (1951) *Theory of Elasticity*. McGraw-Hill, New York, pp. 277-278.
16. RILEM Draft Recommendation (1985) Determination of fracture energy of mortar and concrete by means of three-point bend tests on notched beams. *Materials and Structures*, Vol. 18, No. 106. pp. 285-290.
17. Banthia, N., Trottier, J.-F., Wood, D. and Beaupre, D. (1992) Steel fibre reinforced dry-mix shotcrete. Effect of fibre geometry on fibre rebound and mechanical properties. *Fibre Reinforced Cement and Concrete: Proceedings of the Fourth RILEM International Symposium*, (ed. R.N. Swamy), E &FN Spon, London, pp. 277-295.
18. Barr, B., Evans, W.T. and Dowers, R.C. (1981) Fracture toughness of polypropylene fibre concrete. *The International Journal of Cement Composites and Lightweight Concrete*, Vol. 3, No. 2. pp. 115-122.

16 DEVELOPMENT AND TESTING OF SELF-COMPACTING LOW STRENGTH SLURRIES FOR SIFCON

D.L. MARRS and P.J.M. BARTOS
Advanced Concrete Technology Group, Department of Civil, Structural and Environmental Engineering, University of Paisley, Paisley, Scotland, UK

Abstract
Slurry infiltrated fibre concrete, or SIFCON, is a high performance composite material produced by infiltrating a preplaced steel fibre bed with a cement based slurry. The resulting material possesses outstanding properties which suggest that it will have significant applications in the field of seismic resistant structures. Current production methods, which include the use of vibration to aid the infiltration process and which employ the use of expensive cement rich slurries, have hindered the development of possible applications.

This paper outlines the first part of a research project to investigate the use of SIFCON in structural applications, and in particular, seismic resistant structures. A study of the fluid cement based slurries is presented and test procedures are proposed which measure the effects of varying material contents on fluidity, cohesion and infiltration. Low strength slurries consisting of cement, fine sand, superplasticiser and a viscosity agent are proposed. These slurries have excellent fluidity and can easily infiltrate fibres without the aid of vibration, and at the same time, posses excellent stability with regard to segregation and bleeding.
Keywords: Cohesion, steel fibres, infiltration, Marsh cone, SIFCON, slurries

1 Background

1.1 Introduction

Slurry Infiltrated Fibre Concrete, (SIFCON), is a high performance cement based composite which contains a high content of steel fibres. The development and applications of conventional steel fibre reinforced concretes have been hindered by mixing and placing difficulties which limit the maximum fibre contents to between 2 to 3 percent by volume. SIFCON removes the need to include the fibres at

Production Methods and Workability of Concrete. Edited by P.J.M. Bartos, D.L. Marrs and D.J. Cleland.
Published in 1996 by E & FN Spon, 2–6 Boundary Row, London SE1 8HN. ISBN 0 419 22070 4.

the mixing stage and thus the fibre content is limited only by the geometric properties of both the fibres and the mould or form. Fibres are first preplaced, by hand or by mechanical means. The fibre bed is then infiltrated with a cement based slurry, usually with the aid of vibration. The result is a material containing between 4 and 27 percent by volume of fibres, which has outstanding material properties with respect to strength, ductility and energy absorption capacity.

1.2 Applications
The applications of SIFCON to date have generally made use of its excellent blast resistant nature and high strength. Schneider [1] reported successful applications such as pavement overlays, impact resistant panels and bridge repairs to name but a few. Since first reported by Lankard and Newell [2] the material properties of SIFCON have been extensively investigated. Naaman et al [3-4] have taken this research further and have considered the use of SIFCON in conjunction with conventional steel reinforcementand in seismic resistant structures . This research was the first step towards developing SIFCON as a truly structural material. Important conclusions   include the need   for a low strength SIFCON, which displays similar shrinkage properties to a typical structural concrete.

1.3 Production
Laboratory practice has been to aid the infiltration of the fibre bed by the use of vibration. This process is not only difficult to reproduce in practice, but has a significant effect on fibre alignment, especially when steel reinforcement is present. Vandengerghe [5] concluded that the optimum fibre for use in most applications is in the region of 50-60mm in length, and that the improved performance obtained when using shorter fibres did not justify the cost implications of the increased fibre content. Fig. 1 shows photographs of fibre beds consisting of two of the most commonly used fibres to make SIFCON, Dramix ZL 30/0.5 and ZL 60/0.8

ZL 30/0.5   ZL60/0.8

Fig. 1 Examples of fibre beds showing relative magnitude of interstitial spaces

The photographs show the approximate magnitude of the interstitial spaces between fibres. These interstitial spaces are a key parameter affecting the infiltration of the slurry. The use of the longer fibres (i.e. ZL 60/0.8) present the real opportunity for producing SIFCON using a self compacting, fluid slurry.

2 Research significance

To date little research has been carried out which specifically looks at the rheology of the cementitious slurries which are used to make SIFCON. The cement rich slurries which have been used are costly in comparison to conventional concretes and this together with awkward production techniques has hindered the development of the material. The aim of this research is to develop reliable site test methods which will aid the development of low strength fluid slurries which are capable of infiltrating fibre beds without the need for vibration. The use of cost effective materials can therefore be investigated and it is hoped that developments in this area, together with more efficient production techniques, will lead to increased acceptance of SIFCON.

3 Test procedures

3.1 Requirements
Like most concretes, the materials used to make SIFCON, are local in nature. This applies to aggregates, cements, cement replacements and additives. Testing procedures are therefore required if the effects on infiltration properties of different materials, and material contents, are to be investigated. Simple tests are preferred, which can quickly provide information on slurry rheology and which can also act as a site quality control measure.

3.2 Flow cones
Mondragon [6] used a flow cone to look at the fluidity of various slurry mixes containing both cement and flyash. Other researchers have used flow cones of varying dimensions to look at the fluidity of cement pastes and these include Al-Manaseer and Keil [7], de Larrard [8] and Aitcin [9]. Although there is no definitive standard in Europe, the Marsh cone (Fig. 2) is probably the best known, and has been used extensively in the petrochemical industry.

Results using the Marsh cone are obtained by placing 1.5 litres of slurry in the cone with the outlet sealed. The time taken for 1.0 litre of slurry to flow into a container is measured to the nearest 0.5 seconds.

The parameter measured by a flow cone is often loosely described as the fluidity. In reality, flow cones measure a combination of the rheological properties of the slurry and the roughness of the cone walls. Lombardi [10] recognised this fact using the term 'apparent viscosity' and went on to show that, for grouts which have low values of cohesion, flow cones do actually represent the viscosity of the grout.

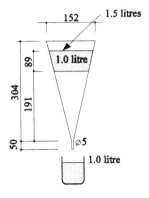

Fig.2   Marsh flow cone dimensions

3.3 Plate Cohesion Meter

It has been shown that the Marsh cone provides a measure of the viscosity of cohesionless grouts which behave approximately as newtonian fluids. Slurries for use in the production of SIFCON do not all fall into this category. It is therefore necessary to include a test which will in some way measure the cohesiveness of the slurry which is related to the yield stress of the material. The simple plate cohesion meter developed by Lombardi [10] was chosen. This simple and quick test consists of a plate (100mm x 100mm x 1mm thick) which, for this application, was manufactured from similar material to the steel fibres used (Fig. 3).

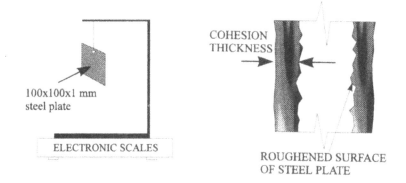

Fig. 3 Plate cohesion meter as developed by Lombardi

The plate is weighed before and after being submerged in the slurry, thus enabling a mass of material retained on the plate to be calculated. If the specific weight of the slurry is known then a cohesive thickness may be calculated. The plate is abraded to ensure adhesion between the steel and the slurry. The thickness calculated is a measure of cohesion, which is essentially the ability of the material to adhere to itself, and is directly related to the yield stress of the slurry.

Lombardi [10] developed a system whereby plastic viscosity could be measured using a Marsh cone in conjunction with the plate cohesion meter. The results compared well with those obtained using a rotary viscometer and suggest that the accuracy is sufficient for field purposes.

For the purposes of this investigation, the value of the cohesion thickness is sufficient to monitor the effect of cohesion, and hence yield stress, on infiltration.

3.4 Direct infiltration tests

To gauge the effectiveness of the flow cone test and the plate cohesion meter, with respect to the prediction of infiltration properties, a direct test was devised. This test consists of a column of fibres (60mm diameter x 500mm high) as shown in Fig. 4

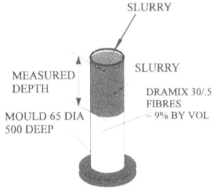

Fig. 4 Direct infiltration test apparatus

The fibres used to pack the column were chosen so as to represent approximately the fibre content when using 60mm long fibres. The combination of 30mm long fibres and column edge effects provided a suitable content. The results from the test are in the form of a measured depth as shown in Fig. 4.

4 Materials

4.1 Water/cement (w/c)

The w/c is the most influential parameter with regard to the strength of the slurry. Mondragon [6] suggests that the upper practical limit for the w/c is 0.4, otherwise segregation of the cement particles will occur. If however the slurries are to have comparable strength characteristics to conventional structural concrete, then higher values of w/c must be considered and some method of stopping the segregation found.

4.2 Superplasticisers

Four commercially available superplasticisers were considered in this project and these are detailed in Table 1.

Table 1. Superplasticisers

Reference	Active ingredient	Specific gravity kg/m³	Form
S1	Melamine Formaldehyde	1110	liquid
S2	Lignosulphate	1150	liquid
S3	Melamine Formaldehyde	1240	liquid
S4	vinyl copolymer	1110	liquid

The flow cone was used to determine which of the four admixtures was the most compatible with the cement being used (ordinary portland cement BS12:1991 class 42.4N). The slurry used was a simple cement paste with a w/c of 0.4. Fig. 2 shows clearly that additive S4 is the most efficient.

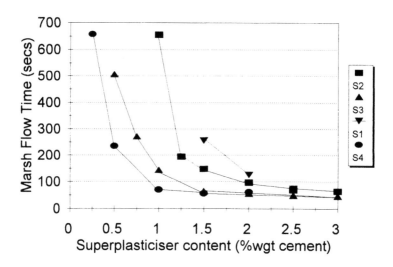

Fig. 5 Effect of superplasticiser content on Marsh flow time

The choice of superplasticiser in practice depends not just on this compatibility. Other factors such as cost and availability must be considered. The information provided by the test is useful, however, as it also allows the quantity of superplasticiser required for saturation to be determined.

4.3 Fine sand

Balaguru [11] found that sand may be added to the cement up to a ratio of 1:1.5 (cement:sand) before the flexural strength is adversely affected. The use of fine sand to create what is essentially a fluid mortar was used by Naaman [3] to create a low strength SIFCON. The problem of using large amounts of sand is that fluidity is lost if the sand content is increased and the w/c is not. If adequate fluidity is provided by increasing the w/c, the problem changes to one of bleeding and segregation. The use of very fine sands overcomes the problem of bleeding to some extent, however the increased specific area tends to decrease the fluidity of the mix. The use of high values of w/c will significantly increase the yield however this will have a detrimental effect with regard to durability.

There has been a tendency to assume that, because the slurry is required to infiltrate a fibre bed, the aggregate used should be very fine. As Fig. 1. shows the interstitial spaces between the fibres are of the order of 1.5-2.5mm. This would suggest that the aggregates do not need to be ultrafine. Segregation of large sand particles can result in the fibres causing a sieving effect which will be detrimental to both infiltration and quality. In practical terms the larger the grain size which can be used the better. This is because the larger sand particles are more effective in reducing shrinkage and also that the cost of very fine sand tends to be prohibitive. The Marsh cone was used to investigate the effect of sand grading on the apparent viscosity of slurries. Fine sand obtained from local sources was used. The material obtained from the quarry was sieved to provide the graded sands shown in Fig. 6.

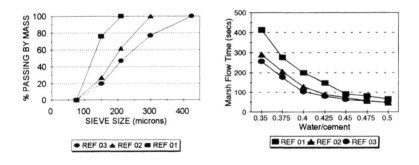

Fig. 6 Effect of sand particle size on Marsh flow time

A sample of the results obtained is also shown in Fig. 6 and this suggests that the fluidity is improved by using a larger grading. This is not altogether unexpected as the larger particles will have a smaller specific surface. The problem with using larger particle sizes is that fluid slurries are unable to hold the grains in suspension and segregation will occur.

4.4 Pulverised fuel ash (flyash)
Pulverized fuel ash (flyash) conforming to BS 3892 Part 1 : 1993, provided by Scottish power ash sales, was used as a cement replacement material. Flyash has been extensively used in SIFCON research and, as a waste product from the coal burning power industry, it is now readily available.

The fine particles (<12% retained in 45μm sieve) help to reduce bleeding and segregation, while the pozzolanic properties contribute to the strength of the material. The latter property, an advantage in many cases, proves to be a hinderance when considering the specific application of SIFCON as a ductile material for seismic applications. This is due to the fact that the use of flyash results in a 'late' strength which may cause maximum strength criterion to be exceeded. Strength reduction may be achieved by replacing cement by flyash however final ultimate strengths are difficult to predict.

4.5 Viscosity agents
Viscosity agents are now used in the field of compaction-free or flowing concrete. The aim is to avoid detrimental segregation of aggregate whilst maintaining fluidity and the ability to flow. The idea of using viscosity agents in SIFCON is a new one. The purpose is to allow increased w/c and increased sand particle size without harmful segregation, and at the same time allowing the cohesiveness of the slurry to be controlled.

5 Infiltration tests

5.1 Test procedure
A total of 20 different mix designs were created using the materials described above and direct infiltration tests were carried out for each. The results recorded were the Marsh flow time, the cohesion thickness and the depth of infiltration.

The aim was to determine the relevance of the Marsh flow cone and plate cohesion meter to the prediction of infiltration properties.

5.2 Marsh flow cone v infiltration depth
It quickly became apparent that despite the flow cones ability to predict trends in situations were material contents were varied, it was unable to compare slurries containing different materials. Indeed for the 20 tests carried out there was no distinguishable relationship between flow time and infiltration depth. The most obvious reason for this is that the flow cone measurements represent a combination of rheological properties of the slurry. It is therefore impossible to distinguish between a cohesive slurry and a purely viscous one.

5.3 Cohesion thickness v infiltration depth
The results obtained using the plate cohesion meter are plotted against infiltration depth in Fig. 7. A relationship is clearly visible, which is consistent for all of the mixes. This suggests that the ability of the slurry to infiltrate the fibre bed is directly related to the cohesiveness, and hence yield stress, of the material.

Using Fig. 7 it is possible to establish a value of cohesion thickness below which complete infiltration of the 500mm high column will be achieved. It is this type of result which is significant and which will enable new mixes to be designed using the test.

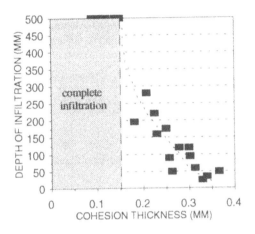

Fig. 7   Effect of cohesion thickness on infiltration depth

6 Conclusions

The flow cone is not a reliable test for predicting the infiltration properties of slurries. Using a plate cohesion meter, a relationship has been established between cohesiveness and infiltration. The test procedures adopted allow the infiltration properties of slurries containing new materials to be investigated. As a result, fluid low strength slurries which contain cement, fine sand, superplasticiser and a carefully controlled quantity of viscosity agent have been successfully developed. These slurries are free from segregation and do not require vibration to aid infiltration of the fibre bed.

Further work is required to establish the effects of different fibre types on infiltration and to relate such factors as fibre content, aspect ratio and specific surface.

7 References

1.   Schneider, B. (1992) Development of SIFCON Through Applications, *High Performance Fiber Reinforced Cement Composites.* Ed. Reinhardt, H.W., and Naaman, A.E., 1992 RILEM, pp. 177-194.

2.   Lankard, D.R. and Newell, J.K. (1984) Preparation of highly reinforced concrete composites, *Fibre reinforced concrete - International Symposium*, SP-81, American Concrete Institute, Detroit, 1984, pp 286-306

3.   Naaman, A.E., Wight, J.K. and Abdou, H.M. (1986) SIFCON connections for seismic resistant frames, *Concrete International*, vol. 9, No. 11, Nov. 1987, pp 34-39.

4.   Naaman, A.E., Reinhardt, H.W., Fritz, C. and Alwan, J. (1993) Non-linear analysis of RC beams using a SIFCON matrix, *Materials and Structures*, vol. 26, pp 522-531

5.   Vandengerghe, M.P. (1992) A Practical Look at SIFCON, *High Performance Fiber Reinforced Cement Composites*. Ed. Reinhardt, H.W., and Naaman, A.E., 1992 RILEM, pp. 226-234.

6.   Mondragon, R. (1987) SIFCON in Compression, *Fiber Reinforced Concrete Properties and Applications,* SP-105, American Concrete Institute, Detroit, 1987, pp 269-281.

7.   Al-Manaseer, A.A., and Keil, L.D. (1992), "Physical properties of cement grout containing silica fume and superplasticizer." *A.C.I. Materials Journal*, 1992, vol. 89, No.2, pp 154-160.

8.   De Larrard, F. (1992) Ultrafine particles for making very high performance concretes, *High Performance Concrete: From material to structure*, 1992, ed Malier,Y., E & F.N. Spon.

9.   Aitcin,P-C. (1992) The use of superplasticizers in high performance concrete, *High Performance Concrete: From material to structure,* 1992, ed Malier,Y., E & F.N. Spon.

10. Lombardi, G. (1985) The role of cohesion in cement grouting of rock, Fifteenth congress on large dams. International Commission on large dams, Lausanne, vol.III pp 235-261.

11. Balaguru, P., and Kendzulak, J. (1987) Mechanical Properties of Slurry Infiltrated Fibre Concrete (SIFCON), *Fiber Reinforced Concrete Properties and Applications*, SP-105, American Concrete Institute, Detroit, 1987, pp 247-268.

17 INTERACTION BETWEEN METAL–METALLOID GLASSY ALLOY RIBBON REINFORCEMENT AND FRESH AND HARDENED CONCRETE

K. FORKEL
Technische Fachhochschule Wildau, Germany
D. VOLLHARDT, H. ZASTROW and H. LICHTENFELD
Max-Planck-Institut für Lolloid- und Grenzlächenforschung, Berlin, Germany
P.J.M. BARTOS
Advanced Concrete Technology Group, Department of Civil, Structural and Environmental Engineering, University of Paisley, Paisley, Scotland, UK

Abstract
In composites of $(Fe,Cr)_{80}(P,C,Si)_{20}$ and $Fe_{75}Cr_5P_8C_{10}Si_2$ respectively, containing ribbons and ordinary Portland cement (OPC) or concrete matrices, oxygen-rich layers exist on the glass surface and can strongly influence the workability of fresh concrete and the composite properties.

Zeta (ζ) potential measurements demonstrate only a small effect of the pH value of the aqueous solution on the electrokinetic potential determined for the finely dispersed powders of the chemically resistant glassy alloy.

Already low concentrations of the anionic surfactant sodium dodecyl sulphate (SDS) affect the ζ potential of the finely dispersed particles. At higher concentrations it approaches to a limiting value.
Keywords: metal-metalloid glassy alloy, finely dispersed glass powder, electrokinetic potential measurements, anionic surfactant sodium dodecylsulphate.

1 Background and introduction

Since the 1960s it is known that glasses from melts of alloys can be obtained easily if they have approximately the chemical composition $M_{80}m_{20}$, where 'M' is a transition or noble metal and 'm' represents a metalloid [1,2]. The metal-metalloid glass containing the transition metals Fe and Cr can be made in an economic relevant way and is characterised by an excellent chemical resistance [3,4]. Nevertheless the question arises concerning the structure and the properties of this new glass [5]. With regard to the physical chemistry of the glass surface a special field is the application of surfactants which can influence the corrosion behaviour of steel [6] and glassy alloy [7] in different aqueous solutions. Surfactants are also important for the production methods

Production Methods and Workability of Concrete. Edited by P.J.M. Bartos, D.L. Marrs and D.J. Cleland.
Published in 1996 by E & FN Spon, 2–6 Boundary Row, London SE1 8HN. ISBN 0 419 22070 4.

and the workability of fibre reinforced cements and concretes.

In the fabrication of the majority of glass fibre reinforced cementitious composites the starting components [8,9] are cement, aggregate, water and fibres in combination with admixtures and additives such as highly active plasticizers. Now it is widely accepted that the microstructure which develops during the hardening process of the cementitious matrix is determined largely by the characteristics of the starting materials including the influence of the additives.

The typical fabrication of GRC involves processing components in the liquid water. The successful processing of composites in liquids generally requires the achievement of stable particle dispersions. Surfactants are often used to render dispersions stable; their adsorption at the solid/liquid interface promotes the wetting process and contributes to stability by the formation of surface charge and/or adsorbed layers. In particular, the glass surface state of ribbons of the metal-metalloid glassy alloy $(Fe,Cr)_{80}$ $(P,C,Si)_{20}$ and $Fe_{75}Cr_5P_8C_{10}Si_2$ respectively is of great interest with regard to a high-strength reinforced composite and to eliminate defective structures in the morphology of the ribbon reinforced concrete [10] for the production of fresh concrete which is responsible for the quality of the final product.

An increased use of surfactants in many industrial applications, not only as conventional type dispersants, but also as coupling agents, dopants, binders and wetting agents is expected in future. It is therefore of scientific and technological interest to investigate the potential of composites using surface modified ribbons of the glassy alloy.

This paper focuses on the characterisation of the glass surface which is responsible for microstructural phenomena at the interface with Portland cement pastes, mortars or concrete. Furthermore we tried to study and to simulate the physical and chemical processes which occur in the period of the workability of fresh concrete and during the cement hydration.

Zeta potential measurements are suitable to characterise the mineral building materials such as cement pastes, mortars, additives, natural minerals and oxidic materials [11]. Based on the so-called Dorn effect (sedimentation potential) it seems to be possible to measure exactly the zeta potentials of solid particles dispersed in water [12]. The knowledge of the ζ potential is important for electrochemical characterisation and for the assessment of the stability of colloidal particles in fresh mixes. To measure this value the microelectrophoresis, i.e. the movement of colloids placed in an electric field, is a usual method. It can be used to determine the sign of the particle charge as well as their electrophoretic mobility, which is related to the zeta potential.

2 Experimental

2.1 Zeta potential measurements

For electrochemical characterisation and for the assessment of the stability of colloidal particles the knowledge of the zeta potential is an important quantity. The microelectrophoresis, i.e. the movement of colloids placed in an electric field, is a usual method to determine the zeta potention. Then it follows the sign of the particle charge as well as the electrophoretic mobility.

This measurement was carried out using the Zeta Sizer 4 from Malvern Instruments Ltd. This instrument is a cross beam laser anemometer. As the velocity has to be measured in the stationary layer, a small volume is defined by the crossing point of two laser beams. Where the two beams cross they produce interference fringes. Particles inside the scattering volume interact with these fringes to produce scattered light which oscillates in time in such a way which depends on the velocity of the particle. If the zeta potential is near zero the measurement of a particle can be improved by a mirror situated in one light beam with an oscillating module, which modulates the response signal. The electrophoretic mobility has to be changed into the zeta potential. If $\kappa a > 100$ (κ - Debye-Hückel parameter and a - particle radius) the calculation can be done with the formula of v. Smoluchowski:

$$v_e = E \, \varepsilon \, \varepsilon_o \, \zeta / \eta$$

where v_e is the velocity of the particle under the influence of an electric field of intensity E in a solution of viscosity, relative permittivity ε and permittivity of a vacuum ε_o.

2.2 Materials and procedures

The ribbons of the metal-metalloid glassy alloy $(Fe,Cr)_{80}(P,C,Si)_{20}$ and $Fe_{75}Cr_5P_8C_{10}Si_2$ respectively were broken up with a vibrating cup mill ®pulverisette 9 from Fritsch GmbH Laborgerätebau, Idar-Oberstein, Germany, using a tungsten carbide (WC + Co) cup. The ®pulverisette 9 operates on a vibration grinding principle. The grinding set is placed on a vibrating disc and the grinding elements (disc and rings) are accelerated by centrifugal force and grind the sample using both impact and friction.

The ground particles of the glassy alloy were fractionated with a 40 μm sieve and the finely dispersed particle fraction was used for the investigation.

Specimens of (0.5 g) powder of the glassy alloy were suspended in 50 ml distilled water and sodium dodecylsulphate (SDS) solution respectively, of different concentration (1.10^{-3}, 2.10^{-3} or 3.10^{-3}) and 2 times for 2 minutes dispersed in an ultrasonic bath. After leaving the samples over one night at room temperature, the solution containing smallest particles was decanted from the sediment, divided into three equal parts and each aprt adjusted to the pH 3.7 or 5 by adding a few drops 0.1 M HCl and 0.1 M NaOH.

The electrophoretic measurements were performed by adding of 10^{-2} M KCl as the ion-background to determine the adsorption behaviour at different pH values. Parallel measurements were carried out to determine the zeta potential in dependence of the electrolyte concentration after adsorption of the anionic surfactant sodium dodecyl sulphate (2.10^{-3} M).

3 Results and discussion

The zeta potential at the interface of the metal-metalloid glass particles suspended in aqueous solution with and without an anionic SDS surfactant was determined. The zeta potential values were calculated automatically by the instrument from electrophoretic velocities according to the v. Smoluchowski equation. For example, Fig. 1 shows the zeta potential for the metal-metalloid glass particle surface in aqueous

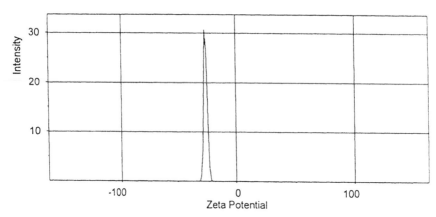

Fig. 1 Zeta potential

2.10^{-3} M SDS and 1.10^{-2} M KCl solution. The results given in Tables 1 and 2 are average values which were determined from ten singular measurements. Deviations of approximately ± mV are typical for this measuring method.

Measurements in water with and without surfactant have given an average value of the medium zeta potential from 10 mV. In presence of electrolyte (5.10^{-3} M KCl) this value decreases to 1 mV. This means that no potential can be measured because the positive charge carriers at the glass surface are compensated by the counterions (Table 1). In the neutral medium the zeta potential of the metal-metalloid glass particles modified by SDS is independent of the electrolyte concentration. This result has been measured only for ≤ 5.10^{-3} M SDS concentrations.

Table 2 shows the effect of the surfactant concentration on the adsorption behaviour, and thus on the zeta potential. In all cases, the adsorption of the anionic surfactants gives rise to a change of the surface charge owing to the negative charged sulphate groups. In the acid pH range, an increase in the surfactant concentration corresponds to an increase in the zeta potential. In the pH range > 7 no change in the zeta potential can be observed.

Table 1 Zeta potentials after adsorption of SDS
of the same concentration but with a
different electrolyte content

c_{SDS} (M)	c_{KCl} (M)	ζ (mV)
2.10^{-3}	without	- 21
2.10^{-3}	1.10^{-3}	- 19
2.10^{-3}	5.10^{-3}	- 19
2.10^{-3}	1.10^{-2}	- 25

Table 2 Zeta potentials after adsorption of SDS of different
concentration and different pH values in 1.10^{-2} M KCl

c_{SDS} (M)	ζ (mV)		
	pH 3	pH 7	pH 10
1.10^{-3}	- 14	- 29	- 26
2.10^{-3}	- 20	- 25	- 27
5.10^{-3}	- 29	- 25	- 25

4 Conclusion

Zeta potential measurements can be used to characterise the surface properties of powdered glassy alloys. New and interesting surface behaviour has been found. The pH dependence of the zeta potential of the powdered glassy allow is rather small and exists only for acid solutions, whereas in alkaline solutions differences do not occur. This suggests that apart from the oxide layer formed at the particle surface of powdered glassy alloys, the metallic character of the particles also determines their surface behaviour. The oxide layer on the metal-metalloid glass surface clearly affects the surface properties in an acid medium, whereas its effect is strongly reduced in an alkaline medium. Then the metallic character of the particles becomes dominant and causes conductivity effects on the surface.

In acid solutions, adsorption of the anionic SDS affects the surface properties of the glassy alloy particles. At higher concentrations the zeta potential quickly approaches a limiting value.

Further investigations will have to be done to study the influence of cationic and non-ionic surfactants on the surface behaviour of the glassy alloy in an alkaline environment such as cement or concrete.

Acknowledgement

The authors are indebted to the Fachvereinigung Faserbeton e. V., Düsseldorf (Germany) for supplying the ribbons of the metal-metalloid glassy $(Fe,Cr)_{80}(P,C,Si)_{20}$ and $Fe_{75}Cr_5P_8C_{10}Si_2$ respectively.

5 References

1. Cahn, R.W. (1982) Metallic glasses - some current issues. *Journal des Physique*, Vol. 43, pp. C9-55 - C9-66.
2. Feltz, A. (1993) *Amorphous Inorganic Materials and Glasses*, VCH, Weinheim.

3. de Guillebon, B. (1987) Resistance a la corrosion de la fibre de fonte, Centre de Recherches de Pont-A-Mousson, France.

4. Burghoff, M., Wihsmann, F.G., Forkel, K., Kästner, R., and Born, D. (1995) Electrochemical study to the corrosion behaviour of metallic glass ribbons in an aqueous alkaline environment at higher temperature, in *Proceedings XVII International Congress on Glass,* The Chinese Ceramic Society, Beijing, P.R. of China, Vol. 3: Glass Properties, pp. 66-71.

5. Vogel, W. (1992) *Glaschemie* Springer-Verlag, Berlin Heidelberg.

6. Morsi, M.S., Barakat, Y.F., El-Sheikh, R., Hassan, A.M. and Baraka, A. (1993) Corrosion inhibition of mild steel by amphoteric surfactants derived from aspartic acid, *Werkstoffe und Korrosion,* Vol. 44, pp. 304-308.

7. Gomaa, N.G., Khamis, E., Ahmed, A and Abaza, S. (1993) Effect of surfactants on the corrosion of amorphous 76Ni-24P alloy in neutral solutions, *Werkstoffe und Korrosion,* Vol. 44, pp. 461-466.

8. Mayer, A. (1986) Zusammensetzung und Eigenshaften der Faserbeton-Matrix/Composition and properties of the fibre concrete matrix, *Betonwerk + Fertigteil-Technik,* No. 1.

9. Majumdar, A.J. and Laws, V. (1991) *Glass Fibre Reinforced Cement,* BSP Professional Books, London Edinburgh Boston Melbourne Paris Berlin Vienna.

10. Bartos, P.J.M. (1995) unpublished results.

11. Nägele, E. and Schneider, U. (1988) Das Zeta-Potential mineralischer Baustoffe-Theorie, Eigenshaften und Anwendungen, *TIZ internat.,* Vol. 112, pp. 458-467.

12. Jacobasch, H.-J. and Kaden, H. (1983) Elektrokinetische Vorgänge - Grundlagen, Meßmethoden, Anwendungen, *Z. Chem.,* Vol. 23, pp. 81-91.

18 INFLUENCE OF DIFFERENT STEEL FIBRES ON WORKABILITY OF FRESH CONCRETE

K. TRTIK and J. VODICKA
Department of Concrete Structures and Bridges, Faculty of Civil
Engineering, Czech Technical University, Prague, Czech Republic

Abstract
This paper deals with results of workability tests of fresh fibre concrete. The tests
were carried out for six different kinds of steel fibre. The concretes have fulfilled the
conditions of having a relatively low volume of fibres and a good pumpability. The
data are complemented by values of strengths of the hardened concretes.
Keywords: Cube test, flexural test, flow table test, setting of flow table test cone,
splitting tensile test, steel fibre concrete, workability.

1 Introduction

The experimental programme, the results of which are included in this paper, was carried out at the Faculty of Civil Engineering of the Czech Technical University in Prague. The research was carried out in co-operation with FATEK Ltd, a company which
provided financial resources needed for this programme.

This paper presents the results of the part of the programme which concerned the
workability of fresh fibre concrete. The presented data were collected for this conference and have not been published before.

The main goal of the experimental programme was to verify the influence of the
steel fibre of the "FATEK" fibre on basic properties of concrete, i.e. its cube strength,
splitting tensile strength and tensile strength in bending. Moreover, a load-deformation
diagram was determined under a deformation controlled regime. Consequently, a
toughness index was calculated on the basis of that diagram.

Production Methods and Workability of Concrete. Edited by P.J.M. Bartos, D.L. Marrs and D.J. Cleland.
Published in 1996 by E & FN Spon, 2–6 Boundary Row, London SE1 8HN. ISBN 0 419 22070 4.

2 Programme of experiments

The authors tested concretes which were reinforced by four types of "FATEK" fibres, each in combinations of two lengths and two kinds of end form. These results were compared with the results of testing of plain concrete and two other concretes reinforced by different types of steel fibres from other sources ("Bohumin", "Dramix"). The basic parameters are provided in Fig. 1.

Serie	A	C	D	G1	G2	G3
Fibre						
l/d	35/0.98	35/0.98	45/0.45	50/1.0	55/0.98	55/0.98
Label	FT 35	FT 35 H	Bohumín	Dramix	TC 55	TC 55 H
Type of fibre	1	2	6	3	4	5

Fig. 1. Types of fibres used and their shapes and sizes.

A relatively low content of steel fibres was chosen because of two main reasons. The first was the demand for a sufficient pumpability of fresh concrete. The second was the idea to test a concrete suitable for the design of industrial floors.

All the fresh concretes had the same mix proportions (see Table 1) and the same method of mixing and placing was used. Also, the method of treatment of the young concrete was kept identical. As for the method of compacting the fresh concrete, punching with a steel bar (diameter of 14 mm) was used in all cases. Effects of this method of compaction were checked by measuring of the fresh concrete density.

Table 1. Mix proportions of the basic concrete

Constituent	kg/m^3
Portland cement PZ 35F	320
Water	182.4
Aggregate 0 - 2 mm	635
2 - 8 mm	255
8 - 16 mm	380
16 - 32 mm	540
Plasticizer	3.2
Steel fibres	50

3 Assessment of workability

The workability tests on the fresh concrete were carried out for all concrete mix-variations examined. The authors decided to use the method of Flow Table Test (FTT). The results of this test were complemented by the measurement of the settlement of a cone, which was made for the FTT. The results of this non-standard measurement have given very good correlation with the FTT results (see Fig. 2) with respect to the pumpability of fresh concrete.

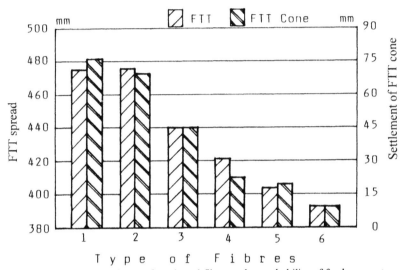

Fig. 2. The influence of type of used steel fibre on the workability of fresh concrete.

4 Tests on hardened concrete

The following specimens were made of every kind of concrete:
- cubes 150 x 150 x 150 mm (3 pieces for series "G1", "G2" and "G3", 9 pieces for the series "A", "C" and "D",
- beams 150 x 150 x 700 mm (the same numbers for beams as for the cubes).

These characteristics of hardened concrete were measured for every type of concrete:
- cube strength,
- splitting tensile strength on cubes,
- bending tensile strength,
- compressive strength on the rest of beams,
- splitting tensile strength on the rest of beams,
- load - deformation diagram,
- toughness index.

All kinds of tests were carried out at the age of 28 days and the results are shown in Fig. 3. More details can be found in [1], [2], [3].

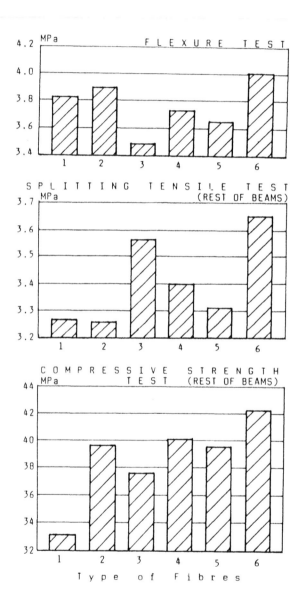

Fig. 3. Average values of the strength of concretes with different fibres at the age of 28 days.

5 Conclusions

The test results support the following conclusions:

- geometrical characteristics of used fibres steel have a significant influence on workability of the fresh concrete,
- workability of the fresh concrete decreases with an increasing length of the fibres,
- geometrical characteristics of the end form of the fibres have practically no influence on workability of fresh concrete when using the low content of fibres,
- having the same volumetric content of fibres and the same length of fibres the increase of the number of fibres (i.e. a decreasing diameter of the fibres) decreases the workability of the fresh concrete.

Considering the results of tests for strength it is possible to claim:

If a proper compaction of concrete is ensured, it is possible to expect the higher values of the strength for concretes, whose workability has been more difficult.

6 References

1. Abramowicz M., Kratky J., Trtik K., Vodicka J. (1994) *Strains of steel fibre reinforced concrete under a long-term load.* Proceedings of the fourth international symposium on brittle matrix composites, pp. 408 - 414.
2. Goldau E., Trtik K., Vitek J., Vodicka J. (1994) *Bending of fibre reinforced concrete beams.* Proceedings of the fourth international symposium on brittle matrix composites, pp. 561 - 568.
3. Kratky J., Trtik K., Vodicka J. (1984) *Concrete with dispersed reinforcement.* Publication of Building Research Institute Prague (in Czech).

7 Acknowledgement

The expenditure connected with the evaluation of the test results was covered by means of a grant GA CR 103/95/0731 and grant GA CR 103/96/0841.

PART FIVE
FLOWING AND SUPERFLUID MIXES

19 DESIGN AND TESTING OF SELF-COMPACTING CONCRETE

P.L. DOMONE and H.-W. CHAI
Department of Civil and Environmental Engineering, University College London, London, UK

Abstract
Recent developments in high performance concrete have included self-compacting concrete, that is concrete which, without segregation, can flow into place and compact under its own weight. Published examples of uses, mix designs and testing programmes are reviewed and important features highlighted. The initial results of an experimental programme aimed at producing and evaluating self-compacting concrete made from readily available UK materials are presented and discussed. This includes evaluating the rheology using a two-point workability apparatus.
Keywords: High performance concrete, self-compacting concrete, literature review, mix design, rheology, two-point workability testing.

1 Introduction

In recent years, the use of cement replacement materials (CRMs) and admixtures in concrete for other than purely economic reasons has led to the development of mixes for increasingly demanding circumstances. Improvements in the quality and uniformity of the CRMs and the attention given to admixture formulations have greatly assisted progress. Many of the results can be considered as "high performance concrete" in the general sense of a concrete which has one or more property or properties significantly enhanced. Examples include low heat, high strength, high durability, underwater and foamed concrete.

Super workable or self-compacting concrete (SCC) is a particularly interesting example. As its name implies, it is a concrete that can achieve full compaction without vibration, i.e. under self-weight only. For structural use in reinforced sections, high fluidity is therefore a prime requirement. It is also important that the resulting concrete is uniform and homogeneous, and therefore high cohesion or segregation resistance during flow is equally

Production Methods and Workability of Concrete. Edited by P.J.M. Bartos, D.L. Marrs and D.J. Cleland. Published in 1996 by E & FN Spon, 2–6 Boundary Row, London SE1 8HN. ISBN 0 419 22070 4.

Table 1 Uses of self compacting concrete

reference	date	structure	details	concrete volume (m³)
Hasan et al [1]	1993	power plant water intake	reinforced concrete velocity caps, tremied underwater concrete	1500
Hayakawa et al [2]	1993	high rise building	heavily reinforced structural concrete	3000
Kawai et al [3]	1993	cable stayed bridge	lightweight reinforced concrete main span	93
Kuroiwa et al [4]	1993	20 storey building	lower part of central heavily reinforced core	1600
Matsuo et al [5]	1993	stadium	reinforced concrete guide track for retractabe roof	10,000
Miura et al [6]	1993	research laboratory	heavily reinforced structural wall, 6-8m high, 200mm thick	80
Miura et al [6]	1993	LNG storage tank	heavily reinforced wall base junction, flowing distance up to 10m	800
Sakamoto et al [7]	1993	high rise building	infill of steel tubular columns - 40m fill height	885
Sakamoto et al [7]	1993	cable stayed bridge	upper part of central column into permanent formwork	650
Furnya et al [8]	1994	suspension bridge (main span 1990m)	cable anchorages	380,000
Kubota et al [9]	1994	tunnel culvert	heavily reinforced upper slab	1091
Nagayama et al [10]	1994	arch bridge	heavily reinforced section	1500
Umehara et al [11]	1994	precast panels	lightweight concrete in thin sections	

important. The terms "passing ability" and "filling capacity" can also be used to describe these two properties.

Most developments and uses of SCC have been in Japan. This paper briefly reviews examples of these, discusses some of the important factors governing material selection and mix design, and presents initial results from a laboratory programme aimed at producing and evaluating SCC made from UK materials.

2 Applications

Table 1 shows some of the typical applications. The advantages claimed for SCC include:
- reduced need for skilled labour during construction;
- shorter concrete placing times;

- improved compaction in areas in high reinforcement density, with difficult access for vibrators;
- improved compaction and hence enhanced durability of the critical cover zone of structural members;
- the possibility of altering construction procedures to advantage e.g. by higher single lifts in narrow columns, or the use of alternative composite construction;

3 Materials and mix proportions

Many authors [e.g. 1-19] have reported details of the mix constituents and proportions resulting from laboratory investigations and/or used in construction. Although, perhaps not surprisingly, these show considerable variation, several factors common to a majority of mixes are apparent.

3.1 Water content
This is typically within the range 160-185kg/m^3, i.e. similar to that for conventional medium workability concrete. It is not clear from the literature whether this is a necessary requirement for SCC, or if it results from other considerations such as a durability [27]. It is clear, however, that it is an important criterion. If it is too low, there is insufficient fluidity, even with admixtures; if it is too high, the segregation resistance reduces and the paste or mortar will easily separate from the larger aggregate particles which can then form arches across gaps, e.g. between reinforcing bars, hence blocking the flow.

3.2 Admixtures
The fluidity is provided by superplasticizers (high range water reducers), most being based on either a naphthalene or melamine formaldehyde, often modified to provide extended retention of fluidity and set times - an important requirement for some applications. These often also provide air entrainment, but it is not clear whether the prime reason for the air is to assist the rheology, or for durability considerations.

3.3 Binders
Binder contents are high, typically 450-550kg/m^3, but it would seem that, at the required high fluidity, the use of Portland cement alone will result in inadequate cohesion or segregation resistance. Ground granulated blast furnace slag (ggbs) is beneficial in this respect, typically replacing 40% by weight of the Portland cement. In some mixes fly ash (pfa) and/or microsilica (csf) have also been used, and limestone powder as an inert filler seems particularly useful in enhancing cohesion.

3.4 Water/binder ratios
It follows from the above that water/binder ratios are normally in the range 0.30-0.36. Sufficient segregation resistance has been achieved at higher values by including small quantities of a "viscosity agent" e.g. a cellulose or a polysaccharide "biopolymer", in addition to the plasticizer or superplasticizer. It is suggested that this works by restricting the movement of the free water in the mix, hence ensuring that it is fully available to provide lubrication between the particles. It also increases the viscosity of the free water, and so high doses are detrimental. It is also claimed that it reduces the sensitivity of the self-

compacting properties to variations in mix proportions, an important criterion for large scale production.

3.5 Aggregates
SCC has been successfully produced with both gravels and crushed rocks and gravels, normally with a maximum size of 20mm. Lightweight SCC has been produced using an artificial lightweight coarse aggregate [2,11].

3.6 Workability measurement
All the mixes have collapsed slumps, and a conventional slump value is therefore inappropriate. A "slump flow" value is normally used, which is the diameter of the concrete after a standard slump test. Values at least in excess of 500mm, and normally more than 600mm are required. Interestingly, there does not seem to be any use of the flow table test, which is commonly used in the UK and elsewhere for superplasticized high flow mixes.

The slump flow is a single value, and, as with all such measurements, it is insufficient to fully describe the rheology of a near Bingham material such as fresh concrete, which has a yield stress and a plastic viscosity. Several authors have recognised this and have consequently also measured a second value. For example, Kuroiwa et al [4] have measured the time to reach a slump flow value of 500mm. They have called this the flow speed, and have indicated that a value of between 4 and 10secs is required. They claim that this is related to the viscosity of the mix, and that the slump flow is related to the yield stress.

Sakomoto et al [7] have used a similar approach, but have defined their flow speed as the time to reach a slump flow of 600mm. A value between 10 and 20secs is required.

At least two groups [6,26] have used the efflux time of concrete from a funnel as a measure of segregation resistance. The funnel capacity is typically 10 litres.

Mitsui et al [12] have developed an "L-flow test", shown in figure 1. The vertical leg is filled with concrete and, on lifting the dividing plate, this flows into the horizontal leg. Mitsui claims that the final fall of the horizontal surface (L_L) is related to slump, the total horizontal flow (L_h) is related to slump flow, and the average speed of flow of the concrete over a distance from 50 to 100mm from its inital vertical face i.e. A to B in fig. 1 (called the flow velocity) is related to the plastic viscosity of the mix.

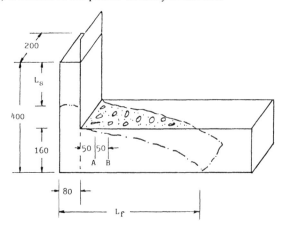

Fig. 1   L-flow test used by Mitsui et al [12]

There is little published information on measurements using a rigorous rheological test method such as the two-point workability apparatus. In a notable exception, Kawai and Hashida [20] have tested SCC in a two point workability test apparatus with a helical impeller i.e. similar to the Tattersall apparatus. They related the resulting yield stress and plastic viscosity to slump flow, V-funnel flow time and the passing ability through a mesh with a gap sizes of 34 and 58mm. They suggest that sufficient passing ability can be achieved with a yield stress of 50Pa, and a plastic viscosity of 20-80Pas. A minimum plastic viscosity of 30Pas is required for segregation resistance.

3.7 Strength
A range of strengths have been achieved, either by varying the relative proportions of the CRM's or the overall water/binder ratio (incorporating a viscosity agent at higher values). In one case at least, high strength has been achieved by using a water/binder ratio of 0.2 and including microsilica, but the mix did require some vibration during placing [12].

3.8 Heat output
Low heat mixes can be produced by variation of the binder constituents. Pfa and limestone powder may be particularly useful in this respect [22].

4 Demonstration tests

Many authors have demonstrated the effectiveness of SCC using tests representing placing conditions. These vary in scale from laboratory measurements of flow through meshes of reinforcing bars, to near full size mock-ups of structural elements.

The smaller scale tests are used as both as a qualitative demonstration and for quantitative development and assessment of mixes. A variety of arrangements have been used, assessing such factors as flowability, filling capacity, segregation resistance and passing ability through gaps.

Larger scale mock-ups which demonstrate that all aspects of concrete production, handling and placing can be translated from laboratory to site are extremely important.

5 Mix design

Some authors give little information on the process by which they have arrived at or near their final mix proportions. Others give the results of numerous tests to assess a variety of properties of the binder pastes, mortars and concrete, and the effect of the material types and their relative proportions etc.

From such studies, some mix design guidelines can be formulated. These include:

- For a full understanding of the mix and its design it is important to consider the volume fractions of the various components.
- For a 20mm gravel, the volume fraction of the coarse aggregate should not exceed 50% of that obtained from a dry rodded test [23].
- The sand content of the mortar fraction should be about 40% by volume [23].
- An adequate free water content is required. Ozawa et al [13] have defined the free

water as the total water less that retained (i.e. absorbed or adsorbed) by the powder materials (cement, CRM's, inert filler) or aggregates. The adsorption on to the particle surfaces becomes more significant for the smaller particles, and they defined and measured a water retaining factor (β) as the ratio of the weight of water retained to the weight of powder. This will be different for each material. Any fine aggregate with a particle size of less than 60µm should be considered as binder powder, although this can be ignored if it is less than 2% of the total fine aggregate.

The minimum shear resistance of the concrete is obtained at an optimum free water content, which will be different for different binder/aggregate ratios, but can be considered as independent of the binder constituents and admixtures. The retained water, and hence the total water content, will vary with binder constituents, admixtures etc.

The effect of plasticizers is to release water from particle flocs, thereby increasing the amount of free water at any given total water content [13].

- For mixes without a viscosity agent, the water/powder ratio by volume should be approximately $k_p.\beta_p$, where k_p is a factor between 0.8 and 0.9, and β_p is the water retaining factor for the combined powders, as defined above.
- The minimum spacing of reinforcing bars should be 1.8 times the maximum size of the coarse aggregate [19]

A detailed analysis of the reported mixes indicates that two important criteria are:

- the ratio of the water content to the total fines content, defined as the total weight of binder, inert filler and fine aggregate; and
- the coarse aggregate content, expressed as a proportion of the total aggregate content.

Figure 2 shows the values of these quantities for a variety of concrete types, including self compacting concrete with and without a viscosity agent, normal structural concrete and high strength concrete.

Self compacting concrete mixes without a viscosity agent fall in a relatively small area with a maximum water/fines ratio of 0.14 and coarse aggregate content of 40% of the total concrete weight. With a viscosity agent, water/fines ratios of up to 0.16 can be used. High strength concretes have higher aggregate contents, between 0.4 and 0.5, and a wider range of water/fines ratios, from 0.095 to 0.155; normal strength concretes have similar amounts of coarse aggregate at higher water/fines ratios ranging from 0.15 to about 0.19.

All of the above values relate to concrete containing normal coarse aggregates of 20mm maximum particle size. Data for limited number of lightweight aggregate mixes show that these have similar water/fines ratios, but, as might be expected, lower coarse aggregate proportion. Some limited data (not shown) on mixes with a larger coarse aggregate size indicate that these will tend to have higher coarse aggregate contents.

6 UCL programme

Initial studies in the current programme have concentrated on producing an SCC with materials readily available in the London area. Tests have been carried out to determine he concrete's properties and effectiveness, and to compare it with a conventional flowing concrete, a high slump concrete without any admixtures, and an underwater concrete.

The tests used were considered to be the most effective of those used by others, so that comparisons could be made with reported data. Two-point workability testing was used

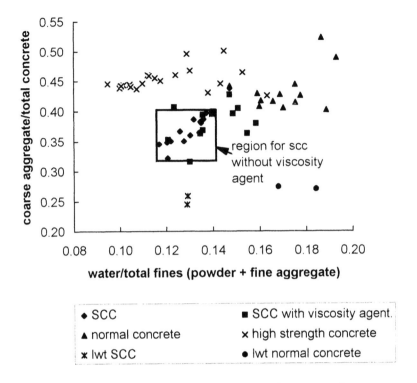

Fig. 2   Relative mix proportions (by weight) for various types of concrete

to provide some fundamental rheological data.

6.1 Materials
Details of the Portland cement (class 42.5N), the ggbs and the pfa are given in table 2.
   The admixtures were:
- a sulphonated naphthalene formaldehyde superplasticizer with a 40% solids content.
- a water reducing air entraining agent
- a two-component underwater concrete admixture

   The aggregates were 20-5 mm Thames Valley gravel and Thames Valley sand with a fineness modulus of 2.47

6.2 Mixes
The guidelines for mix proportioning SCC as outlined above were used. The underwater concrete was designed according using the recommendations of the Japan Society of Civil Engineers [24] and the admixture suppliers [25]. Since this concrete was used for comparison only, extensive trial mixes were not carried out, and the resulting mix may not be the optimum for underwater use.
   The mix proportions are given in table 3.

Table 2 Composition of binder materials

	Portland cement	ggbs	pfa
oxides	percent by wt		
SiO_2	19.6	33.7	51.4
Al_2O_3	5.5	11.5	25.0
Fe_2O_3	3.0	1.8	9.4
CaO	64.2	41.3	1.4
MgO	0.7	9.0	1.4
SO_3	2.9		0.7
K_2O	0.8		3.7
Na_2O	0.1		1.3
LOI	1.3		4.8
free CaO	1.9		
compounds	(Bogue)		
C_3S	59		
C_2S	12		
C_3A	9.6		
C_4AF	9.0		
SSA (m^2/kg)	376		

6.3 Tests and results

Each mix was tested for slump or slump flow, and values of the yield (g) and plastic viscosity (h) terms were measured in a Tattersall two-point workability apparatus fitted with the helical impeller [21]

The flow and filling characteristics were assessed in a L-shaped flow trough, the principle

Table 3 Concrete mix proportions

type of concrete	pc	pfa	ggbs	water	aggregate		admixture		
					coarse	fine	sp	aea	u/w(2 pt)
				kg/m³					
self compacting	218	125	280	179	785	686	8.2	0.8	
flowing	465			195	1155	545	4.65		
high workability	465			195	1155	545			
underwater	207		169	188	1044	703			3.7+1.85

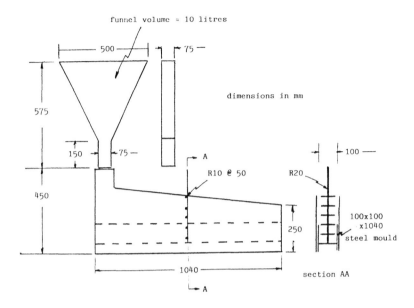

Fig. 3   Funnel and trough arrangement for the assessment of the flow, passing and filling ability of the concrete

dimensions of which are shown in fig. 3. The V-shaped funnel was filled with the concrete, and the trap door pulled back. The time of flow out the funnel was measured to give an assessment of the viscosity. The concrete then flowed into the metre long trough. A reinforcing mesh was fixed half way along this so that the passing ability of the concrete could be assessed.

The flow curves from the two-point test are shown in fig. 4, and all the results in table 4.

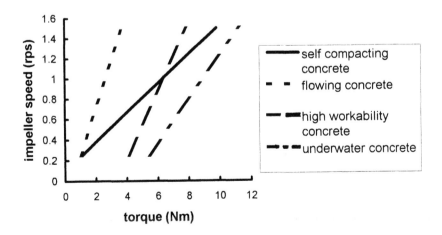

Fig. 4 Flow curves from two-point workability tests

Table 4 Results of workability and flow tests

type of concrete	slumpflow or (slump)	two-point workability		V-funnel time (secs)	filling ability	passing ability
		g (Nm)	h (Nms)			
self compacting	>700 mm	[-0.74]	7.02	9	rank B	pass
flowing	640 mm	0.46	2.08	8	rank B	blocked
high workability	(190 mm)	3.45	2.77	blocked	n.a.	blocked
underwater	420 mm	4.83	4.27	70		

All mixes had high slump flow or slump values, but the differences in the rheology are apparent from the two point workability test results. The self compacting concrete had, in effect, a zero yield. (The extrapolation of best fit straight line through the data points had a small negative intercept on the torque axis - indicating the flow curve probably becomes nonlinear near the origin.) This is consistent with flow at very low shear stresses, e.g. those due self weight. The flowing concrete had a low, but positive yield term, with the 'normal' high workability concrete and the underwater concrete much higher values. The flowing and high workability concretes had similar low values of plastic viscosity; the underwater concrete had a higher value, indicating greater cohesion, and the self compacting concrete the highest value of all.

The self compacting and flowing concrete had similar short efflux times from the V-funnel, with the underwater concrete taking much longer. The ranking for filling ability is based on the criteria suggested by Ozawa et al [26] shown in figure 4. The two factors that are calculated from the slump flow and efflux time values are:
1. Relative flow area (A_f) = (slump flow/600)^2
2. Relative flow speed (R_f) = 5/V-funnel time.

The ranking criteria were obtained from subjective assessment of the performance of a large number of mixes. On this basis, both the self compacting and flowing concretes tested would   be classified as rank B, i.e. good filling ability. However, they differed in their ability to pass through the reinforcing mesh. The self compacting concrete readily passed through, with the free surface on the downstream side nearly reaching the level of that on the upstream side after about one minute. Conversely, very little of the flowing concrete passed through the mesh before blocking occurred. Filling of the trough was only possible after removal of the mesh.

Throughout the testing, the cohesive and non-segregating nature of the self compacting concrete was visually apparent.

The inference is that a low yield and a high plastic viscosity value is therefore a requirement for self compacting concrete. Further testing is necessary to determine the limitations of the required values, and the sufficiency of this approach.

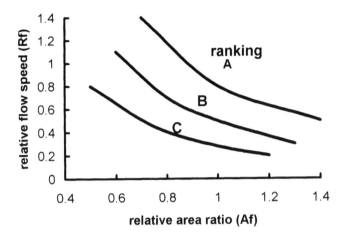

Fig. 5 Ranking criteria for self-compacting concret mixes suggested by Ozawa et al [26]

Quantitative data on the resulting properties of the concrete was obtained by allowing the concrete in the flow trough to harden, and the dividing this into two prisms, each 500 x 100 x 100mm i.e. consisting of the concrete either upstream or downstream of the reinforcing mesh. These were cured in water for 7 days, and tested for density, dynamic elastic modulus and ultrasonic pulse velocity. For comparison, specimens were also obtained from a second similar test in which the concrete was vibrated to ensure full compaction. The results are shown in table 5.

All show little difference between the upstream and downstream beams i.e. the concrete has flowed through the reinforcing mesh with little or no segregation. The vibration has had some effect; the density is increased by about 1%, the dynamic elastic modulus by about 4% and the ultrasonic pulse velocity by about 0.5%. However, all of these increases are small and indicate successful self compacting properties of the concrete.

Table 5 Results of tests on hardened self compacting concrete

	unvibrated		vibrated	
	beam 1	beam 2	beam 1	beam 2
density (kg/m³)	2289	2293	2321	2329
dynamic elastic modulus (kN/mm²)	37.5	36.7	38.7	38.9
ultrasonic pulse velocity (km/sec)	4.55	4.56	4.60	4.55

7 Conclusions

The tests reported have indicated that self compacting concrete can be produced on a laboratory scale from UK materials using the mix design 'rules' resulting from extensive work in Japan. Two point workability tests show that SCC has a very low or zero yield value and a high plastic viscosity. This combination is consistent with the requirements of high deformability and high segregation resistance, but acceptable limits for the yield and plastic viscosity values have not yet been defined. It has not yet been shown if the specification of these two values is a sufficient requirement to also indicate passing ability and filling capacity.

The work is being extended to include a study of the properties of the cement paste component of SCC, to assessment of the merits of including limestone powder and a viscosity agent, and to the measurement of other hardened properties such as strength.

Acknowledgements

The authors would like to thank the technical staff of the Elvery Concrete Technology Laboratory, University College London, for their help during the experimental programme, and Rugby Cement Ltd, Civil and Marine Ltd, Ash Resources Ltd, Fosroc Expandite Ltd and Cormix Construction Chemicals for the supply of materials.

8 References

1. Hasan, N., Faerman, E. and Berner D. (1993). Advances in underwater concreting: St Lucie plant intake velocity cap rehabilitation, in *ACI SP-140 High Performance Concrete in Severe Environments*, (ed. P. Zia) American Concrete Institute, Detroit, USA, pp 187-215.
2. Hayakawa, M., Matsuoka, Y. and Shindoh, T. (1993). Development and application of superworkable concrete, in *"Special Concretes - Workability and Mixing"*, *University of Paisley*, ed. P. J. M. Bartos. London: E&FN Spon.
3. Kawai , T. et al (1993) An application of highly flowable concrete to lightweight concrete with over 300 kgf/cm^2 at the age of three days (in Japanese), in *Cement and Concrete*, Japan Cement Association, No 558, pp 67-70
4. Kuroiwa, S., Matsuoka, Y., Hayakawa, M. and Shindoh, T. (1993). Application of super workable concrete to construction of a 20-Story Building, in *ACI SP-140 High Performance Concrete in Severe Environments* (ed. P. Zia), American Concrete Institute, Detroit, USA, pp 147-161
5. Matsuo, K. et al (1993) Placing of 10,000 m^3 of super workable concrete for a guide track structure of a retractable roof (in Japanese) in *Cement and Concrete*, Japan Cement Association, No 558, pp 15-21
6. Miura, N., Takeda, N., Chikamatsu, R. and Sogo, S. (1993). Application of super workable concrete to reinforced concrete structures with difficult construction conditions, in *ACI SP-140 High Performance Concrete in Severe Environments* (ed. P. Zia) American Concrete Institute, Detriot, USA, pp163-186.

7. Sakamoto, J., Matsuoka, Y., Shindoh, T. and Tangtermsirikul, S.. (1993) Application of super-workable concrete to actual construction, in *Concrete 2000* (ed R K Dhir and M R Jones), E&FN Spon, London, pp 891-902.
8. Furuya, N., Itohiya, T. and Arima, I (1994) Development and application of highly flowing concrete for mass concrete anchorages of Akashi-Kaikyo Bridge, in *Proceedings of International Conference on High Performance Concrete (supplementary papers)*, Singapore, American Concrete Institute, Detroit, USA, pp 371-396.
9. Kubota, T., (1995) Adoption of highly workable concrete to construction of the Asukayama Tunnel (in Japanese), in *Concrete Journal*, Vol 33, No 5, Japan Concrete Institute, Tokyo pp 52-60
10. Nagayama, I., (1995) Application of superworkable concrete to arch bridges for labour saving (in Japanese), in *Concrete Journal*, Vol 33, No 2, Japan Concrete Institute, Tokyo pp 47-53
11. Umehara, H., Uehara, T., Enomoto, Y. and Oka, S. (1994). Development and usage of lightweight high performance concrete, in *Proceedings of International Conference on High Performance Concrete (supplementary papers)*, Singapore, American Concrete Institute, Detroit, USA, pp 339-353.
12. Mitsui, K., T. Yonezawa, M. Kinoshita, and T. Shimono (1994) Application of new superplasticizer to ultra high strength concrete, in *ACI SP-148 Superplasticizers in Concrete*, American Concrete Institute, Detroit, USA, pp 27-45.
13. Ozawa, K., Maekawa, K., and Okamura, H., (1992) Development of high performance concrete, in *Journal of Faculty of Engineering, Univ of Tokyo* XLI(3):pp 381-439.
14. Okazawa, S., Umezawa, K., and Tanaka, Y., (1993). A new polycarbonate based polymer: physical properties of concrete, in *Concrete 2000* (ed R K Dhir amd M R Jones), E&FN Spon, London, pp 1813-1824
15. Imai, M., (1993) Production of percast concrete beams of super workable concrete (in Japanese) in *Cement and Concrete*, Japan Cement Association, No 558, pp 34-38.
16. Tanaka, M (1993) Achievement of 900 kgf/cm^2 super workable high strength concrete with belite Portland cement (in Japanese), in *Cement and Concrete*, Japan Cement Association, No 558, pp 52-59.
17. Takashi (1993) Using 2000 m^3 of highly workable concrete for box-culverts in the Meishin Expressway (in Japanese), in *Cement and Concrete*, Japan Cement Association, No 558, pp 91-98.
18. Hirano, T (1993) Use for the construction of liquid natural gas tanks (in Japanese) in *Cement and Concrete*, Japan Cement Association, No 558, pp 84-90.
19. Yoshino, A. et al (1993) A study of the estimation of the flow behaviour of high flowability concrete, in *Transactions of the Japan Concrete Institute*, Vol 15 pp 7-14
20. Kawai, T. and Hashida, H., (1994) Fundamental research on the rheological properties of super workable concrete (in Japanese), in *Proceedings of the Japan Concrete Institute*, Vol 16. No 1 pp 125-130.
21. Tattersall, G.H., (1991) *Workability and Quality Control of Concrete* E&FN Spon, London
22. Ayano, T., Sakata, K., Ogawa, A. and Kaneko, T., (1993) Study of the mix proportions of highly flowable concrete with limestone powder, in *Transactions of the Japan Concrete Institute*, Vol 15, pp 1-6
23. Okamura, H. and Ozawa, K. (1994) Mix design method for self-compactable concrete

(in Japanese), in *Proceedings of the Japan Society of Civil Engineers*, No 490, Vol 24 pp 1-8

24. Japan Society of Civil Engineers (1992) *Recommendations for Design and Construction of Antiwashout Underwater Concrete*, Concrete Library International of JSCE, No 19, Tokyo

25. Cormix Construction Chemicals (1994) *Data sheet on Cormix UCS Liquid Underwater Concreting System*, Warrington

26. Ozawa, K., Sakata, N. and Okamura, H. (1994) Evaluation of self compactability of fresh concrete using the funnel test (in Japanese), in *Proceedings of the Japan Society of Civil Engineers*, No 490, Vol 23 pp 71-80

27. Japanese Architectural Standard (1986) *Specification for reinforced concrete work*, JASS 5

20 BASIC PROPERTIES AND EFFECTS OF WELAN GUM ON SELF-CONSOLIDATING CONCRETE

N. SAKATA
Kajima Corporation, Niigata, Japan
K. MARUYAMA
Nagaoka University of Technology, Niigata, Japan
M. MINAMI
Research Center, Sansho Co. Ltd, Osaka, Japan

Abstract
It has been reported that suitable quantity of the viscosity agent Welan gum, a kind of natural water soluble polysaccharide, is very effective for stabilizing the rheological property of self-consolidating concrete. In this report, we have just studied basic properties of Welan gum experimentally and effects of Welan gum for fresh concrete such as self-consolidating property. As a result, we understand that viscosity of Welan gum's water sol. is not affected by calcium concentration and pH, except for a little bit high viscosity in alkaline solution. And we learned that it stabilizes the rheological property and mobility in small space of self-consolidating concrete to apply suitable quantity of Welan gum for concrete, in addition, the product provides long-term stability and improves self-consolidating property in wide range of slump-flow. We made clear partially how the product provides these effects.
Keywords: polysaccharide, rheological property, self-consolidating concrete, slump-flow, welan gum

1 Introduction

Self-consolidating concrete has been studied and developed in lots of laboratory institutions, since it was advocated by Professor Okamura at Tokyo University in Japan[1]. Now this concrete is applied for not only general constructs but huge scale constructs and very important constructs. This self-consolidating concrete is defined as

Production Methods and Workability of Concrete. Edited by P.J.M. Bartos, D.L. Marrs and D.J. Cleland. Published in 1996 by E & FN Spon, 2–6 Boundary Row, London SE1 8HN. ISBN 0 419 22070 4.

a concrete, which has excellent self-consolidating and transfiguring properties, which can be cast into every nock and corner in forms without using vibrators, which is easy to avoid defects such as shrinkage at the time of hardening at early-age step, and which has a long-term durability. The report by Dr.K.Ozawa indicates that these properties can be estimated through two tests, slump flow test and V type funnel test[2].

In general, the self-consolidating property of self-consolidating concrete depends on mix proportion and uneven quality of materials. These causes troublesomeness for producing stable self-consolidating concrete which has stable fluidity and outstanding self-consolidating property. To stabilize fluidity which greatly affects self-consolidating property, the use of a viscosity agent has been investigated. It was reported with the results of indoor-tests at the laboratory institution and outdoor-tests at the actual plant that Welan gum, a kind of natural polysaccharide, is extremely effective for stabilizing fluidity of concrete[3][4].

In this report, typical properties of Welan gum, the visocosity agent for self-consolidating concrete was estimated experimentally. And this report describes effects of Welan gum on fluidity, ease to flow in small space and self-consolidating property of concrete by slump flow test, V type funnel test and U type casting test.

2 Typical properties of the viscosity agent Welan gum

2.1 Chemical structure
Welan gum is a kind of natural polysaccharide and its structure is shown in Figure 1. The backbone is the repeat unit of four saccharides and it is composed of D-glucose, D-glucuronic acid, D-glucose and L-rhamnose units and the side chain is either a L-mannose or a L-rhamnose unit[5].

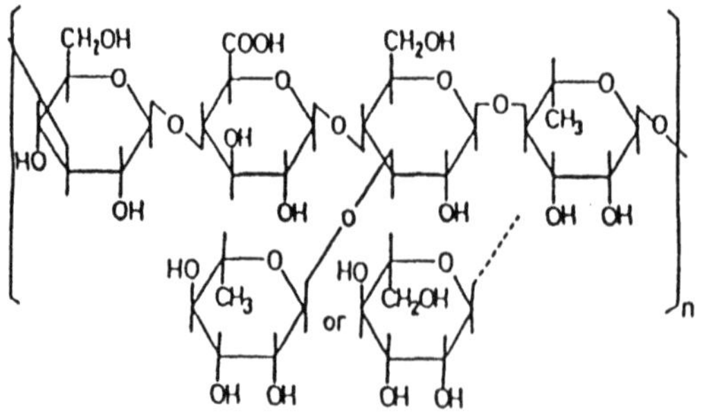

Fig. 1. Chemical structure of Welan gum.

2.2 Production process

Welan gum is produced by a carefully controlled aerobic fermentation using an alcaligenes strain ATCC31555 and the production process is shown in Figure 2. This process is almost identical to that of other biopolymers such as Xanthan gum.

2.3 Rheological property

There are a lot of natural polysaccharide based viscosity agents such as Xanthan gum, Guar gum and Curdlan and so on except for Welan gum. In this clause, the rheological properties of the concrete including Welan gum were examined in order to investigate the mechanism of effects which Welan gum brings to the concrete, comparing with Xanthan gum and Guar gum as a typical natural polysaccharide-based viscosity agent, MC(Methyl Cellulose) and HEC(Hydroxy Ethyl Cellulose) as a typical cellulose based viscosity agent which is used for concreting under water, and Polyacrylate. Chemical structures of each viscosity agent are shown in Figure 3.

【Carbon resource 】 ⇨ 【Aerobic fermentation】 ⇨ 【Dilution】 ⇨
【Sterillization】 ⇨ 【Filtering 】 ⇨ 【Precipitation 】 ⇨ 【Filtering 】
 ⇨ 【Drying】 ⇨ 【Milling 】 ⇨ 【Product 】

Fig. 2. Production process of Welan gum

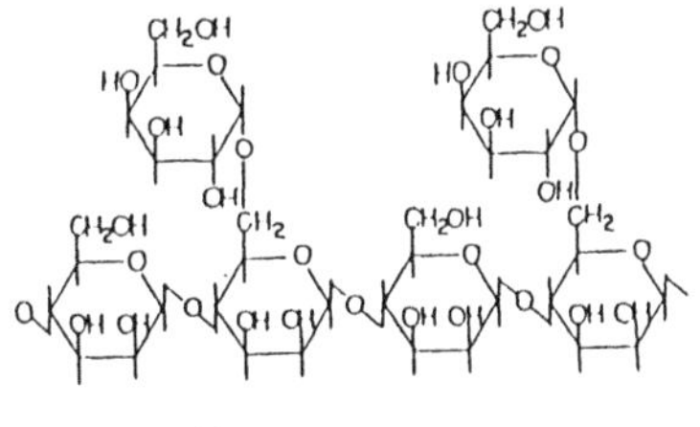

Guar gum

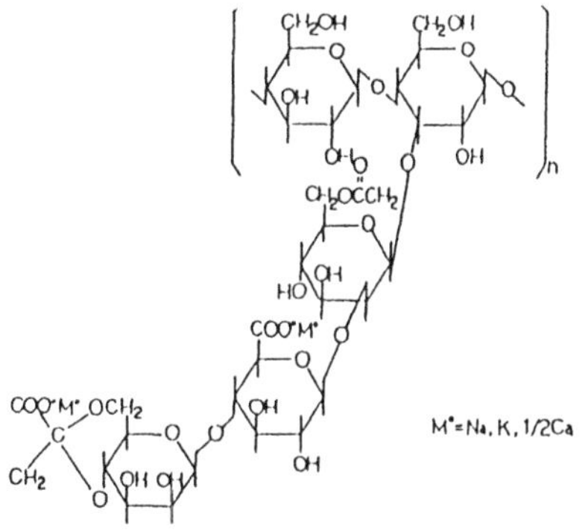

Xanthan gum

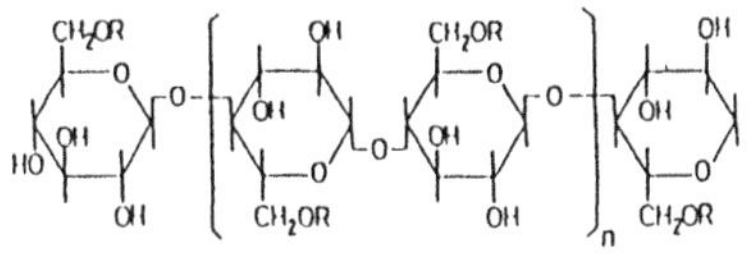

$M^* = Na, K, 1/2Ca$

R= - CH₃ MC(Methyl Cellulose)
R= - C₂H₄OH HEC(Hydroxy Ethyl Cellulose)
Cellulose based viscosity agent

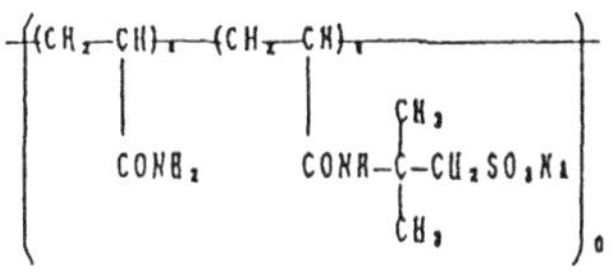

Polyacrylate

Fig. 3. Chemical structure of each viscosity agent.

2.3.1 Effects of water

The object of this study is to understand effects of viscosity agents on viscosity in some different type of water. Solutions of the above mentioned viscosity agents were made, with using 4 different types of water, deionized water, alkaline solution(2% sodium hydroxide and saturated calcium hydroxide), filtered cement water dispersion A(100g of cement in 1000g of deionized water), and filtered cement water dispersion B(300g of cement in 1000g of deionized water). Viscosity was measured at 60 rpm by BM type viscometer manufactured by Tokimeck Co. Concentration of viscosity agents was chosen so as to give the almost same viscosity to that of 1% Welan gum solution in deionized water.

In the past actual use, the standard usage of Welan gum in highly fuidize concrete is 0.05% to the water content. But Welan gum concentration to free water is estimated at 1%. And so we chose 1% as concentration of Welan gum to free water and tested. Free water is assumed as the water which was squeezed out of the paste through 10 minutes-centrifuge at 3000 rpm. The paste contains 2% superplasticizer (β-naphthalene sulfonate) to the cement content in the system of 30% water/cement ratio. As shown by Figure 4, the proportion of free water in the water content is approx.5%, in the case that high level concentration of superplasticizer is added.

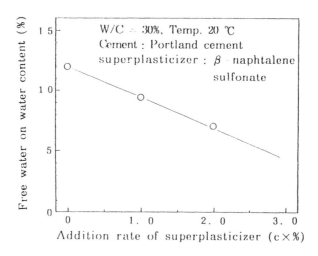

Fig. 4. Relation between SP addition & free water

Figure 5 shows the viscosity of each visocosity agents'solution under various types of water comparatively. Welan gum provides the almost same viscosity under 4 kinds of different water, except that it tends to give a little higher viscosity under the alkaline solution and the filtered cement water dispersion than that under deionized water. Xanthan Gum provides higher viscosity than Welan gum under the alkaline solution. But in the case under the filtered cement water dispersion, the viscosity of Xanthan'solution was lower under higher cement concentration than under lower cement concentration. It is assumed that Xanthan gum interacts strongly with the calcium in cement, leading gelation and providing higher viscosity under low cement concentration. It is also supposed that viscosity decreased because of the aggregates of Xanthan and the precipitation with the calcium under high cement concentration.

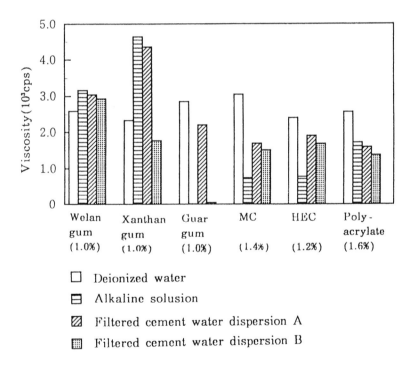

Fig. 5. Each viscosity of viscosity agent'solution under various types of water.

Figure 6 shows the relationship between slump flow and gum concentration in the case that Welan gum or Xanthan gum is added in self-consolidating concrete. Materials and mix proportion in this test will be seen in Table 1 & 2. The slump flow of self-consolidating concrete containing Xanthan gum decreased greatly, even if Xanthan concentration is extremely low. Because it might be assumed that Xanthan gum aggregated in concrete. As Guar gum is insoluble in the strong alkaline system, Guar gum provided hardly viscosity under the alkaline solution and the high concentration cement water.

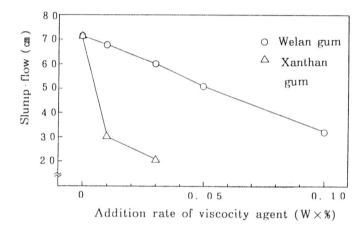

Fig. 6. Relation between viscocity agent addition & slump-flow

Table 1. Property of materials

Notation	Name	Specific gravity	Blaine value(cm²/g)etc.
C	Ordinary Portland cement	3.16	Blaine 3790
SD	Limestone dust	2.70	Blaine 3210
S	Sand	2.57	River sand, Absorption 1.53(%) F.M. 2.62
G	Gravel	2.65	Crushed stone, Max.agg.size 20(mm) Solid Volume 60(%) F.M. 6.65
SP	β-naphtalene sulfonate	—	—
VA	Welan gum or Xanthan gum	—	—

Table 2. Mix proportion

Water cement ratio (%)	Water powder ratio (%)	Unit weight(Kg/m³)					SP * (%)	VA ** (%)
		W	C	SD	S	G		
52.9	32.0	175	331	216	691	878	1.5	0 - 0.10

* Caluculated on the basis of (C+SD) weight
** Caluculated on the basis of water weight

In the case of cellulose based derivatives (MC & HEC) and polyacrylate viscosity agent, the viscosity of solutions under the alkaline solution and filtered cement water dispersion were lower than that under deionized water and the viscosity made a big differnce under some kinds of various water. From these results, it might be expected that these viscosity agents are affected by the differnce of cement concentration and it might be possible for these agents to vary the flow property and/or the casting property.

It was thought that Welan gum has a property to provide a little higher viscosity in alkaline solution than that in deionized water. To confirm this thought, the viscosity vs pH and vs calcium concentration regarding Welan gum and MC solution were measured at 25℃ by BM type viscometer manufactured by Tokimeck Co.. Figure 7 shows the viscosity of 2 kinds of viscosity agent's solution containing 1%, 5%, or 10% of calcium chroride. As this graph shows,the visocosity of both solutions of Welan gum and MC increased as calcium chroride concentration increased. Fugire 8 shows visocosity vs pH(2 to 12), using hydrogen chloride and sodium hydroxide to deionized water. The viscosity of solutions containig Welan gum is not changed substantially over a wide pH range, as shown in this graph. On the other hand, the viscosity of solutions containing MC reduced as acid or alkali was stronger. From these tests, it is expected that the visocosity of the filtered cement water dispersion containing Welan gum increases a little with calcium concentration increase and that the visocosity of the filtered cement water dispersion containing MC decreases as pH is higher or alkali is stronger.

2.3.2 Effects of temperature

For understanding effects of temperature on the viscosity of voscosity agents, the viscosity of 1% of each viscosity agent's solution were measured at 5, 10, 20 & 30 ℃ by BM type viscometer manufactured by Tokimeck Co.. Alkaline solution was used as solution. In this test, Guar gum was not examined because Guar is insoluble in alkaline solution. Figure 9 shows the relation between the percentage of the viscosity which is calculated the viscosity at 5℃ as 100% and the

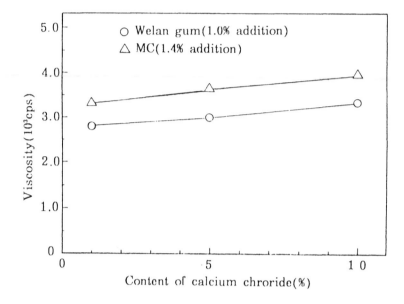

Fig. 7. Relation between content of calcium
chroride and viscosity

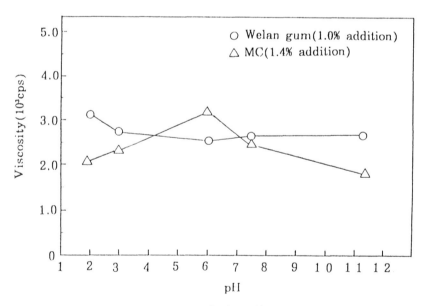

Fig. 8. Relation between pH and viscosity

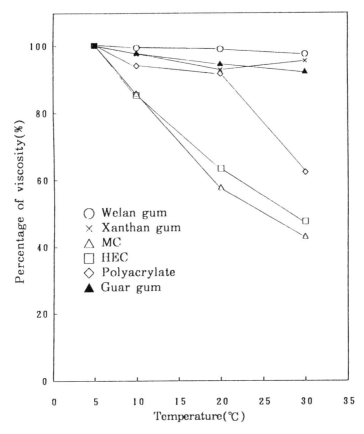

Fig. 9. Relation between % of vis.
and temperature

temperature. The viscosity of solutions containing Welan gum and that containing Xanthan gum are independent with respect to temperature and indicate the identical viscosity within 5 to 30 ℃. This is the reason why the slump flow of self-consolidating concrete containing Welan gum is not changed substantially at various temperatures. Compared to this, the viscosity of solutions containing cellulose derivatives or Polyacrylate viscosity agent decreased as temperature increased. With regard to the degree of decrease, cellulose derivatives was bigger than polyacrylate. The viscosity of water soluble polymer's solution usually decreases as temperature increase but the viscosity cahange with respect to Welan gum or Xanthan gum did not occur.

2.3.3 Relationship between shear rate and viscosity.

For understanding the relationship between shear rate and viscosity, 1% solution of each viscosity agent's were examined by Rheometer, using alkaline solution. Figure 10 shows the shear rate vs the viscosity. With regard to Welan gum and Xanthan gum, the visosity on the solution decreases sharply as the shear rate increases(pseudoplastic property). This property is assumed to the one reason why Welan gum stabilizes the flow property of self-consolidating concrete. In the slump flow test in the case of highly fuidized concrete containing Welan gum, the flow speed is getting slowly gradually from the point which the corn is given a lift and the transfiguring and the viscosity favorably balance each other with the pseudoplastic property's appearance, resulting the stable flow. This is applicable in actual plants. It is assumed that the transfuiguring property is rich while it is flowing and the suitable visocosity appears at the part of the flow-end, preventing materials in concrete to separate.

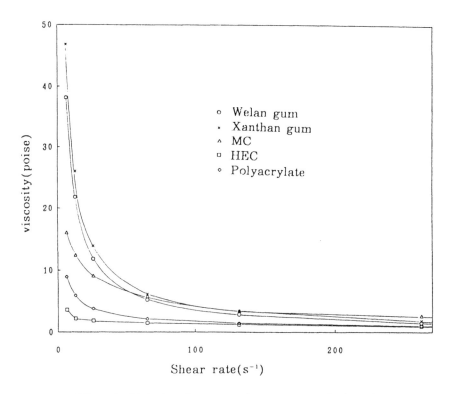

Fig. 10. Relation between shear rate and viscosity.

3 Flow property, ease to flow in small space and self-consolidating property

It has been reported that a suitable addition of Welan gum to self-consolidating concrete provides stability on the flow property which can be examined by slump flow test. In this clause, the flow property, the ease to flow in small space and the self-consolidating property of the concrete containing Welan gum were investigated experimentally, investigating a change in the time.

In the test regarding the stability of the flow property, 5 kinds of water content per unit volume of concrete to the basic mix proportion, -10, -5, ± 0, $+5$ & $+10 \text{kg/m}^3$ were used, imaging the differnces of the surface moisture ratio of fine aggregates. This 5 kinds means 155 to 175kg/m^3 of water content per unit volume of concrete. Materials used in the test and the mix proportion are shown in Table 3 & 4. This proportion was decided so as to satisfy the properties as self-consolidating concrete. The same proportion were prepared, except for the dosage of Welan gum and superplasticizer. 0.5% and 0.0% of Welan gum to the water content were chosen and examined on each tests. The dosage of superplasticizer was chosen so as to let the slump flow be $65 \pm 1 \text{cm}$ just after mixing. And so the dosage of superplasticizer was decided to 2.5% to the powder content in the case of the concrete containing Welan gum and it was decided to 1.7% in the case of that without Welan gum. The air content was adjusted so as to be desirable by controlling the content of the air entraining agent. To see effects of Welan gum accuratly, a simple of β - naphtalene sulfonate without containing slump - retentive agents such as retarders was used. And the weight of fine aggregate was increased or decreased as the water content was increased or decreased.

80 ℓ of concrete on each tests was mixed by 100 ℓ - pan type of mixer for 90 sec. and measured at the initial, 30 minutes and 60 minutes after mixing.

3.1 Slump flow test

The slump flow test was done to grasp the flow property of the concrete. The results is shown in Figure 11. In the case of the concrete system containing Welan gum, the range of the slump flow just after the mixing was 60.5 to 71.0cm and the difference was 10.5cm in response to the change of the water content per unit volume of concrete, -10 to $+10 \text{kg/m}^3$. In case of the concrete system without containing Welan gum, the range of the slump flow was 51.0 to 75.0cm and the difference was 24.0 cm, which was bigger than the system containing Welan gum. With regard to the condition of the concrete sysytem, no segregation was found on the concrete system containing Welan gum over all cases. By contrast with this result, the concrete system without containing Welan gum provided segregation for example

Table 3. Property of materials

Notation	Name	Specific gravity	Blaine value(cm^2/g)etc.
C	Ordinary Portland cement	3.16	Blaine 3790
SD	Limestone dust	2.70	Blaine 3370
S	Sand	2.69	River sand, Absorption 1.31(%) F.M. 2.69
G	Gravel	2.69	Crushed stone, Max.agg.size 20(mm) Solid Volume 63(%) F.M. 6.61
SP	β - naphtalene sulfonate	—	—
VA	Welan gum	—	
AE	Alkylaryl sulfonate	—	—

Table 4. Mix proportion

Mix No.	Water powder ratio (%)	Slump -flow (cm)	Air content (%)	Unit weight(Kg/m^3)					SP * (%)	VA ** (%)	AE * (%)
				W	C	SD	S	G			
1	30.2	65 ±1	4.5±1	165	331	216	713	888	1.7	0.00	0.05
2	30.2	65 ±1	4.5±1	165	331	216	713	888	2.5	0.05	0.02

* Calculated on the basis of (C+SD) weight
** Calculated on the basis of water weight

the floating paste on the surface of the concrete system when the water content per unit volume of concrete was +5 and +10 kg/m³. in the case containing Welan gum, the slump flow at 30min. and at 60min. after the mixing were the almost same under ±0, +5 and +10kg/m³ of the water content per unit volume of concrete and its flow property did not decrease. The slump flow of the concrete system containing Welan gum has a tendency to decrease under -5 and -10kg/m³. But even under the case of -10kg/m³ of the water content per unit, the slump flow was more than 50cm.

The slump flow of the concrete sysem without containing Welan gum did not decrease under the case of +5 and +10kg/m³ of the water

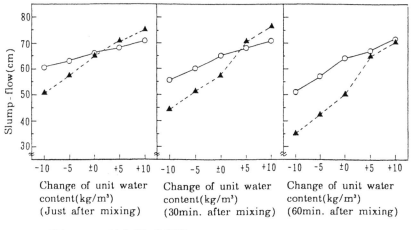

Fig. 11. Change of slump-flow of fresh concrete

content per unit volume of concrete. But these were segregated just after mixing. The concrete system at 60 min. after mixing has no segregation and became the excellent flowable concrete. Under the case of -5 and -10kg/m³ of the water content per unit volume of concrete,the slump flow resulted in great degradation as time-passing.

From these results, it is assumed that the addition of Welan gum minimizes the change of the flow property by the change of the water content per unit volume of concrete and provides the effect of maintaining the stable flowability on the concrete system.

3.2 V type funnel test

The measurement of the time for flowing down through V type funnel, which was shown in Figuer 12, was done to grasp the ease to flow in small space. V type funnel is a equipment which was developed by Prof. Okamura and his staff at Tokyo University to measure the flow property of self-consolidating concrete. Figure 13 shows the average time for flowing down through V type funnel. The average time was calculated from some measured time for flowing down. The less the water content and the longer time-passing, the longer the average time for flowing down of the concrete containing Welan gum tend to be. The change was comparatively small, 7.0 to 16.9cm/s. The change of the concrete system without conctaining Welan gum was big, 6.5 to 27.5cm/s in the case that the slump flow is more than 70cm. This is because low viscosity and segregation cause the concrete to flow with

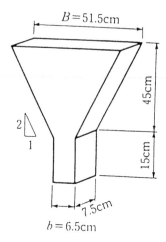

Fig. 12. Shape and size of
V type funnel

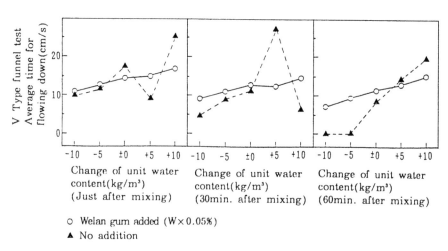

○ Welan gum added (W × 0.05%)
▲ No addition

Fig. 13. Change of average time for flowing down of fresh concrete

a thud or flow slowly owing to coarse aggregates' rocking each other.

In the case of that the slump flow is less than 50cm, the time for flowing down was extremely long or the concrete was not able to flow because of a blockade in the funnel.

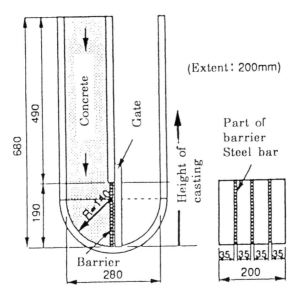

Fig. 14. Shape and size of U type
casting test equipment

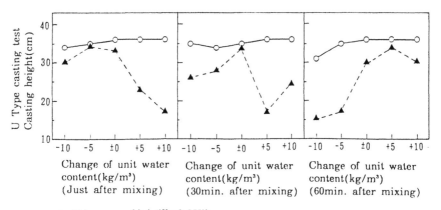

Change of unit water   Change of unit water   Change of unit water
content(kg/m³)   content(kg/m³)   content(kg/m³)
(Just after mixing)   (30min. after mixing)   (60min. after mixing)

○ Welan gum added (W×0.05%)
▲ No addition

Fig. 15. Change of casting height of fresh concrete

3.3 U type casting test

U type casting test was done to understand self-consolidating property of the concrete. The outline of the equipment of U type casting test shown in Figure 14 and the test results are shown in Figure 15[6]. The height of the casting of the concrete system containing Welan gum was more than 30cm in all cases of the range of the water content per unit volume of concrete, -10 to +10kg/m³, and in all cases of time passing, initial, 30 & 60 minutes after mixing. And This results mean the concrete system containing Welan gum has an excellent casting property. In this results, the height of casting was 31.0cm in the case of the system of which the water content per unit volume of concrete was -10kg/m³ and which was left for 60 minutes after mixing in spite that the slump flow was small, 51.0cm. This means Welan gum provide the suitable viscosity in the concrete system and improve the casting property, even in the case that the transfiguring property is small. The height of casting of the concrete system without containing Welan gum was small, because of coarse aggregates' rocking each other in case of too big slump flow and because of small transfiguring property in case of too small slump flow. In Figure 15, the peak of the height of the concrete system without containing Welan gum is moving to the higher water content per unit volume of concrete as time-passing. Because the change of the slump flow by time-passing was big and the most suitable casting property was performed at that time accidentally.

From three results just after mixing, it can be said that the height of the casting of the concrete system containing Welan gum is higher than that of the system without containing Welan gum in spite that the slump flow is the same and so the use of Welan gum make a contribution to an improvement on the self-consolidating property.

Through some tests by us, it was understood that the addition of suitable dosage of Welan gum stabilizes the fluidity, makes it easy to flow in small space and provides self-consolidating property over wide range of slump flow.

4 Conclusion

With experimentally estimating typical properties of the visocosity agent Welan gum, effects of Welan gum on the fluidity of self-consolidating concrete, the ease to flow in small space and the self-consolidaing property were investigated in detail.
The summary are as follows.

1. Welan gum solution purovides same viscosity without depending on the calcium content or pH. This property is unique and it cannot be seen in cellulose based derivatives and Polyacrylate viscosity agents.

2. The viscosity of Welan gum solution is unchanged substantially with in range of 5 to 30℃. This property is one of reasons why the change of the slump fow of the self-consolidatiug concrete contain- ing Welan gum is small at various temperature.
3. Welan gum solution indicates very high viscosity at rest and very low viscosity at high shear-rate(pseudoplastic property). This property is one of the effects to avoid segregation in self-consol- idating concrete.
4. The addition of a suitable dosage of Welan gum can not only gives the stable fluidity and the stable ease to flow in small space to self-consolidating concrete but provides excellent self-consolidat- ing property over the wide range of slump flow.

5 References

1. K.Ozawa,K.Maekawa,M.Kunishima and H.Okamura:High-Performance Concre te based on the durability design of concrete structures, Proc. of the Second East Asia-Pacific Conference on Structural Engineering and Construction,1989
2. K.Ozawa,N.Sakata and II.Okamura:):Evaluation of self-compactability of fresh concrete using the funnel test" Proc. of JSCE,No.490,V-23, 1994
3. M.Yurugi,N.Sakata and M.Iwai:Annual meeting of the Japan Concrete Institute,Vol.14-1,pp51-56,1992
4. N.Sakata,et al:Annual meeting of the Japan Concrete Institute,Vol.1 5-1,1993
5. P.E.Jansson,B.Lindberg and P.A.Sandford:Structural Studies of an Ex tracellular Polysaccharide(S-130)Elaborated by Alcaligenes ATCC3155 5,Carbohyd.Res.,Vol.139,pp.217-223,1985
6. T.Shindoh:Fundamenrtal Study on Properties of Super Workable concre te, Proc. Annual meeting of Japan Concrete Institute,Vol.18,No.1,pp 179-184,1991 .

21 EFFECT OF PROPERTIES OF MIX CONSTITUENTS ON RHEOLOGICAL CONSTANTS OF SELF-COMPACTING CONCRETE

S. NISHIBAYASHI, A. YOSHINO, S. INOUE and
T. KURODA
Department of Civil Engineering, Tottori University, Tottori, Japan

Abstract
In this study we consider whether the rheological constants of self-compacting concrete can be calculated from the physical properties of the materials and mix proportions. Firstly, we investigate the effects of the properties of the materials on the rheological constants (plastic viscosity and yield value) of mortar and concrete. Secondly, we study a method for calculating the plastic viscosity of concrete from the physical properties of the mix constituents and mixture proportions.
Key words: Self-compacting concrete, rheological constants, thickness of excess paste, thickness of excess mortar

1 Introduction

In recent years there has been greater demand for the workability of fresh concrete to be estimated from physical constants (rheological constants) instead of values derived from empirical tests. Many believe that the estimation of workability in terms of rheological constants will promote both the systemization and automation of concrete construction work [1]. Nevertheless, there are problems associated with concrete construction based on the rheological constants of concrete. One is the establishment of a method to calculate proportional mix designs that will produce concrete with the required rheological constants. In pursuing a solution to this problem, it is helpful to think of mortars and concretes as highly concentrated suspensions and to consider the rheological constants of the mixture in terms of both the concentration of the suspended particles and the properties of the suspended particles and matrix.

In this study, we studied the effects of physical properties of aggregate and rheological properties of paste on the rheological constants of mortar and concrete. We

Production Methods and Workability of Concrete. Edited by P.J.M. Bartos, D.L. Marrs and D.J. Cleland. Published in 1996 by E & FN Spon, 2–6 Boundary Row, London SE1 8HN. ISBN 0 419 22070 4.

Table 1. Physical properties of binding materials

Mark	Constituents	Specific gravity	Blaine fineness (cm²/g)	Replacement percentage (%)
PC	Portland cement	3.15	3220	-
SG	Blast-furnace slag	2.89	6020	50
LS	Limestone powder	2.73	5010	30

Table 2. Physical properties of fine aggregates

Mark	Kind	Specific gravity	F.M.	Solid volume percentage (%)	Specific surface area (cm²/cm³)
a	Crushed sand + Fine sand	2.67	2.81	67.5	341
b	River sand + Fine sand	2.58	2.81	65.5	291
c	Crushed sand	2.68	3.01	66.8	288
d	Crushed sand	2.66	3.01	66.5	255
e	Crushed sand	2.68	3.45	63.8	232

Table 3. Physical properties of coarse aggregates

Mark	Kind	Max. size (mm)	Specific gravity	F.M.	Solid volume percentage (%)	Specific surface area (cm²/cm³)
A	Crushed stone	20	2.69	6.81	58.0	6.1
B	Crushed stone	15	2.69	6.71	57.9	6.2
C	Crushed stone	10	2.69	6.52	56.9	6.5

examined a method to calculate rheological constants (especially plastic viscosity) based on the properties of mix constituents and mixture proportions.

2 Experimental summary

2.1 Materials
Ordinary portland cement, blast-furnace slag and limestone powder were used in the experiments. The physical properties and the replacement percentage of these materials are shown in Table 1. The physical properties of the aggregates used are given in Tables 2 and 3. The specific surface area of fine aggregate was calculated using Loudon's equation [2]

$$\log (kS^2) = 1.365 + 5.15n \tag{1}$$

where S is the specific surface area (cm²/cm³), k is the coefficient of permeability (cm³/s·cm²) and n is the void ratio.

The specific surface area of the coarse aggregate was calculated using the projection area method. This method is based on the fact that the mean projection area is a quarter of the full surface area. In this study, the projection areas in three directions were determined.

Four kinds of chemical admixtures were used in the mortar testing, and an AE high-range water-reducing admixture (HAE2) was used in the concrete testing. The main components of the chemical admixtures are given in Table 4.

Table 4.   Main components of chemical admixtures

Mark	Kind	Main components
WR	AE water-reducing admixture	Lignosulfonic acid compound and hydroxycarboxylic acid complex
SP	Sperplasticizer	Naphtalensulfonic acid compound
HAE1	AE high-range water-reducing admixture	Complex of denatured lignin, Alkyarylsulfonic acid and continuous activation polymer
HAE2	AE high-range water-reducing admixture	Polycarboxylic ether complex

Table 5.   Condition of mixture proportions of mortar

Series No.	Binding material	W/C	Fine aggregate	Volume ratio of fine ag.	Chemical admixture
1	PC	0.40	a, b, c, d, e	4 levels	WR, SP, HAE1
2	PC, PC+SG	0.35, 0.40, 0.45	a, b	4 levels	SP, HAE1, HAE2

Table 6.   Condition of mixture proportions of concrete

Binding material	W/C	Fine ag.	Coarse ag.	Volume ratio of coarse ag.	Chemical admixture
PC, PC+SG, PC+LS	0.30, 0.35	a	A, B, C	3 levels	HAE2

Table 7. Size of pull-up viscometer

Sample	Diameter of sphere	Container
Paste and mortar	3.14 cm	ϕ 15 $\times$ 30 cm
Concrete	9.98 cm	ϕ 50 $\times$ 50 cm

2.2 Mixture proportions and test procedure

In this study, we regarded mortar and paste in the following manner to design the mix proportions for mortar under the conditions given in Table 5: mortar as a suspension of fine aggregate particles; and a paste as matrix. The volume ratio of fine aggregate ranged from 0.328 to 0.480. In series 1, we investigated the effects of properties of fine aggregate on the rheological constants of mortar. In series 2, we investigated the effects of the properties and volume of binding material on the rheological constants of mortar.

We considered self-compacting concrete to be a suspension in which the coarse aggregates are suspended particles and the mortar is a matrix. The mixture proportion for the concrete was designed under the conditions in Table 6. The volume ratio of coarse aggregate ranged from 0.257 to 0.322, the slump flow values ranged from 50 to 65 cm. The rheological constants of paste, mortar and concrete were measured using a pull-up viscometer [3]. The dimensions of the pull-up viscometer are shown in Table 7.

3   Results and discussions

3.1 Mortar

Examples of the relationship between the volume ratio of fine aggregate and the rheological constants of mortar are shown in Fig. 1. This figure indicates that for each aggregate and each chemical admixture, as the volume ratio of fine aggregate increased, the rheological constants grew exponentially. The results can be expressed by the equation

$$\log R = a_1 Vs + b_1 \tag{2}$$

where R is the rheological constants of mortar (plastic viscosity or yield value), Vs is the volume ratio of fine aggregate, a_1 and b_1 are constants depending on kinds of fine aggregate and paste.

This equation does not appear to be an ordinary rheological one, since the constants change for each aggregate and each chemical admixture.

We studied the effects of properties of fine aggregate on rheological constants of mortar. Using the same rheological constants, we studied the relationship between the kinds of fine aggregate and the volume ratio calculated from equation (2). Fig. 2 shows the volume ratio of each fine aggregate at the same plastic viscosity or yield value. This figure indicates that the larger the specific surface area of the aggregate (Table 2), the smaller the volume ratio at the same rheological constants. Therefore, we can presume that the specific surface area of fine aggregate affected the rheological constants as well as the volume ratio.

We studied the relationship between the rheological constants and the total surface area, calculated from following equation

$$St = Ss \times Vs \times 100 \tag{3}$$

where St is the total surface area (m²/m³), Ss is the specific surface area (cm²/cm³) and

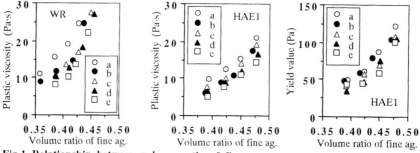

Fig.1 Relationship between volume ratio of fine aggregate and rheological constants

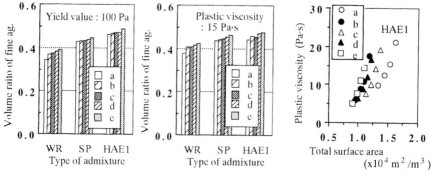

Fig.2 Effect of kind of fine aggregate

Fig.3 Relationship between total surface area and plastic viscosity

Vs is the volume ratio of fine aggregate.

An example of the relationship between the total surface area of the fine aggregate and the rheological constants of mortar is shown in Fig. 3. This figure indicates that for each aggregate, the rheological constants grew exponentially as the total surface area of fine aggregate increased. Therefore, it seems that there is another property, other than the specific surface and volume ratio of fine aggregate, which affects the rheological constants.

According to the Excess Paste Theory [4], the consistency of concrete depends of two factors: the thickness of excess paste and the consistency of the paste itself. The thickness of excess paste is the volume of paste (minus the volume of paste required to fill the voids of the compacted aggregate) divided by the surface area of the aggregate. The thickness of excess paste can be expressed by the following equation

$$Fp = (1- Vs \times 10^2 / Cs) \times 10^4 / (Ss \times Vs) \qquad (4)$$

where Fp is the thickness of excess paste (μm), Cs is the solid volume percentage (%), Ss is the specific surface area (cm^2/cm^3) and Vs is the volume ratio of the aggregate. From this equation the solid volume percentage of the aggregate appears to affect the rheological constants as well as the specific surface and volume ratio.

Examples of the relationship between the thickness of excess paste and the rheological constants of mortar are shown in Fig. 4. This figure indicates that the rheological constants decreased as the thickness of excess paste increased, and that a thickness of the excess paste determined the plastic viscosity or the yield value irrespective of the kind of aggregate. The results may be expressed by the equation

$$\log R = a_2 Fp + b_2 \qquad (5)$$

where R is the rheological constant, Fp is the thickness of excess paste, a_2 and b_2 are constants depending on the kind of paste. It follows that for each paste, equation (5) has different constants.

It is clear that the rheological constants of mortar are affected by the rheological constants of the paste as well as the thickness of the excess paste. The plastic viscosity is particularly affected, and transforms into the relative viscosity used in other suspension fields. If a mortar is considered to be a suspension in which fine aggregates

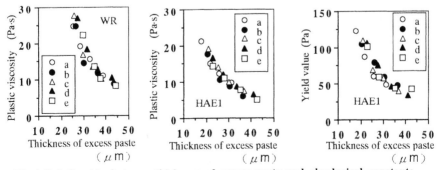

Fig.4 Relationship between thickness of excess paste and rheological constants

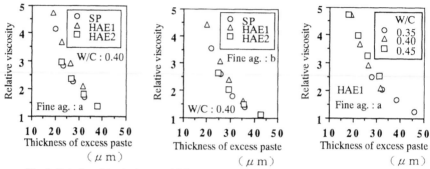

Fig.5 Relationship between thickness of excess paste and relative viscosity

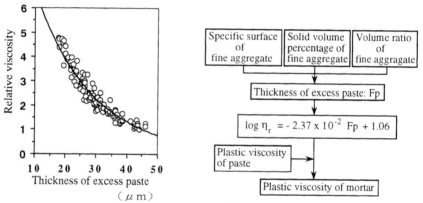

Fig.6 Relationship between thickness of excess paste and relative viscosity (all data)

Fig. 7 Method of prediction of plastic viscosity of mortar

are suspended particles, the relative viscosity is given by the ratio of the plastic viscosity of mortar to the plastic viscosity of paste.

The examples of the relationship between the thickness of excess paste and the relative viscosity are shown in Fig. 5. This figure indicates that the relative viscosity decreased as the thickness of the excess paste increased, and the relationship between the thickness of excess paste and the relative viscosity was unaffected by the kind of paste. The relationship between the thickness of excess paste and the relative viscosity was plotted on a graph for all mixtures (Fig. 6). Based on this graph, it appears that the relationship between the thickness of excess paste and the relative viscosity can be expressed by the following equation

$$\log \eta_r = -2.38 \times 10^{-2} Fp + 1.06 \qquad (6)$$

where η_r is the relative viscosity and Fp is the thickness of excess paste.

The method proposed to calculate the plastic viscosity of mortar is shown in Fig. 7. The plastic viscosity of mortar can be estimated from the plastic viscosity of paste and the thickness of excess paste (the latter being calculated from the specific surface area, solid volume percentage and volume ratio of the fine aggregate).

3.2   Self-compacting concrete

Examples of the relationship between the volume ratio of coarse aggregate and the rheological constants of self-compacting concrete are shown in Fig. 8.   This figure indicates that the rheological constants increased as the volume ratio of the coarse aggregate increased, and that the kind of coarse aggregate affected the rheological constants.   Fig. 9 indicates the effect of the kind of binding material on the plastic viscosity of concrete and suggests that plastic viscosity is affected by the kind of binding material and the ratio of water to binding material.   Based on these results, the rheological constants of concrete seem to be affected by the properties of the coarse aggregate and the rheological constants of the mortar (or paste).

In the same manner as our earlier consideration of mortar, we thought of self-compacting concrete as a suspension in which coarse aggregates are deemed to be suspended particles and mortar a matrix.   We studied the relationship between the thickness of excess mortar and relative viscosity.   The thickness of excess mortar is calculated by the following equation

$$Fm = (1 - Vg \times 10^2 / Cg) \times 10 /(Sg \times Vg) \tag{7}$$

where Fm is the thickness of excess mortar (mm), Cg is the solid volume percentage (%), Sg is the specific surface area (cm^2/cm^3) and Vg is the volume ratio of coarse aggregate.

Furthermore, the relative viscosity is given by the ratio of the plastic viscosity of concrete to the plastic viscosity of mortar.

Fig.10 shows the relationship between the thickness of excess mortar and the relative viscosity.   From this figure, it appears that the relationship is expressed by the following equation

$$\log \eta_r = -0.290 \, Fm + 1.59 \tag{8}$$

where η_r is the relative viscosity and Fm is the thickness of excess mortar.

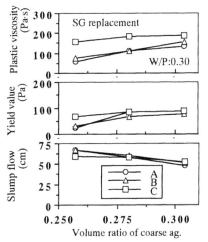

Fig.8 Effect of kind of coarse aggregate

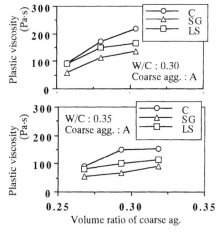

Fig.9 Effect of kind of binding material

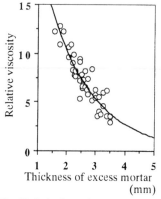

Fig.10 Relationship between thickness of excess mortar and relative viscosity

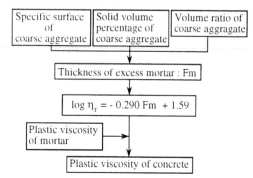

Fig. 11 Method of prediction of plastic viscosity of concrete

The method we propose to calculate the plastic viscosity of self-compacting concrete is shown in Fig.11. We believe that the plastic viscosity of concrete can be calculated from the plastic viscosity of mortar and the thickness of excess mortar (the latter being calculated from specific surface area, solid volume percentage and volume ratio of the coarse aggregate).

4 Conclusion

In this study, we discussed that the rheological constants of mixtures could be calculated based on the physical properties of the materials and their mix proportions.

We confirmed that the plastic viscosity of mortar could be calculated from the plastic viscosity of paste and the thickness of excess paste (the latter being calculated from the specific surface area, solid volume percentage and volume ratio of the fine aggregate). Moreover, we confirmed that plastic viscosity of self-compacting concrete could be calculated from the plastic viscosity of mortar and the thickness of excess mortar (the latter being calculated from specific surface area, solid volume percentage and volume ratio of the coarse aggregate).

5 References

1. Tanigawa,Y., Mori,H. and Watanabe,K. (1990) Analytical study on flow of fresh concrete by suspension element method. *Proc. of RILEM Colloquium: Properties of Fresh Concrete*, Chapman and Hall, pp. 301-8.
2. Loudon, A. G. (1952) The computation of permeability form simple soil tests. *Geotechnique*, pp.165-83.
3. Nishibayashi,S., Yamura,K. and Inoue,S. (1981) Rheological properties of super-plasticized concrete. *SP-68*, ACI, pp. 441-66.
4. Powers, T.C. (1968) *The Properties of Fresh Concrete*, John Wiley & Sons, Inc., pp. 501-3

22 EFFECT OF MIX CONSTITUENTS ON RHEOLOGICAL PROPERTIES OF SUPER WORKABLE CONCRETE

T. SHINDOH, K. YOKOTA and K. YOKOI
Technology Research Center, Taisei Corporation, Yokohama, Japan

Abstract
A super workable concrete is a new type of concrete which can be filled in heavily reinforced area without applying vibration. The authors developed the super workable concrete with utilization of a special kind of viscosity agent derived from process of biotechnology.

This study investigated influence of mix constituent on filling ability of the super workable concrete. Especially , this paper focused on correlation between filling ability of the super workable concrete and rheological properties of its mortar.

It was found from result of the evaluation tests that acceptable range of filling ability was quantified by relationship of yield value and plastic viscosity.

Keywords: Super Workable Concrete , Viscosity Agent , Filling Ability, Deformability, Resistance to segregation, Rheology

1 Introduction

A super workable concrete is defined as the concrete which has excellent deformability and high resistance to segregation , and can be filled in heavily reinforced area without applying vibration. Idea of the super workable concrete was originated by the research team from concrete laboratory of University of Tokyo[1].

For improvement of manufacturing and handling, the authors developed the super workable concrete with utilization of a

Production Methods and Workability of Concrete. Edited by P.J.M. Bartos, D.L. Marrs and D.J. Cleland. Published in 1996 by E & FN Spon, 2–6 Boundary Row, London SE1 8HN. ISBN 0 419 22070 4.

viscosity agent which is a kind of natural polysaccharide polymer derived from process of biotechnology[2][3].

This viscosity agent makes it possible to increase resistance to segregation of super workable concrete without negative influence such as decreasing deformability of the concrete[4].

To increase filling ability of super workable concrete, harmonizing deformability with resistance to segregation is required. Since these abilities are essentially incompatible, they have a tendency to be sensitive to variation of mix constituent such as kinds of powder, solid volume and fluctuation of water content, as compared with conventional concrete.

This study investigated influence of mix constituent on filling ability of the super workable concrete. Especially , this paper focused on relationship between filling ability of the super workable concrete and rheological properties of its mortar.

2   Evaluation test

2.1   Filling ability

The filling ability of concrete is evaluated by using U-shaped apparatus as shown in Fig.1. Deformed bars with nominal diameter of 13mm are installed at the gate with clear spacing of 35mm between bars.

To conduct the test, concrete sample is filled in room R1. The gate is then opened, through the clearance of the reinforcing bars installed at the gate, into the destination R2. The filling height'Uh' of the concrete in R2 is then measured and denoted as the filling ability of the concrete.

When the concrete rises to a height of 30cm or more , it is judged as "good" in filling ability.

Assuming there is no friction between the sample and the inner surface of the vessel, the calculated maximum filling height is 36.5cm. However, from facts of a filling test using a vessel of extremely dense reinforced conditions[1] and from the application to construction of actual structures[5], fairly good filling ability has been known to be achieved if the filling height reaches over 30cm.

2.2   Deformability

Deformability of super workable concrete is evaluated by slump flow test. The slump flow test is conducted just same as a slump test of conventional concrete. In the slump flow test, deformability of concrete is expressed by the average base diameter of the concrete mass after the slump test.

Previous study affirms that slump flows that ensures good deformability range between 60 to 70 cm on average[3].

Together with the measurement of slump flow, the time required

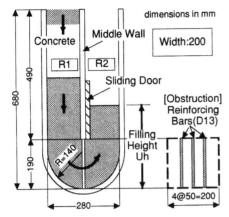

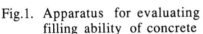

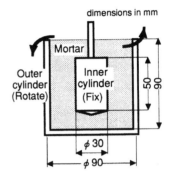

Fig.1. Apparatus for evaluating filling ability of concrete

Fig.2. Details of viscometer for mortar

for slump flow to reach 50cm from stationary, was measured.
It is suggested that the velocity of deformation at slump flow 50cm serve as average velocity of deformation and indicator of resistance to segregation of super workable concrete[6].

2.3 Rheological property
The coaxial cylinders viscometer for mortar is shown in Fig.2. The rotationary speed of outer cylinder is continuously variable up to maximum revolution of 50rpm within 2 minutes. The system of viscometer and test condition are provided to minimize errors which is occurred in a slippage layer under the surface of cylinders. Yield value and plastic viscosity are calculated according to the flow curve of torque against speed.

2.4 Test condition
The evaluation test of Filling ability was given to the super workable concrete in question, while deformability would be measured. At the same time, the concrete sample was wet screened by a 5mm sieve to sample mortar, and rheological properties of the mortar were measured. Deformability of the concrete and rheological properties of its mortar can be quantitatively obtained by investigating correlations between filling height, slump flow, flow time at 50cm flow, plastic viscosity and yield value.
The mix proportions of super workable concrete used in this test are shown in Table 1. Table 2 shows quality of the materials used. The test was divided into 3 series.
Series A was intended to check effect of addition or withholding of the viscosity agent on filling ability of super workable concrete

Table 1. Test series and mix proportions of mixtures

Series	Case	Water/Powder (%)	Water	Op	Bl	Bs	Fa	Ls	S	G	Sp (% of Powder)	Bp (% of Water)
A	C-1	30	165	-	500	-	50	-	792	813	1.95	0.3
A	C-2	29.5	162	-	500	-	50	-	800	813	2.00	0
B	C-3	47	165	350	-	-	-	0	982	793	2.30	0.3
B	C-4	37	165	350	-	-	-	100	884	793	2.00	0.3
B	C-5	30	165	350	-	-	-	200	788	793	1.80	0.3
C	C-6		178	215		215	107		881	663	1.60	0.6
C	C-7	33	165	200	-	200	100	-	819	795	1.60	0.6
C	C-8		157	190		190	96		782	875	1.70	0.6

(Unit content (kg/m³): Water, Powder [Op, Bl, Bs, Fa, Ls], S, G)

Table 2. Properties of materials used in this study

Materials		Type	Properties
Powder	[Op]	Ordinary portland cement	Specific gravity=3.16, Finesse=3270cm²/g
	[Bl]	Belite-based low heat cement	Specific gravity=3.22, Finesse=3440cm²/g
	[Bs]	Blast furnace slag	Specific gravity=2.90, Finesse=4430cm²/g
	[Fa]	Fly ash	Specific gravity=2.23, Finesse=3080cm²/g
	[Ls]	Lime stone powder	Specific gravity=2.70, Finesse=3000cm²/g
Fine aggregate	[S]	Specific gravity=2.59, Finesse modulus=2.64	
Coarse aggregate	[G]	Maximum size=20mm, Specific gravity=2.65, Finesse modulus=6.68, Solid volume=0.61	
Chemical admixture	[Sp]	Superplasticizer	Polycarbonacid-based
	[Bp]	Viscosity agent	Natural polymer(insoluble in water) derived from process of biotechnology

and check rheological properties of its mortar, while varying superplasticizer content.

Series B was intended to check relationship between filling ability and rheological properties, while varying water content.

In series B, variation of powder volume was used as parameter that should be studied.

Series C focused on arching reaction of coarse aggregate, and checked influence of coarse aggregate volume would cause on filling ability and rheologcal properties.

3 Results and Discuss

3.1 Series A

Influence of superplasticizer content on filling height of concrete and rheological properties of its mortar are shown in Fig.3.

In Fig.3, if filling height'Uh'≧30cm is set as acceptable range of filling ability, C-1(with addition of the viscosity agent) shows good filling ability, especially with increase of superplasticizer content. This indicates that addition of the viscosity agent allows a

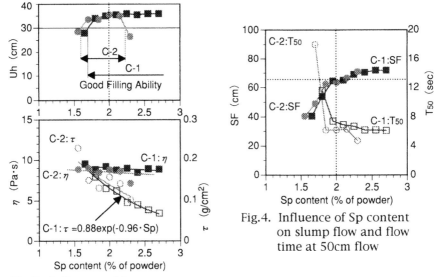

Fig.4. Influence of Sp content
on slump flow and flow
time at 50cm flow

Fig.3. Influence of Sp content
on Filling height and rheological properties

proportion with high resistance to segregation, despite addition of a considerable amount of superplasticizer.

The rheological property findings indicated that, as addition of superplasticizer increases or decreases, yielding value changes linearly, while plastic viscosity shows almost constant values.

This fact, obtained as expected, is generally known. What deserves special note is that the effects of addition of superplasticizer on yield value or plastic viscosity are almost identical, regardless of addition or withholding of the viscosity agent. This result indicates that addition of the viscosity agent in question does not obstruct the effect of superplasticizer on improving deformability.

Fig.4 shows influence of superplasticizer content on slump flow and flow time at 50cm flow. The slump flow ranged from 40cm to 70cm for cases, and the flow time at 50cm flow showed a tendency of drastic deceleration as the amount of superplasticizer added was reduced.

3.2 Series B

Fig.5 shows influence of various water content on filling height'Uh', plastic viscosity'η' and yield value'τ'.

In Fig.5, when yield value and plastic viscosity at the upper and lower value of acceptable range of filling ability are calculated, the ranges of yield value and plastic viscosity to ensure good filling ability can be quantified.

Fig.6 shows relationship between paste/mortar volumetric ratio

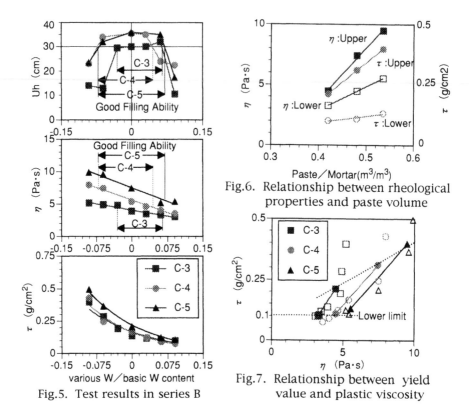

Fig.5. Test results in series B

Fig.6. Relationship between rheological properties and paste volume

Fig.7. Relationship between yield value and plastic viscosity

and upper and lower limit of yield value and plastic viscosity that ensure good filling ability, as obtained earlier.

Yield value and plastic viscosity both increase linearly with the increase of paste volume, simultaneously widening the range between the upper and lower limit.

Fig.7 shows the range of upper and lower limit of yield value and plastic viscosity of acceptable range of filling ability on relationship between yield value and plastic viscosity.

As shown by the figure, the upper limit of yield value widens with the increase of paste volume, but the lower limit of yield value stays at value of $0.1 g/cm^2$ for each case. This result indicates that the lower limit of yield value exists to ensure good filling ability as super workable concrete.

3.3 Series C

Fig.8 shows influence of various water content on filling height'Uh', plastic viscosity'η' and yield value'τ'. The physical arching among coarse aggregate particles is a major factor that drastically changes the filling ability of super workable concrete.

As shown by this figure, rheological properties are almost the

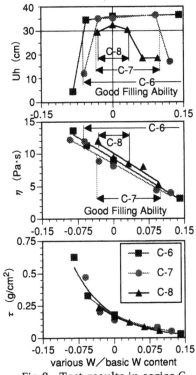

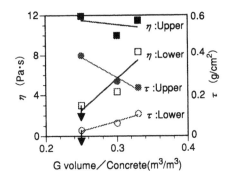

Fig.9. Relationship between rheological properties and gravel volume

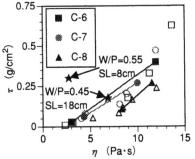

Fig.8. Test results in series C

Fig.10. Relationship between yield value and plastic viscosity

same for varying water content for all cases, while superplasticizer content greatly varies.

Fig.9 focuses on coarse aggregate/concrete volumetric ratio, together with the upper and lower limit of yield value and plastic viscosity, calculated as described in the Series B explanation.

With the increase in coarse aggregate volume, the range between the upper and lower limit of both yield value and plastic viscosity decreases drastically. In this series, a proportion with coarse aggregate content of 0.35 was evaluated and the filling height of the conventional mix proportion, was approximately 20cm.

This result can be explained by the fact that it is almost equal to the coarse aggregate volume at which the upper and lower limit converge, as shown in Fig.9.

Fig.10 shows the range between the upper and lower limit of yield value and plastic viscosity of acceptable range of filling ability.

The figure also shows the results from conventional concrete samples that measured a slump of 8cm with water/powder ratio is 0.55 and a slump of 18cm by adding superplasticizer with water/powder ratio is 0.45. Obviously these conventional concrete samples have no filling ability since their coarse aggregate

volumetric ratio of around 0.4.
As shown in Fig.10, the conventional concrete with a slump of 8cm is quite remarkably off the range of super workable concrete with good filling ability, but concrete with a slump of 18cm and W/P=0.45 is positioned in the rheological range of super workable concrete that ensures acceptable filling ability.
With all its defects that filling ability is sensitive to fluctuation of water content as in Case 3 of Series B, this result suggests that it is possible to accept filling ability by reducing solid volume.

4 Conclusions

This paper provides the evaluation results that focused on the relationship between the acceptable range of filling ability of super workable concrete and rheological properties of its mortar, and discusses the influence of mix constituent on filling ability of super workable concrete.
In this study, quantifying correlation between filling ability of concrete and rheological properties of its mortar for each mix constituent was accomplished. There is considerable validity to the indication that constant lower limit of yield value exists to ensure good filling ability as super workable concrete.
Having got this point firmly established, the study can turn to establishing highly precise methods to select a suitable mix proportion for super workable concrete.

5 References

1. K.Ozawa,et.al, (1989) High performance concrete based on durability of concrete, Proceeding of 2nd East Asia-Pacific conference on Structural Engineering and Construction, Vol.1, pp.445-456
2. T.Shindoh,et.al, (1990) Properties of super workable concrete in fresh state, Proceeding of 45th JSCE conference, pp.228-229
3. M.Hayakawa,et.al, (1994) Development and application of super workable concrete, in Special Concretes:Workability and Mixing, E&FN Spon, London, pp.183-190
4. K.Nara,et.al, (1994) Properties of β-1,3-Glucan(Curdlan) applied for viscosity agent to concrete, Concrete Research and Technology, Vol.5, No.1, pp.23-28 (in Japanese)
5. J.sakamoto,et.al, (1993) Application of super-workable concrete to actual construction, Concrete2000, Scotland, pp.891-902
6. S.Tangtermsirikul,et.al, (1992) A study on velocity of deformation of super workable concrete, Proceeding of 14th JCI annual conference, Vol.1, pp.1161-1166

23 EXPERIMENTAL RESEARCH ON THE MATERIAL PROPERTIES OF SUPER FLOWING CONCRETE

J.K. KIM and S.H. HAN
Korea Advanced Institute of Science and Technology, Taejon, Korea
Y.D. PARK and J.H. NOH
Tongyang Central Laboratories, Suwon, Korea
C.L. PARK and Y.H. KWON
Daewoo Institute of Construciton Technology, Suwon, Korea
S.G. LEE
Chonnam National University, Kwangju, Korea

Abstract
In this study, the properties of super flowing concrete containing fly ash were experimentally investigated and compared with those of ordinary concrete. Tests were carried out on five types of super flowing concrete mixes containing fly ash and three types of ordinary concrete mixes without fly ash. Flow test, O-funnel test, box test, L type test and slump test were carried out to obtain the properties for flowability and workability of fresh concrete. The mechanical properties of hardened concrete were also investigated in terms of compressive strength, splitting tensile strength, modulus of elasticity, creep and drying shrinkage.
    In fresh concrete, it was found that super flowing concrete had excellent workability and flowability compared with ordinary concrete, and the volume ratio of coarse aggregate to concrete greatly influenced flowability and workability. Super flowing concrete also had good mechanical properties at both early and late ages with compressive strength reaching as high as 40 MPa at 28 days. The creep deformation of super flowing concrete investigated was relatively lower than that of ordinary concrete, but drying shrinkage was much higher.
Keywords: compacting, flowability, fly ash, HPC, mechanical property, super flowing concrete, workability,

1 Introduction

Nowadays, the introduction of new admixtures and cementitious materials has allowed the production of high performance concrete(HPC). HPC has three important properties which are high strength, good flowability and high durability.

Production Methods and Workability of Concrete. Edited by P.J.M. Bartos, D.L. Marrs and D.J. Cleland.
Published in 1996 by E & FN Spon, 2–6 Boundary Row, London SE1 8HN. ISBN 0 419 22070 4.

Super flowing concrete is a type of HPC and has very high flowability. The quality and long-term durability of concrete structure are greatly effected by compacting work with vibrators at construction sites. However, it is very difficult to improve the quality of concrete structure greatly due to the lack of skillful labors and other social constraints. This problem could be solved by the use of super flowing concrete which do not need compacting with vibrators. Super flowing concrete could improve the long-term durability greatly, reduce construction cost, and is being considered as revolutional materials in concrete technology. Also, industrial by-products such as fly ash, ground granulated blast furnace slag and silica fume could be used as admixtures of super flowing concrete, and we could save the expense to treat the wastes, recover the resources and protect the environment.[1]

The concept of super flowing concrete and its significance were firstly raised by Okamura(Professor of the University of Tokyo) in 1986. The first prototype was developed and published in 1988 and now being quickly and widely implemented in the Japanese construction. By the way, the first study, in Korea, developing the methods for mixture optimization of super flowing concrete was undertaken in 1993. Recently, in order to utilize practically super flowing concrete, a wide range investigations are being conducted, which are evaluation methods of rheology and workability, investigation of the mechanical properties of super flowing concrete, field studies for guidelines on mixing and placing and so on.[2]

Objectives of this investigation are to find out the properties of fresh concrete and the mechanical properties of hardened concrete, and to compare the results of super flowing concrete with those of ordinary concrete.

2 Experimental programs

2.1 Materials

Material properties are shown in Table 1 and the properties of superplasticizer are also shown in Table 2.

Table 1. Material properties

	Cement	Fly ash	Fine aggregate	Coarse aggregate
Type	Type I portland cement	By-product of power plant	Sea sand	Crushed stone Max. size 19 mm
Specific gravity	3.15	2.21	2.58	2.61
Specific surface	3315 cm^2/g	4201 cm^2/g	–	–
Fineness modulus	–	–	2.73	6.82

Table 2. Superplasticizer

Specific gravity	pH	Solid content(%)	Quantity (Cement weight %)	Main component
1.18	7~9.5	41	0.8~2.0	Sulfonate naphthallene formaldehyde condensate

2.2 Mix proportions of concrete

As shown in Table 3, SF 1-1, SF 1-2, SF 1-3 and SF 1-4 include fly ash equivalent to 30% of the binder content, and SF 2 includes fly ash equivalent to 40% of the binder content. The volume ratios of coarse aggregate to concrete in SF 1-1, SF 1-2, SF 1-3 and SF 1-4 are 0.27, 0.31, 0.35 and 0.39, respectively. The water-cement ratio of SF 1-1, SF 1-2, SF 1-3 and SF 1-4 is 35%, and that of SF 2 is 38%. NC 1, NC 2 and NC 3 were designed to have the target compressive strength of 20MPa, 40MPa and 60MPa, respectively.

Table 3. Basic mix proportions

Specific number	Classification	Water (kg/m^3)	Binder(kg/m^3) Cement	Binder(kg/m^3) Fly ash	Aggregate(kg/m^3) Fine	Aggregate(kg/m^3) Coarse	Superplasticizer (kg/m^3)	Water-Cement ratio (%)	Volume ratio of coarse aggregate to concrete
SF 1-1	Super flowing concrete mixes	195	390	167	814	733	10.0	35	0.27
SF 1-2		185	370	159	782	820	10.0	35	0.31
SF 1-3		175	350	150	739	917	10.0	35	0.35
SF 1-4		165	330	141	678	1032	9.9	35	0.39
SF 2		190	300	200	778	819	8.5	38	0.31
NC 1	Ordinary concrete mixes	185	350	0	775	1035	0.0	53	0.39
NC 2		175	400	0	760	1035	4.0	44	0.39
NC 3		175	500	0	677	1032	7.5	35	0.39

2.3 Test Procedures

2.3.1 Tests of fresh concrete

Flow test, slump test, O-funnel test, L type test and box test were used to estimate flowability and workability of fresh concrete. Flow test and slump test have almost the same test procedures. The former evaluates the flowability and workability of fresh concrete by the bottom radius of the concrete flowed, while the latter by the height of the concrete. O-funnel test measures the time that fresh concrete flows from a funnel. Good flowable concrete would consume short time to flow out. For the evaluation of self-compactable performance, L type apparatus shown in Fig. 1 (b) was used, and four grades which were excellent, good, plain and poor were selected. Box test, as shown in Fig. 1 (c), puts fresh concrete in left box and opens the gate. When flow is stopped, flowability is estimated by the height difference of fresh concrete in left and right boxes.[3]

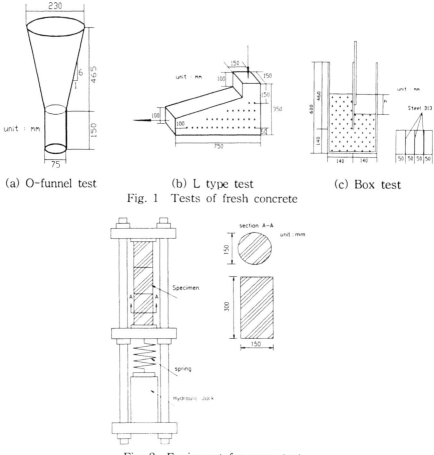

| (a) O-funnel test | (b) L type test | (c) Box test |

Fig. 1   Tests of fresh concrete

Fig. 2   Equipment for creep test

2.3.2   Tests of hardened concrete

The properties of hardened concrete which were compressive strength(f_c'), splitting tensile strength(f_t), modulus of elasticity(E_c), creep and drying shrinkage were investigated. Compressive strength, splitting tensile strength and modulus of elasticity were obtained from $\Phi\,100\times200$ mm cylinders, creep from $\Phi\,150\times300$ mm cylinders, drying shrinkage from $100\times100\times400$ mm hexahedrons specimens. For all properties except drying shrinkage, the average of the experimental results of three identical specimens was used. Drying shrinkage strain was determined from an average of two identical specimens.

Compressive strength was tested according to ASTM C 39, splitting tensile strength according to ASTM C 496 and modulus of elasticity according to ASTM C 469. For all mixes, compressive and splitting tensile strength of concrete were tested at 3, 7, 28, 90 days and modulus of elasticity at 28, 90 days. The ends of specimens were capped with sulfur mortar before testing.

Creep was tested for SF 1-2 and NC 2. Fig. 2 shows the creep tester and specimens. Creep specimens removed from the molds not less than 24 hour after molding and cured in a moist condition at a temperature of $23.0\pm2.0\,\degree C$ until the age of 28 days. To evaluate the basic creep, the specimens were wrapped in plastic film to prevent the gain or loss of water during the storage and test period. The specimens were stored at a temperature of $23.0\pm2.0\,\degree C$ during a test period. The specimens were initially loaded at the age of 28 days and at the 40% intensity of the compressive strength of the loading age.

Shrinkage was tested for all super flowing concrete mixes and NC 2, NC 3 of ordinary concrete. The drying shrinkage specimens were cured in the same condition as the creep specimens during 7 days. After curing, the specimens were stored at a temperature of $20.0\pm1.0\,\degree C$ and relative humidity of $60\pm5\%$. The comparator was used to test drying shrinkage strains.[4][5]

3 Experimental results and discussion

3.1 Experimental results of fresh concrete

SF 1-1 and SF 1-2 are more flowable and workable than SF 1-3 and SF 1-4, as shown in Table 4. Mix proportions of SF 1-1, SF 1-2, SF 1-3, and SF 1-4 were designed to have the volume ratios of coarse aggregate to concrete of 0.27, 0.31, 0.35 and 0.39, respectively. Therefore, it appears that the critical value of the volume ratio of coarse aggregate on flowability and workability exists between 0.31(SF 1-2) and 0.35(SF 1-3). The difference of fly ash contents ranging 30%~40% hardly influenced flowability and workability, as shown in Table 4.

Ordinary concrete, which was NC 1, NC 2 and NC 3, had very poor flowability and workability compared with SF 1-1 and SF 1-2 of super flowing concrete mixes. For flow test, the experimental data of SF 1-1 and SF 1-2 were two times greater than those of ordinary concrete. Insufficient flowability of ordinary concrete made the outlet of a funnel closed, and O-funnel test could not be carried out for ordinary concrete. Due to the same reason as O-funnel test, L type test could not be carried out for ordinary concrete.

Table 4. Experimental data of fresh concrete

Specific number	Flow (cm)	O-funnel (sec)	Box (cm)	L type	Slump (cm)
SF 1-1	65	6.9	2.5	excellent	–
SF 1-2	67	7.8	2.0	excellent	–
SF 1-3	65	closed	45.7	poor	–
SF 1-4	60	31.8	48.5	poor	–
SF 2	65	7.6	22.5	excellent	–
NC 1	40	closed	–	–	18.7
NC 2	36	closed	–	–	18.0
NC 3	36	closed	–	–	18.7

3.2 Experimental results of hardened concrete

3.2.1 Compressive strength(f_c')

The typical compressive strength developments of super flowing concrete and ordinary concrete are shown in Fig. 3 (a). At early ages, there was little difference in the compressive strength between super flowing concrete and ordinary concrete, but the difference is increased with age due to the contribution to the compressive strength of fly ash. Fig. 3 (b) shows the compressive strength of super flowing concrete with fly ash contents of 30% and 40%. As expected, the compressive strength of 30% fly ash content was 30%~40% higher than that of 40% fly ash content.

Fig. 4 (a) and (b) show the effect of curing condition on the compressive strength for SF 1-2 and NC 3 which have the same water-cement ratio. At 3 days, the compressive strength with curing condition was similar. But with age, the compressive strength development of the specimens cured in moist condition was higher than that of the specimens cured in air condition. At 28 days, the difference of the compressive strength for SF 1-2 with curing condition was about 10%, while that for mix NC 3 was more than 18%. It can be concluded that the effect of curing condition on the compressive strength of super flowing concrete is less than that of ordinary concrete.

3.2.2 Splitting tensile strength(f_t)

The 28-day splitting tensile strength of super flowing concrete ranged from 3.6 to 4.0 MPa, as shown in Table 5. These experimental data were about 8% ~10% of the 28-day compressive strength, and these ratios were comparable with those of ordinary concrete of the same compressive strength. Therefore, the splitting tensile strength of super flowing concrete is identical with ordinary

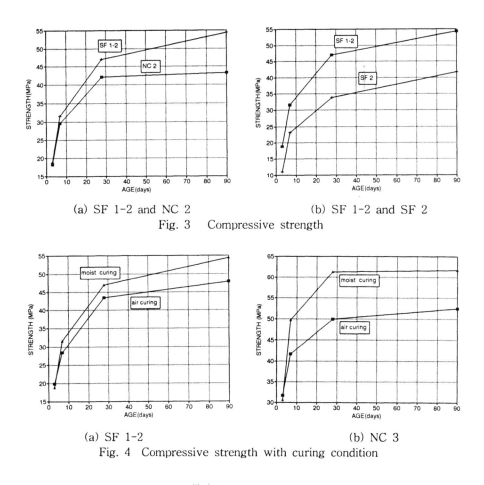

(a) SF 1-2 and NC 2   (b) SF 1-2 and SF 2
Fig. 3   Compressive strength

(a) SF 1-2   (b) NC 3
Fig. 4   Compressive strength with curing condition

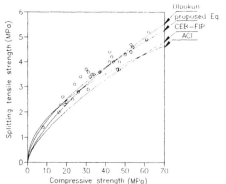

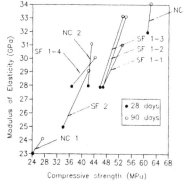

Fig. 5   The relation between splitting tensile strength and compressive strength

Fig. 6   The relation between modulus of elasticity and compressive strength

concrete. Also, the relation between splitting tensile strength and compressive strength is independent of the testing age, as shown in Table 5.

Several relations have been proposed for splitting tensile strength relationship to compressive strength of concrete.

The American Concrete Institute recommended[6]

$$f_t = 0.556 \ (\ f_c' \ )^{0.5} \tag{1}$$

The Committé Euro-International du Beton(CEB-FIP) recommended[7]

$$f_t = 0.273 \ (\ f_c' \ )^{0.67} \tag{2}$$

Oluokun recommended[8]

$$f_t = 0.295 \ (\ f_c' \ )^{0.69} \tag{3}$$

for f_t, f_c' in MPa

The experimental results of splitting tensile strength and compressive strength, as shown in Fig. 5, are compared with those based on American Concrete Institute(ACI), Committé Euro-International du Beton(CEB-FIP), Oluokun recommendations. CEB-FIP model, as shown Fig. 5, always underestimates splitting tensile strength for total compressive strength level, therefore CEB-FIP model can be used as the lowest limit of splitting tensile strength of concrete. ACI model overestimates splitting tensile strength for concrete with compressive strength less than 20 MPa and underestimates for concrete with compressive strength in excess of 30 MPa. the estimation of Oluokun model was generally in closest agreement with the experimental results.

Based on the experimental results of super flowing concrete and ordinary concrete, the relation for estimating splitting tensile strength from compressive strength was proposed.

$$f_t = 0.43 \ (\ f_c' \ )^{0.59} \tag{4}$$

for f_t, f_c' in MPa

3.2.3 Elastic Modulus(E_c)

The elastic modulus of super flowing concrete ranged from 27 to 28 GPa at 28 days and from 30 to 33 GPa at 90 days, as shown in Table 5. For all mix proportions, it was observed that the elastic modulus of concrete tested at 28 days was about 10~20% less than the estimation by ACI model. The relationship between elastic modulus and compressive strength is shown in Fig. 6. For concrete with compressive strength greater than 30 MPa, the relation, as shown in Fig. 6, is not independent of the type of concrete. The slope of super flowing concrete is smaller than that of ordinary concrete. Based on this observation, it can be concluded that the increasing rate of compressive strength, at late ages, is greater than that of elastic modulus for super flowing

Table 5. Experimental data of splitting tensile strength and elastic modulus

	3 days		7 days		28 days				90 days		
	f_c' (MPa)	f_t (MPa)	f_c' (MPa)	f_t (MPa)	f_c' (MPa)	f_t (MPa)	E_c (GPa)	E_c by ACI Code	f_c' (MPa)	f_t (MPa)	E_c (GPa)
SF 1-1	20	2.4	33	3.5	47	4.0	28	32	53	4.7	31
SF 1-2	19	2.3	32	3.4	47	3.7	28	32	54	4.6	33
SF 1-3	16	2.0	31	2.9	46	3.7	28	32	53	4.4	33
SF 1-4	11	1.0	26	2.8	37	3.6	28	29	44	4.0	30
SF 2	11	1.1	23	2.5	34	3.5	25	27	42	4.4	29
NC 1	8	1.4	17	2.3	24	3.1	23	23	27	3.4	24
NC 2	18	2.6	30	3.7	42	4.0	28	31	43	4.2	31
NC 3	31	3.6	50	4.3	61	4.9	32	37	62	5.2	34

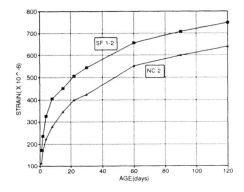

Fig. 7 Experimental creep strains of SF 1-2 and NC 2

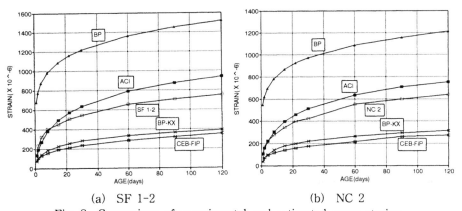

(a) SF 1-2   (b) NC 2

Fig. 8 Comparison of experimental and estimated creep strains

concrete with compressive strength greater than 30 MPa.

3.2.4 Creep

The creep characteristics of super flowing concrete and ordinary concrete are shown in Fig. 7. Until the age of 20 days, the increasing rate of creep strains of super flowing concrete is greater than that of ordinary concrete, but the difference decreases with age. The experimental results of creep strains, as shown in Fig. 8, are compared with those based on American Concrete Institute(ACI), Committé Euro-Internationál du Beton(CEB-FIP), Bazant-Panula (BP) and Bazant-Panula-Kim-Xi(BP-KX) recommendations. BP model overestimates creep strains of super flowing concrete and ordinary concrete, and BP-KX model and CEB-FIP model underestimates. At early ages, ACI model underestimates the creep strains of super flowing concrete, but estimates those of ordinary concrete well. At late ages, ACI model overestimates the creep strains of super flowing concrete and ordinary concrete. It appears that the increasing rate of the creep strains of super flowing concrete is greater than that of ordinary concrete at early ages but decreases with age.[9]~[16]

3.2.5 Drying Shrinkage

Fig. 9 shows the experimental results of drying shrinkage. The drying shrinkage strains of SF 1-1 were the greatest of all super flowing concrete mixes and those of SF 1-4 the smallest, as shown in Fig. 9 (a). This experimental results are the effects of unit water weight and the volume ratio of coarse aggregate to concrete. As shown in Table 3, unit water weights of SF 1-1, SF 1-2, SF 1-3 and SF 1-4 are 195, 185, 175 and 165, respectively, and the volume ratios of coarse aggregate to concrete are 0.27, 0.31, 0.35 and 0.39, respectively. Although the volume ratios of coarse aggregate to concrete were identical with SF 1-2, and unit water weight of SF 2 was a little greater than that of SF 1-2, the drying shrinkage strains of SF 2 were about 20% greater than those of SF 1-2. It can be concluded that the adding of fly ash increases drying shrinkage strains.

The drying shrinkage strains of super flowing concrete, as shown in Fig. 9 (b), were about 30~50% greater than those of ordinary concrete. The cause of this experimental results is that super flowing concrete includes more mortar and less coarse aggregate than ordinary concrete. In order to utilize practically super flowing concrete, the methods that decrease drying shrinkage strains are investigated.

The experimental results of drying shrinkage strains, as shown in Fig. 10, are compared with those based on American Concrete Institute(ACI), Committé Euro-Internationál du Beton(CEB-FIP), Bazant-Panula(BP) and Bazant-Panula -Kim-Xi(BP-KX) recommendations. The models underestimate the drying shrinkage strains of super flowing concrete greatly. It appears that the error is the effects of the adding of fly ash and much water-reducing admixture. Since drying shrinkage is directly associated with the water held by small pores in the

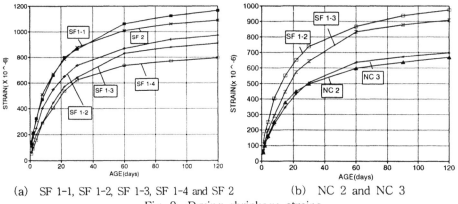

(a)   SF 1-1, SF 1-2, SF 1-3, SF 1-4 and SF 2   (b)   NC 2 and NC 3

Fig. 9   Drying shrinkage strains

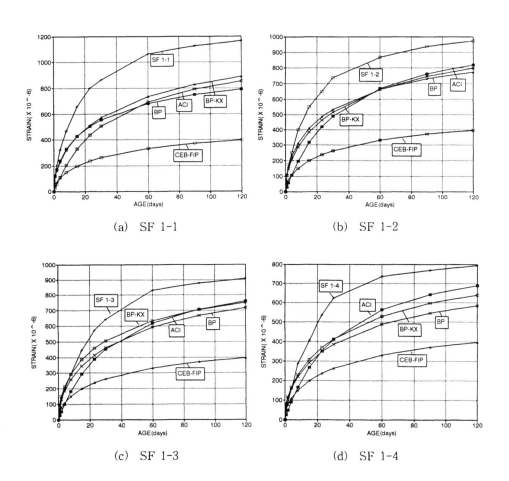

(a)   SF 1-1   (b)   SF 1-2

(c)   SF 1-3   (d)   SF 1-4

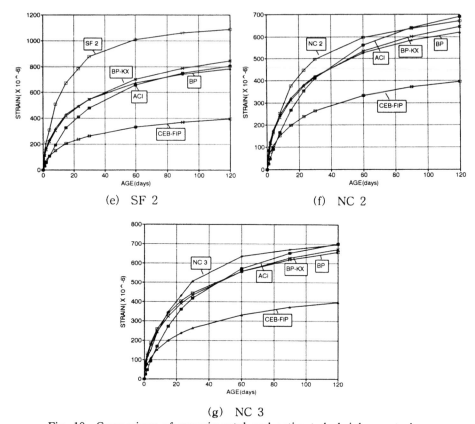

(e) SF 2   (f) NC 2

(g) NC 3

Fig. 10 Comparison of experimental and estimated shrinkage strains

range 3 to 30 nm, concrete containing fly ash capable of pore refinement usually show higher drying shrinkage. Water-reducing admixtures capable of effecting better dispersion of anhydrous cement particles in water, also lead to pore refinement in the hydration product. At early ages, the estimations of BP and BP-KX models were generally in closest agreement with the experimental results, but at late ages, ACI model estimates the experimental results well. The shape of the shrinkage-age curves agrees well with that of the curve estimated by BP model.[17]~[18]

4 Conclusions

Through the investigations and comparisons for super flowing concrete and ordinary concrete in this study, the following conclusions were drawn.
1. Based on the experimental results of fresh concrete, it was found that super flowing concrete had sufficient flowability and workability to obtain

the self compactable performance. The critical value of the volume ratio of coarse aggregate to concrete effecting on flowability and workability existed between 0.31(SF 1-2) and 0.35 (SF 1-3).

2. The effect of fly ash on the compressive strength of super flowing concrete was observed. The increasing rate of the compressive strength of super flowing concrete was lower than that of ordinary concrete at early ages, but at late ages, the increasing rate of the strength of super flowing concrete was higher than that of ordinary concrete. The effect of curing condition on the compressive strength of super flowing concrete was less than that of ordinary concrete.

3. For concrete of similar compressive strength, the splitting tensile strength of super flowing concrete was identical with ordinary concrete. At late ages, the increasing rate of compressive strength was greater than that of elastic modulus for super flowing concrete with compressive strength greater than 30 MPa.

4. The increasing rate of the creep strains of super flowing concrete was greater than that of ordinary concrete at early ages but decreased with age. The drying shrinkage strains of super flowing concrete were about 30~50% greater than those of ordinary concrete.

Acknowledgment

The authors are grateful to the Ministry of Construction and Transportation for providing the financial support for the project.

References

1. Carino, N. J., and Clifton, J. R., (1991) High-Performance Concrete : Research Needs to Enhance Its Use, *Concrete International*, Vol. 13, No. 9, pp.70-76.
2. Ozawa, K., Maekawa, K., and Okamura, H., (1989) Development of the High Performance Concrete, *Proceedings of the Japan Concrete Institute*, Vol. 11, No. 1, pp.699-704.
3. Tatsuo, I., Hotaka, Y., Shin, M., and Tatsuya, M., (1994) Rheological Study on High Flowing Concrete, *Proceedings of the Japan Concrete Institute*, Vol. 16, No. 1, pp.177-182.
4. Tikalsky, P. J., Carrasquillo, P. M., and Carrasquillo, R. L., (1988) Strength and Durability Considerations Affecting Mix Proportioning of Concrete Containing Fly Ash, *ACI Materials Journal*, Vol. 90, No. 6, pp.535-544.
5. Carette, G. G., Bilodeau, A., Chevrier, R. L., and Malhotra, V. M., (1993) Mechanical Properties of Concrete Incorporating High Volumes of Fly Ash from Sources in the U.S, *ACI Materials Journal*, Vol. 90, No. 6, pp.535-544.
6. ACI Committee 318, (1989) Building Code Requirements for Reinforced Concrete and Commentary (ACI 318-89/318R-89), *American Concrete Institute*, Detroit, pp.353.
7. Comité Euro- International du Betón, (1978) CEB-FIP Model Code for Concrete

Structures, Paris, pp.348.

8. Oluokun, F. A., (1991) Prediction of Concrete Tensile Strength from Its Compressive Strength, *ACI Materials Journal*, Vol. 88, No. 3, pp.302-309.

9. Bazant, Z. P., and Panula, L., (1979) Practical Prediction of Time-Dependent Deformation of Concrete, *Materials and Structures*, Vol. 12, No. 69, pp.169-183

10. Bazant, Z. P., and Panula, L., (1982) New Models for Practical Prediction of Creep and Shrinkage, *ACI SP-76*, pp.7-23.

11. ACI Committee 209 (chaired by Rhodes, J. A.) (1982) Prediction of Creep, Shrinkage and Temperature Effects in Concrete Structures, *ACI SP-27 (Designing for Creep & Shrinkage in Concrete Structure)*, pp.193-301.

12. ACI Committee 209/II (chaired by Branson, D. E.) (1971) Prediction of Creep, Shrinkage and Temperature Effects in Concrete Structures, *ACI SP-27 (Designing for Creep & Shrinkage in Concrete Structure)*, pp.51-93.

13. Meyers, B.L., Branson, D.E., Schumann, C.G., and Christiason, M.L., (1970) The Prediction of Creep and Shrinkage Properties of Concrete, *Iowa Highway Commission, Final Report* No. 70-5, pp.140.

14. Bazant, Z.P., and Panula, L., (1978) Practical Prediction of Time- Dependent Deformation of Concrete, *Materials and Structures*, Vol. 11, No. 65, pp.307-328; (1978) Vol. 11, No. 66, pp.415-434; and (1979) Vol. 12, No. 69, pp.169-183

15. Comité Euro-Internationál du Betón, (1990) Evaluation of the Time Dependent Behavior of Concrete, *Bulletin* No. 199, pp.201.

16. Bazant, Z. P., and Kim, J. K., (1991) Improved Prediction Model for Time-Dependent Deformation of Concrete, *Materials and Structures*, Vol. 24, No. 148, pp.219-223.

17. Neville, A. M., (1981) *Properties of Concrete*, pp.268-318.

18. Mehta, P. K., and Monteiro, P. J. M., (1993) *Concrete*, pp. 93-101.

24 FLOWING CONCRETE WITH PACKED POWDER SUPERPLASTICISER

K. GOTO and M. HAYAKAWA
Technology Resaerch Center, Taisei Corporation, Yokohama, Japan
T. UKIGAI and N. TOBORI
Research and Development Headquarters, Lion Corporation, Tokyo, Japan

Abstract

The powder superplasticizer packed in a paper bag that dissolves in alkali solution was developed. With this package, flowing concrete can be easily produced on site in an agitator truck. Consistency of flowing concrete is controlled with selecting number of the packages. Concrete with slump of less than 21cm is produced with only this superplasticizer, and concrete of more than 21cm can be produced with this superplasticizer and a segregation control agent.

To clarify the quality of these concretes, fresh and hardened concrete was investigated in laboratory. Slump, slump flow, air content, bleeding, setting time, compressive strength, drying shrinkage and freezing and thawing resistance of concretes were measured. It was confirmed that compressive strength and durability of the flowing concrete was as good as those of the concrete without the superplasticizer.

The flowing concrete was made by adding this superplasticizer into concrete which was produced at a ready mixed concrete plant. It was confirmed that flowing concrete can be easily produced on site in an agitator truck with the packed powder superplasticizer.

KEY WORDS
Flowing concrete, Production method, Powder superplasticizer, Alkali-decomposable paper, Segregation control agent, Fresh concrete, Compressive strength, Durability

1 Introduction

Flowing concrete produced by adding liquid type superplasticizer to base concrete in an agitator truck are widely employed. The powder superplasticizer packed in a paper bag that dissolves in alkali solution (alkali-decomposable paper) was developed. If the packed powder superplasticizer is added in fresh concrete, the paper bag dissolves for high level alkali solution. Then the superplasticizer is dispersed, and fluidity of concrete become

Production Methods and Workability of Concrete. Edited by P.J.M. Bartos, D.L. Marrs and D.J. Cleland. Published in 1996 by E & FN Spon, 2–6 Boundary Row, London SE1 8HN. ISBN 0 419 22070 4.

higher. Good points of this production method are that the packed powder superplasticizer can be easily added into an agitator truck and that consistency of flowing concrete is controlled with selecting number of the packages. In Japan, concrete is often placed into heavily reinforced concrete members designed for earthquake resistance. The concrete need to have high fluidity and ability of passing through narrow spaces. To accomplish this needs, the flowing concrete with slump flow of 45cm (slump of about 24cm) using the packed powder superplasticizer and powder segregation control agent as well as slump of less than 21 cm using only the packed powder superplasticizer was developed.

In this paper, laboratory tests on the flowing concretes with slump of less than 21cm and slump flow of 45cm were made to determine the characters of fresh concrete and hardened concrete. Field investigations on production of the flowing concrete using the packed powder superplasticizer and the packed powder segregation control agent in an agitator truck were carried out.

2 Flowing concrete with slump of less than 21 cm

2.1 Purpose

Laboratory tests on the flowing concrete with slump of less than 21cm were conducted to determine the quality of the concrete. Field investigation was made to determine the feasibility of producing the flowing concrete by using the packed powder superplasticizer in an agitator truck.

2.1 Laboratory test

2.2.1 Materials

The type of cement is ordinary portland cement (specific gravity of 3.16). Liver sand (specific gravity of 2.62, fineness modulus of 2.74, water absorption of 1.46%) and crushed stone (specific gravity of 2.66, fineness modulus of 6.72, water absorption of 0.56%) were used as the fine aggregate and coarse aggregate respectively. Air entraining agent was used in base concrete. The powder superplasticizer (1) is based on polystyrene sulfonate.

2.2.2 Mix proportion and production method

Table 1 shows the mix proportion of base concrete. Targets of slump and air content of the base concrete are 8cm and 4.5%. Targets of slump and air content of the flowing concrete are 18cm and 4.5%. The concrete was mixed in a tilting mixer. The base concrete was mixed for 3 minutes after materials put in the mixer. The flowing concrete was mixed for 1 minute after the powder superplasticizer of 0.47% of cement weight was added to the base concrete.

2.2.3 Testing of concrete

Table 1 Mix proportion of base concrete

W/C (%)	S/A (%)	W	C	S	G	AE (ml/m³)
		(kg/m³)				
51.7	46.0	166	321	834	993	96

AE : Air entrained agent

Slump, air content, amount of bleeding, setting, compressive strength, drying shrinkage and freezing and thawing resistance were measured in accordance with the methods specified in standards (table 2). Compressive strength of water cured specimen (20℃) was tested at aged 3, 7, 28 days.

2.2.4 Results and discussion
The test results of fresh concrete are shown in Table 3. Comparing with the base concrete and the flowing concrete, amount of bleeding of the flowing concrete increases 0.04 cc/cm². Both initial and final of setting times of the flowing concrete were about an hour longer than those of the base concrete. According to the Standards of the Architectural Institute of Japan (JASS 5T-402), this superplasticizer should be designated as the standard type. The test results of the hardened concrete are shown in Table 4. These results of the flowing concrete and the base concrete were almost equal.

2.3 Field investigation

2.3.1 Materials
The type of cement is ordinary portland cement (specific gravity of 3.15). Land sand (specific gravity of 2.58, fineness modulus of 2.60) and crushed stone (specific gravity of 2.69) were used as the fine aggregate and coarse aggregate respectively. Air entraining and water reducing agent was used in base concrete. The powder superplasticizer with weight of 485g packed in a bag made by alkali-decomposable paper (PACK-A) was used. The alkali-decomposable paper (2) is made by pulp fiber and a kind of cellulose which doesn't dissolve in water but it dissolves easily in alkali solution.

2.3.2 Mix proportion and production method
Table 5 shows the mix proportion of concrete. Targets of slump and air content of the base concrete are 15cm and 4.0%. Targets of slump

Table 2 Tests

Slump	JIS A 1101
Air Content	JIS A 1128
Amount of Bleeding	JIS A 1123
Setting	ASTM C 403
Compressive Strength	JIS A 1108
Drying Shrinkage	JIS A 1129
Freezing and Thawing	JIS A 6204

JIS: Japanese Industrial Standerd

Table 3 Results of fresh concrete tests

		Concretes	
		Base	flowing
Slump (cm)		7.8	17.8
Air content (%)		4.2	4.6
Amount of Bleeding (cc/cm³)		0.16	0.20
Setting (hr-min)	Initial	5-30	6-30
	Final	7-35	8-30

Table 4 Results of hardened concrete tests

		Concretes	
		Base	flowing
Compressive Strength (MPa)	3days	19.0	18.7
	7days	31.6	31.3
	28days	44.3	42.8
Drying Shrinkage 26weeks (μm)		684	676
Durability Factor 200 cycles		96	96

Table 5 Mix proportion of base concrete

W/C (%)	S/A (%)	W	C	S	G	AEW
				(kg/m³)		
55.2	49.5	171	310	882	939	0.775

AEW: Air entrained water reducing agent

and air content of the flowing concrete are 21cm and 4.0%. The base concrete produced in a ready mixed concrete plant was loaded into an agitator truck (5.0m³). The flowing concrete was mixed for 60 seconds in the agitator truck after 4 packs (0.125% of cement weight) of PACK-A was added to the base concrete.

2.3.3 Testing of concrete
Slump, air content and compressive strength were measured in accordance with the methods specified in standards (table 2). Compressive strength of the water cured specimen (20°C) was tested at aged 7, 28, 91 days. Concretes (5m³) in every three of the 30 agitator trucks were tested.

2.3.4 Results and discussion
The results of slump and air content are shown in Fig. 1. Slumps of the base concretes were nearly target of 15cm. Slump of the flowing concrete could be controlled in the narrow range around the target slump of 21cm. Air contents of the flowing concretes were similar to those of the base concretes. Results of average compressive strength at aged 7, 28 and 91days are shown in Fig. 2. Compressive strength of the flowing concrete was as same as that of the base concrete. Flowing concrete with slump of 21cm could be produced easily and stably in an agitator truck to use the packed powder superplasticizer.

3 Flowing concrete with slump flow of 45 cm

3.1 Purpose
Laboratory tests on the flowing concrete with slump flow of 45 cm were made to determine the quality of concrete. Field investigation was made to determine the feasibility of producing it by using the packed powder superplasticizer and the packed powder segregation control agent in an agitator truck.

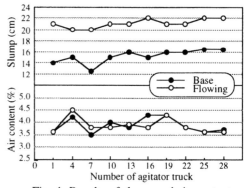

Fig. 1 Results of slump and air content

3.2 Laboratory test

3.2.1 Materials
The type of cement is ordinary portland cement (specific gravity of 3.16). Land sand (specific gravity of 2.59, fineness modulus of 2.68, water absorption of 1.96%) and crushed stone (specific gravity of 2.65, fineness modulus of 6.63, water absorption of 0.63%) were used as the fine aggregate and the

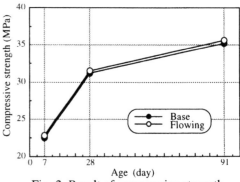

Fig. 2 Result of compressive strength

coarse aggregate respectively. Air entraining and water reducing agent was used in base concrete. The powder superplasticizer was used as same as foregoing. The segregation control agent based on a kind of cellulose (2% water solution of 4000cP of viscosity) as the soluble powder was used.

3.2.2 Mix proportion and Production method

Table 6 shows the mix proportion of three concretes. All concretes are with water content of 175kg/m³ and with water-cement ratio of 50%. A target of slump of these base concretes is 18cm. The powder superplasticizer and the powder segregation control agent is added to base concrete of F45. Targets of slump flow and air content of the flowing concrete (F45) are 45cm and 4.5%. Only the powder superplasticizer are added to base concretes of F21 and N21. Targets of slump and air content of those flowing concretes(F21, N21) are 21cm and 4.5%. The sand percentage of F45 is set high for passing through narrow space. F21 and N21 are designed to compare with F45. Mix proportion of F21 is as same as F45 besides kinds and amount of admixtures. Sand percentage of N21 is 45% like average concrete. Concrete was mixed in a pan type mixer. Base concrete was mixed for 120 seconds after materials put into the mixer. Flowing concrete of F45 was mixed for 90 seconds after adding the powder superplasticizer and the segregation control agent into the base concrete. Flowing concretes of F21 and N21 were mixed for 30 seconds after adding only the powder superplasticizer into the base concrete.

3.2.3 Testing of concrete

The concretes were tested for slump, air content, amount of bleeding, setting, compressive strength, drying shrinkage and freezing and thawing resistance in accordance with the methods specified in standards (table 2) and L-shape flow test (3) and N-shape flow test. The L-shape flow test is that the distance (LF) of flowed concrete is measured after the gate (height of 16cm) is opened, and concrete is filled into the part of long length of the apparatus (Fig. 3). The N-shape flow test is that the distance (NF) of flowed concrete is measured as same as the L-shape test besides height of the gate (5cm). If NF is much lower than LF of the same concrete, the concrete is going to block up the space between bars. Compressive strength of water cured specimen (20℃) was tested at aged 3, 7 and 28 days.

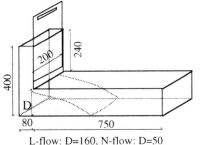

L-flow: D=160, N-flow: D=50

Fig. 3 System of L-Shape and N-shape flow test

Table 6 Mix proportion of concrete

Code	W/C (%)	S/A (%)	Target SL or SF (cm)	W	C	S	G	AEW	SP	SR
								(kg/m³)		
F21	50.0	55.2	SL21	175	350	956	795	0.525	0.70	-
F45			SF45					0.525	1.75	0.06
N	50.0	45.0	SL21	175	350	780	975	0.525	0.35	-

AEW : Air entrained water reducing agent, SP : Superplasticizer, SR : Segregation resistance agent

3.2.4 Results and discussion

The test results of fresh concrete is shown in Table 7. Slump and air content of all the flowing concretes were close to the targets. Air content of the flowing concretes of F45 and F21 with more fine aggregate was 2%-3% lower than that of the base concretes. NF of F45 and F21 was larger than LF, while NF of N21 was much less than LF. According to these results, F45 and F21 with less coarse aggregate were easy to pass through the narrow space. Amount of bleeding of F45 was more than those of F21 and N21, but amount of bleeding of each concrete was small. Initial and final of setting times of F45 were about 150 minutes and 180 minutes longer than those of F21. The test results of hardened concrete are shown in Table 8. Compressive strength of F45 was similar to those of F21 and N21. Drying shrinkage of F45 was as same as that of F21 and was more than that of N21. The result of freezing and thawing resistance were not much difference among F45, F21 and N21. It was determined that the flowing concrete with slump flow of 45 cm (F45) gave the same strength and durability as the flowing concrete with slump of 21cm (F21).

3.3 Field investigation

3.3.1 Materials

The type of cement is portland blast furnace slag cement (type B by Japanese standard, specific gravity of 3.04). Mixed sand (fineness modulus of 2.65) with Land sand (specific gravity of 2.60) and crushed sand (specific gravity of 2.60) was used as fine aggregate. Crushed stone (specific gravity of 2.67, fineness modulus of 6.53) was used as coarse aggregate. Air entraining and

Table 7 Results of fresh concrete tests

		F45	F21	N21
Base	Slump (cm)	19.0	19.0	20.0
	Air content (%)	7.5	6.4	4.5
	SL or SF (cm)	SF46.5	SL20.0	SL21.0
	Air content (%)	4.5	4.4	4.0
	L-Flow (cm)	38.0	32.0	31.5
	N-flow (cm)	42.5	35.0	16.0
	Amount of Bleeding (cc/cm³)	0.26	0.16	0.18
Setting (hr-min)	Initial	8-45	6-03	6-20
	Final	10-47	7-37	7-49

Table 8 Results of hardened concrete tests

		F45	F21	N21
Compressive Strength (MPa)	3days	22.2	22.1	19.6
	7days	34.6	32.6	30.8
	28days	45.2	43.0	43.1
Drying Shrinkage 26weeks (μm)		740	750	690
Durability Factor 200 cycles		88	92	93

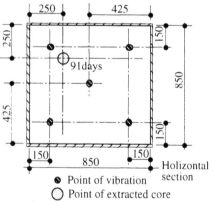

Fig. 4 Large scale specimen

● Point of vibration
○ Point of extracted core

Table 9 Mix proportion of base concrete

W/C (%)	S/A (%)	W	C	S	G	AEW
		(kg/m³)				
50.0	50.0	175	350	871	900	0.875

AEW : Air entrained water reducing agent

water reducing agent was used in base concrete. The powder superplasticizer and the segregation control agent were packed in a bag made by alkali-decomposable paper (PACK-B). PACK-B was a pack with the powder superplasticizer of 485g and the segregation control agent of 30g. PACK-A and PACK-B were used in this investigation.

3.3.2 Mix proportion and Production method

Table 9 shows the mix proportion of concrete. Targets of slump and air content of the base concrete are 18cm and 4.5%. Targets of slump flow and air content of the flowing concrete are 45cm and 4.5%. Base concrete produced in a ready mixed concrete plant was loaded into an agitator truck (5.0m³). Flowing concrete was mixed for 120 seconds in the agitator truck after 5 packs of PACK-B and 6-9 packs of PACK-A were added to the base concrete. Amount of the powder superplasticizer and the segregation control agent become 0.30%-0.39% of cement weight and 30g/m³ respectively.

3.3.3 Testing of concrete

Slump and air content of concretes in 20 agitator trucks were tested in accordance with the methods specified in standards (table 2). Compressive strength of water cured specimen (20 ℃) was measured at aged 28 days in every other agitator trucks.

Further large scale specimens (□ 80cm × 80cm, h=110cm) were made with the base concrete and flowing concrete. Every concrete was placed in two layers. A vibrator was inserted vertically at 5points of the specimen shown in Fig. 4 of each layer for 10 seconds. At aged 28 days, concrete core was extracted vertically from a point shown in Fig. 4 of each specimen and divided into 5 pieces. Ratio of coarse aggregate area and compressive strength were measured from these core specimens.

3.3.4 Results and discussion

The results of slump and slump flow is shown in Fig. 5. Slump of the base concretes were changed from 12cm to 20cm but slump flow of the flowing concretes could be controlled in the range of 45cm ± 8cm. The results of air content are shown in Fig. 6. Air content of the flowing concretes was similar to that of the

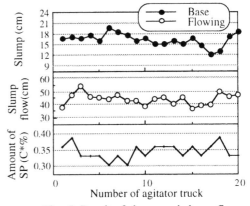

Fig. 5 Result of slump and slump flow

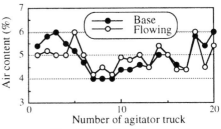

Fig. 6 Result of air content

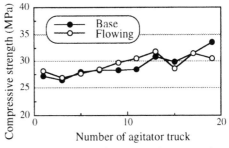

Fig. 7 Result of compressive strength

base concretes. Unlike the laboratory test, drop of air content was not seen. The results of compressive strength are shown in Fig. 7. Compressive strength of the flowing concrete was as same as that of the base concrete.

The results of the ratio of coarse aggregate area and compressive strength in large scale specimens are shown in Fig. 8. Slump and air content of this base concrete were 18.5cm and 4.1%. Slump flow and air content of this flowing concrete were 46cm and 4.2%.

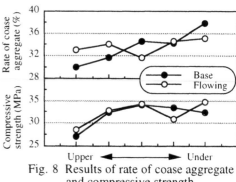

Fig. 8 Results of rate of coase aggregate and compressive strength

Ratio of coarse aggregate area of the upper part was lower than the lower part of the large specimen. But the difference of ratio of coarse aggregate area between upper part and lower pert of the specimen of the flowing concrete was less than that of the base concrete. It was understood that the flowing concrete with slump flow of 45cm has good segregation resistance of coarse aggregate. Compressive strength of the core specimen of the flowing concrete was similar to that of the base concrete. Flowing concrete with slump flow of 45cm could be produced easily and stably in an agitator truck using the packed of the powder superplasticizer and the segregation control agent.

4 Conclusion

The following results were obtained regarding flowing concretes with slump of 21cm and with slump flow of 45cm.

Flowing concrete with slump of 21cm or slump flow of 45cm could be produced easily and stably in an agitator truck with the packed powder superplasticizer or the packed powder superplasticizer and segregation control agent. It was confirmed that the flowing concretes have as same compressive strength and durability as the base concretes.

5 References

1. Y.Sekiguchi, T.Okada, and T.Ukigai (1989), Relative Effects of Ca-Polystyrene Sulfonate and Na-Sulphonate-Based Superplasticizers on Properties of Flowing Concrete, Superplasticizers and Other Chemical Admixtures in Concrete, *ACI, SP-119*, pp157-pp171

2. Y.Sekiguchi (1991), Study on the Properties of Hybrid Mortar added with Carbon Fiber Packaged in Alkali-Decomposable Paper, *Japan Society for Finishings Technology*, 1991

3. T.Yonezawa (1989), L-Shape Flow Test Used for the Quality Control of High Strength Concrete, *Summaries of Technical Papers of Aural Meeting Architectural of Japan* 1989, pp263-pp264

PART SIX
RHEOLOGY

25 NEW GENERATION OF SUPERPLASTICISERS

P. BILLBERG, Ö. PETERSSON and J. NORBERG
Swedish Cement and Concrete Research Institute, Stockholm, Sweden

Abstract
Different superplasticizers have been tested in order to evaluate their effect on concrete properties and to express concrete's rheology in fundamental parameters such as yield stress and plastic viscosity. Measurements were made on concrete's rheology with two different viscometers, one for mortar and one for concrete. Tests were also made using the more traditional slump cone. By using results from slump tests at different times after mixing, slump losses by means of time for losing 30 mm of slump and time for losing half the initial slump value was calculated. The time necessary before trowelling was also measured. Results show that viscosity-measurements on mortar give useful information of the fresh concrete's properties, information that cannot be attained with the traditional methods, such as slump, spread etc. The method of evaluating trowelling hardness is very efficient to measure the admixture's effects on retardation. Only those superplasticizers containing new types of polymers such as acrylic and vinyl shows effects efficient enough to be said to belong to the second generation.
Keywords: Mortar, rheology, slump, slump loss, superplasticizers, trowelling hardness, viscometry.

1 Introduction

The aims of this study were both to test some of the second generation of super-plasticizers and to express the concrete's rheology in fundamental parameters, such as yield stress and plastic viscosity. The second generation of superplasticizers is said to have smaller slump losses and higher short-term and final strength. An other aim of the

Production Methods and Workability of Concrete. Edited by P.J.M. Bartos, D.L. Marrs and D.J. Cleland.
Published in 1996 by E & FN Spon, 2–6 Boundary Row, London SE1 8HN. ISBN 0 419 22070 4.

development of new superplasticizers was to make them more innocuous from a health point of view by minimising or completely removing the formaldehyde content.

2 Methods and materials

Rheology parameters according to Bingham model, τ_0 for yield stress and μ for plastic viscosity, were measured using two different viscometers. On concrete, with D_{max}=16 mm, measurements were made with a BML viscometer and on mortar, with D_{max}=0,25 mm, a Bohlin Controlled Stress Rheometer was used. Both viscometers use a measuring system of coaxial cylinders, see Fig 1 and Table 1. The rheological results from viscometers were compared with more traditional methods such as slump test.

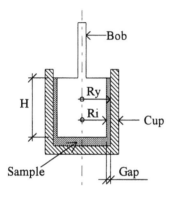

Table 1. Geometry of coaxial cylinders in mm.

Measure	C25	BML
H	37,5	170
Ri	12,50	100
Ry	13,75	150
Gap	1,25	50

Fig 1. Coaxial cylinders, principle.

Consistency was measured on concrete, 5, 30, 60 and 90 minutes after mixing, by means of the slump and BML. Slump losses were evaluated in two ways. The time for slump loss by 30 mm (-30 mm) and the time for slump loss by half the initial value (-50%) was interpolated between slump tests at different times after mixing.

Tests were also performed to determine whether the superplasticizers retard the concrete's hardening. This was done by measuring the time needed before the concrete was hard enough for power floating and power trowelling.

As reference concrete a typical Swedish building concrete was selected with a water to cement ratio of 0.60, see Table 2. The concrete had an initial slump of 70 mm. Batches of 70 litres of concrete were mixed in a paddle mixer of Sandby type.

Table 2. Dry mix quantities of reference concrete.

Component	Amount (kg/m³)
Cement	320.0
Sand 0-8 mm	1035.4
Gravel 8-16 mm	781.1
Water	192.0

Mortar reference mix had a maximum aggregate size of 0.25 mm. The composition of this mix was the same as the concrete, i.e. the amount of aggregate smaller than 0.25 mm in relation to the cement. Batches of 2 litres were mixed.

The aggregates used, are shown in Fig 2.

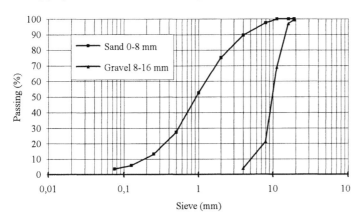

Fig 2. Aggregate particle size distribution curves.

Chemical and physical composition and other characteristics of the cement used, are shown in Table 3. The cement is of standard Portland type and was used throughout the study.

Table 3. Characteristics of cement.

Chemical and physical data	Amount (%)
C_3S	57
C_2S	13
C_3A	8
C_4AF	7
SO_3	3.4
MgO	3.6
Alkali content as equiv. Na_2O	1.10
Density (kg/m^3)	3120
Blaine (m^2/kg)	360

The five superplasticizers investigated in this study are listed in Table 4. They are commonly available on the Swedish market. Their active base differs by means of the type of polymer and the solid content.

Table 4. Superplasticizers used in the test.

Designation	Base	Solid content (%)
M-1	Sulphonated melamine formaldehyde condensate	35
M-2	Sulphonated melamine block polymer	35
N-1	Sulphonated naphtalene condensate	42
MC-1	Sulphonated melamine polymer/vinyl copolymer	20
P-1	Modified acrylic polymer	40

3 Results

The superplasticizer dosages were adapted so that semi-full flow, SFF, and full flow, FF, were achieved. These consistencies correspond to slumps of 180 and 230 mm respectively. Actual dosages are shown in Table 5.

Table 5. Dosages of superplasticizer to achieve SFF and FF.

Superplasticizer	Dosages for SFF		Dosages for FF	
	(%)*	Slump (mm)**	(%)*	Slump (mm)**
M-1	1.05	180	1.40	230
M-2	1.11	180	1.67	225
N-1	0.80	180	1.03	220
MC-1	0.85	185	1.10	225
P-1	0.55	190	0.97	230

* In % of cement weight (as liquid). ** Measured 5 minutes after mixing.

3.1 Rheology of mortar
A comparison between the dosage required to achieve SFF and FF respectively made for all products did not exhibit any large differences in yield stress between the different superplasticizers, see Fig 3.

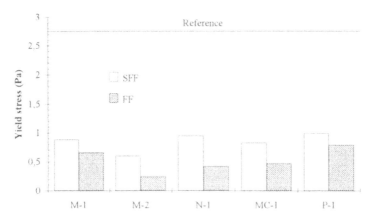

Fig 3. Yield stress measured on mortar at semi-full flow and full flow dosages.

The results obtained using the paste viscometer show good agreement with corresponding concretes whose dosages of superplasticizers had been adjusted to achieve the same slumps.

3.2 Rheology of concrete
For concretes with SFF the yield stress dropped from the reference concrete's 600 Pa to 180-300 Pa. The acrylic polymer based product P-1 exhibited considerably lower values than the rest of the superplasticizers and also exhibited a slower increase with time. The values for the other superplasticizers are relatively well collected. Most products appear to exhibit linear growth in yield stress with time. With dosages to achieve FF consistency, the yield stress dropped to 100 - 200 Pa, see Fig 4.

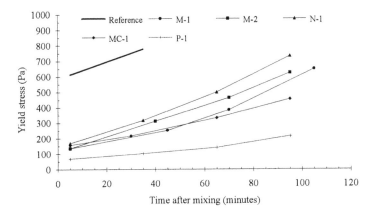

Fig 4. Yield stress growth with time at dosages to achieve full flow.

All superplasticizers exhibited an accelerated increase of yield stress with increasing time after mixing. P-1 exhibited considerably lower values than the other superplasticizers. The melamine/vinylcopolymer-based product MC-1 together with the acrylic polymer product P-1 exhibited less acceleration with time. N-1 and M-1 exhibited a great acceleration with time.

For concrete with SFF the plastic viscosity dropped from 30 Pa s for the reference concrete to 20–25 Pa s for concretes containing superplasticizer. MC-1 exhibits the lowest plastic viscosity value. P-1 and M-1 exhibit high plastic viscosity values. With dosages to FF the plastic viscosity dropped from 30 Pa s for the reference concrete to 15–20 Pa s for concretes containing superplasticizer, see Fig 5. MC-1 had the lowest plastic viscosity and also showed a slow increase with time after mixing. M-2 had the largest plastic viscosity and also showed the greatest increase in viscosity.

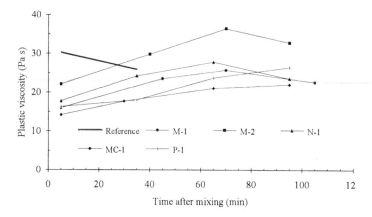

Fig 5. Change in plastic viscosity with time at dosages to achieve FF.

The reason why some curves turn downwards after approximately 70 minutes can be that the concrete gets stiffer (approximately at slump = 100 mm) and slip-surface is formed round the inner cylinder in the BML viscometer. In other words the measurement is more or less made on this slip-surface, not on the concrete. For concretes it should be reasonable to assume that the plastic viscosity increases with increasing time after mixing.

It is interesting to note how P-1, 5 minutes after mixing, achieve a low yield stress but a relatively high plastic viscosity in relation to the other products. Despite the fact that all mixes were adjusted to achieve comparable slumps. The higher value of P-1 after 95 minutes is due to the fact that this concrete mix still has a slump of 150 mm and therefore could be correctly measured with the BML viscometer (no slip-surface). The other mixes have slumps under 100 mm at this time, see Fig 5.

3.3 Slump measurements
Changes in slump with time follow roughly the same pattern for concretes dosed to achieve SFF as to achieve FF. In both cases, P-1 (acrylic polymer) keep the highest slump level with increasing time and N-1 (naphthalene) the lowest. N-1 lost most slump in the first 30 minutes approximately, while the change of slump in time for P-1 was somewhat flatter than the others. In Fig 6 the slumps of FF consistency is shown.

The various products are much more collected for dosages to SFF than FF. In the case of FF, M-1 (melamine based) exhibited basically the same development as P-1 up to about 40 minutes. Here M-2 and N-1 show inferior results compared to the others. The products MC-1 and P-1, based on new types of polymers, show better development with time than the other, traditional superplasticizers.

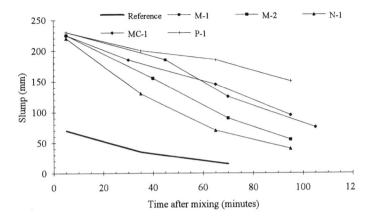

Fig 6. Slump development with time of concrete dosed to achieve FF.

Slump losses were calculated by interpolation between different slump values at different times after mixing, see Fig 7.

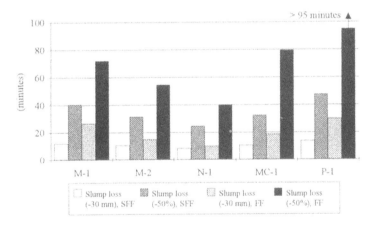

Fig 7. Slump losses for concrete dosed to semi-full flow and full flow.

The calculations were based on the reduction in slump by 30 mm, slump loss (-30 mm), and also on the reduction in slump to half the original value, slump loss (-50%). The acrylic polymer P-1 showed in all cases to have the slowest slump loss, while the naphthalene-based N-1 had the fastest.

3.4 Trowelling hardness
Trowelling hardness was determined by letting a circular drop-weight with a conical point fall onto the concrete at different times after casting, see Fig 8.

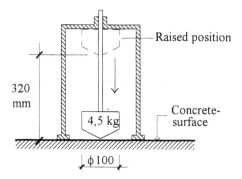

Fig 8. General arrangement of the power trowelling hardness test.

The diameter of the impression made by the cone was measured and compared with measurements made earlier for different types of trowelling: power floating and power trowelling. From experience, the size of the impression for power trowelling should be 60 mm diameter, and for power floating, 45 mm diameter, see [1]. The times required to achieve these impressions are shown in Figure 9.

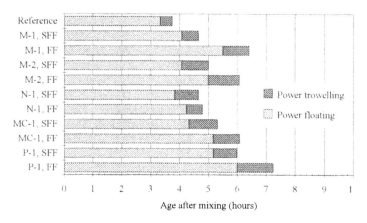

Fig 9. Times to achieve required trowelling hardness for concretes with different
superplasticizers and consistency.

Times required before power trowelling and power floating also indicates, what was stated earlier, that the superplasticizers based on new polymers providing a certain delay in stiffening. Concrete containing large dosage of the melamine-based M-1 also indicates the same tendency.

3.5 Comparison between different consistency tests
The results from measuring on mortar were compared with those measured on concrete five minutes after mixing. There is a relationship between the yield stress measured on mortar and the yield stress measured on concrete, see Fig 10. The upper limit for the range of measurement of the BML viscometer gives too low and incorrect values for

stiff consistencies. Regression analysis gave the relationship between the mortar yield stress and the BML yield stress measured on concrete shown in Fig 10. Results from ref [2] are also shown in the figure. In the calculation, only the mortar mixes that have a yield stress of less than 6 Pa were included.

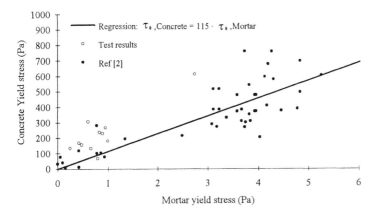

Fig 10. Relationship between yield stress measured on mortar and yield stress measured on concrete.

Plastic viscosity measured on mortar varies between 0.1 and 0.9 Pa, thus, by a factor of 9. Plastic viscosity measured on concrete varies between 10 and 30 Pa, thus by a factor of 3, see Fig 11.

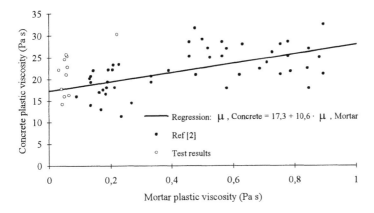

Fig 11. The relationship between the plastic viscosity measured on concrete and plastic viscosity measured on mortar.

The smaller variation of the concrete may be because the plastic viscosity being mainly affected bye the larger particles (stones). The paste is not affecting the concrete viscosity as much as it affects the mortar viscosity. Even if the scatter is large, there is a relationship between the plastic viscosity measured on concrete and the plastic viscosity measured on mortar.

In two-point measurements, differences in the various superplasticizers can be measured that are not detectable in traditional consistency measurements. In Fig 12, testing on mortar, and Fig 13, testing on concrete, the reference concrete is at the extreme top right-hand "point." The curves then show how the properties change with increasing dosage of superplasticizer.

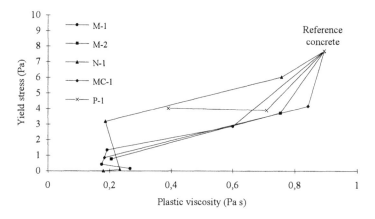

Fig 12. Relationship between yield stress and plastic viscosity in measurements on mortar with $D_{max} = 0.25$ mm.

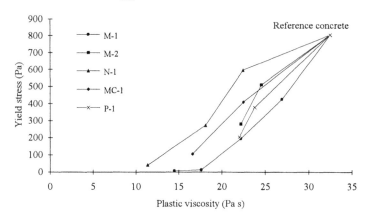

Fig 13. Relationship between yield stress and plastic viscosity in measurements on concrete with $D_{max} = 16$ mm.

M-1 and N-1 exhibit different changes. At a small dosage, N-1 reduces the plastic viscosity, while M-1 reduces the yield stress. With increasing dosage, M-1 provides a lower yield stress than N-1. The properties of the other products lie between these two.

At higher dosages, the yield stress and plastic viscosity provided by the various products are the same.

3.8 Comparison between Bingham model and slump test

A relationship exists between slump measured on concrete and yield stress measured on mortar, see Fig 14.

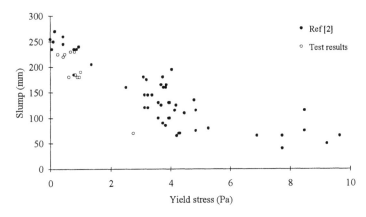

Fig 14. Relationship between slump test on concrete and yield stress measured on mortar.

The results obtained using the BML viscometer show agreement with the results from slump tests, see Fig 15. Earlier investigations also showed a good relationship between yield stress and slump, see [3] and [4].

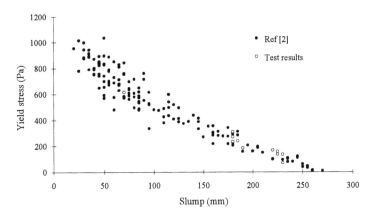

Fig 15. Relationship between the yield stress and slump for concrete.

4 Conclusions

The results show that the superplasticizers based on new types of polymers, P-1 with modified acrylic and MC-1 partly based on vinylcopolymer, are more effective and have less slump losses than the other admixtures. Furthermore the P-1 is completely free of formaldehyde. We consider these two superplasticizers to be the only ones who could be said to belong to the second generation.

Plastic viscosity of concrete is difficult to relate to any of our traditional methods of consistency measurements. Only two-point measurements can provide results that describe the yield stress and the plastic viscosity.

Measurements on mortar using a viscometer can give a relationship to the fresh concrete's properties such as slump. Results that agree with the properties of fresh concrete can be obtained by measuring on mortar that contains the fine fractions of the concrete aggregate.

The method to measure the required time before trowelling is an efficient way to evaluate the superplasticizers effect on retardation.

5 References

1   Ö. Petersson, A. Johansson, *Styrning av glättningshårdhet (Controlling trowelling hardness)*, Swedish Cement and Concrete Research Institute, CBI report 1:91, Stockholm 1991 (in Swedish).

2   Ö. Petersson, P. Billberg, J. Norberg, A. Larsson, *Effects of the second generation of superplasticizers on concrete properties*, Swedish Cement and Concrete Research Institute, CBI report 2:95, Stockholm 1995

3   J. Norberg, *Färsk betongs reologi- teori och mätmetodik (The rheology of fresh concrete-theory and measuring methods)*, Swedish Cement and Concrete Research Institute, CBI report 2:94, Stockholm 1994 (in Swedish).

4   A. Johansson, J. Norberg, *Färsk betongs reologi - mätningar på fabriksbetong (The rheology of fresh concrete-tests on ready mixed concrete)*, Swedish Cement and Concrete Research Institute, CBI report 5:94, Stockholm 1994 (in Swedish).

26 EFFECT OF SOME PLASTICISERS ON THE RHEOLOGICAL BEHAVIOUR OF FRESH CEMENT PASTE

O. WALLEVIK and T. SIMMERMAN
Icelandic Building Research Institute, Iceland

Abstract
A coaxial cylinder viscometer was used to evaluate the dispersing effect of several admixtures in cement pastes with low w/c ratio (0.3). The results indicated that naphthalene based plasticizers had a different dispersing effect depending on their origin, melamine admixtures were not suitable under these special conditions and specially refined lignosulphates could be added to naphthalene based admixtures on a substitute basis and an additional plasticizing effect obtained.
Keywords: admixtures, cement paste, plasticizers, rheology, viscosity, yield stress.

1 Introduction

The objective of the research programme described in this paper was to find a suitable plasticizer for use in a cement paste for high strength light weight aggregate concrete. The light weight aggregate is very porous natural pumice and easily absorbs water from the cement paste. Therefore it was important to obtain a flowable cement paste with an appropriate plastic viscosity that would yield a stable mix in regard to segregation as the yield value had to be as low as possible due to workability requirements.

Preliminary testing indicated that a cement paste of w/c-ratio of about 0.3 would be desirable for obtaining the required compressive strength of the paste.

2 Materials

Throughout the experiments, some parameters remained the same in order to compare the effect of different admixtures. The reference mix had the following composition:

Production Methods and Workability of Concrete. Edited by P.J.M. Bartos, D.L. Marrs and D.J. Cleland.
Published in 1996 by E & FN Spon, 2–6 Boundary Row, London SE1 8HN. ISBN 0 419 22070 4.

- Icelandic Ordinary Portland cement (1000 g)
- water/cement-ratio of 0.30 (300 g)
- 1% of plasticizer (dry powder in % of cement weight)

Plasticizers or dispersing agents are water-reducing admixtures where one of their main functions is to produce concrete of a given workability at a lower water/cement ratio than that of a control concrete containing no admixture. There are several different plasticizers on the market. In the experiments, *superplasticizers* were tested. Traditional superplasticizers are divided mainly into three groups:

- modified sodium lignosulphonate (quoted as lignosulphonate)
- salt of formaldehyde naphthalene (quoted as naphthalene)
- salt of formaldehyde melamine sulphonate (quoted as melamine)

As a reference plasticizer one of the oldest superplasticizers "Mighty 150" was used. To characterise the rheological behaviour of the fresh cement paste, a coaxial cylinder viscometer was used (see part 3).

The plasticizer was mixed with water and the water content of this mixture was taken into account to get the 300 g of total water. The plasticizers tested are listed in Table 1.

Table 1 Commercial names and codes of the admixtures used

Name	Chemical component	Symbol
Mighty 150	naphthalene	M
Woermann FM 420	naphthalene	FM
Woermann FM 27	melamine	-
Woermann BV 13	lignosulphonate	wl
NA	lignosulphonate	NA
Peramin F	melamine	pf
Melment L10 (SP 40)	melamine	sp
Sikament P25	new generation plasticizer	p2
Mapeifluid X	new generation plasticizer	mx

3 Rheology and testing procedure

A coaxial cylinder viscometer *"Bohlin visco 88"* (system set-up C30 DIN53019) with a cross-formed mill or vane of 30 mm diameter (instead of a massive inner cylinder) and a shell were used during the tests.

The testing conditions were:

- all mixes were mixed in a mixer which operates according to the American standard ASTM C 305
- the mixing was carried out in a standard procedure of 3½ minutes, namely:

a. add the water and use speed 1
b. add the cement (30 sec.)
c. stop mixer and handmix (30 sec.)
d. use speed 2 (30 sec.)
e. add plasticizer
f. stop mixer and handmix (30 sec.)
g. mix at speed 2 (60 sec.)

*   the mixes were tested in a viscometer at 6 and 18 minutes after the addition of water, respectively

The total testing time in the viscometer was 308 seconds with 9 measurement points. The changes in rate of shear were carried out stepwise to obtain an equilibrium in shear stress at each rate of shear. Measurement intervals were 45 sec. divided into strain delay time (20 sec.) and integration time (25 sec.). The first point was measured at the rate of shear of 20 Hz, followed stepwise to the 5th point at 75 Hz (the upward curve) followed by reduction in rate of shear to 20 Hz (the downward curve).

The yield value and plastic viscosity was calculated from the downward curve as the time to obtain equilibrium will always be shorter by going from higher to lower rate of shear, the downward curve. If the up- and downward curves are not joined together one can assume that the material tested has a significant thixotropic behaviour (as the measurement time is as long as 45 sec.).

4 Test Programme

Altogether 33 mixes of cement paste were made with various types and contents of superplasticizers. The characteristics of the mixes are given in Table 2.

Three of the mixes were made with different cements (see test No. 3) and three mixes were made with s/w-ratio = 0.25 (see tests Nos. 2 and 9) with the two superplasticizers which were the most effective. As the Bohlin V88 viscometer has a limited range of shear stresses, few of the mixes (in tests Nos. 7 and 8) were too stiff to be tested.

5 Results and discussion

5.1 Test 2: variable dosage of Mighty 150

To establish the sensitivity to an addition of plasticizers for a given cement and w/c-ratios the reference additive Mighty 150 was tested in three dosages. As expected, the shear stresses were higher when the percentage of Mighty was lower. Also the yield value went up but the plastic viscosity was lower. Two mixes were made to see if it was possible to gain the same flowability at w/s-ratio as low as 0.25 by using up to a double dose of the plasticizer. Both these mixes were too stiff; this means that the shear stress was higher than the maximum stress the viscometer could take, or >160 Pa. The results can be used for comparison with other plasticizers to see how much of the Mighty is needed to get the same results. For example, 0.6% of Mighty gives almost the same result as 1% of FM 420, see part 5.2.

Table 2 Mix compositions

Test No.		w/c ratio	Cement (g)	Added water (g)	Plasticizer	% of cement	% of Plast. in water	Amount of plast. (g)
1	a	0.30	1000	284	Mighty	1	39.13	25.6
	b	0.30	1000	284	Mighty	1	39.13	25.6
	c	0.30	1000	284	Mighty	1	39.13	25.6
	d	0.30	1000	284	Mighty	1	39.13	25.6
2	a	0.30	1000	288	Mighty	0.8	39.13	20.4
	b	0.30	1000	291	Mighty	0.6	39.13	15.3
	c	0.25	1000	227	Mighty	1.5	39.13	38.3
	d	0.25	1000	219	Mighty	2	39.13	51.1
3	a	0.30	1000 DK*	284	Mighty	1	39.13	25.6
	b	0.30	1000 DK*	288	Mighty	0.8	39.13	20.4
	c	0.30	1000 NO*	284	Mighty	1	39.13	25.6
4	a	0.30	1000	281	FM 420	1	35.08	28.5
	b	0.30	1000	281	FM 420	1	35.08	28.5
	c	0.30	1000	281	FM 420	1	35.08	28.5
	d	0.30	1000	281	FM 420	1	35.08	28.5
5	a	0.30	1000	279	Mighty/NA	0.8/0.2	39.13/24.9	20.4/8.4
	b	0.30	1000	272	Mighty/NA	0.5/0.5	39.13/24.9	12.8/20.1
	c	0.30	1000	265	Mighty/NA	0.2/0.8	39.13/24.9	5.1/32.1
	d	0.30	1000	260	NA	1	24.9	40.2
6	a	0.30	1000	277	FM 420/NA	0.8/0.2	35.08/24.9	22.8/8.7
	b	0.30	1000	277	FM420/BV13	0.8/0.2	35.08/23	22.8/8.7
7	a	0.30	1000	281	FM 27	1	35.08	28.5
	b	0.30	1000	263	FM 27	2	35.08	57.0
	c	0.30	1000	277	FM 27/NA	0.8/0.2	23.7	8.4
8	a	0.30	1000	279	Peramin	1	32.04	31.2
	b	0.30	1000	268	Peramin	1.5	32.04	46.8
	c	0.30	1000	263	Peramin	1.75	32.04	54.6
	d	0.30	1000	258	Peramin	2	32.04	62.4
	e	0.35	1000	329	Peramin	1	32.04	31.2
9	a	0.30	1000	285	Mel. 1.10	1	40.19	24.9
	b	0.30	1000	277	Sikament P25	1	30.52	32.8
	c	0.30	1000	276	Mapeifluid X	1	29.45	34.0
	d	0.25	1000	202	Mapeifluid X	2	29.45	67.9

* DK and NO stands for Danish and Norwegian Rapid Portland Cement (RPC) respectively

All tests were compared with representative test results of Mighty from test 1.

Note: Paste with w/c-ratio 0.3 has about 48% water by volume (no air included). With reduced w/s-ratio down to 0.25 the volume of water will go down to 43%. One may expect that there is a certain threshold where a reduction of water will have a crucial influence on the flowability of the cement paste. Also, an addition of 2% plasticizer by weight of cement will lead to a concentration of 8% in the water. This high concentration of polymer in the water will increase its (Newtonian) viscosity from about 1 mPas to about 6-9 mPas (depending on rate of shear). This increase of vis-

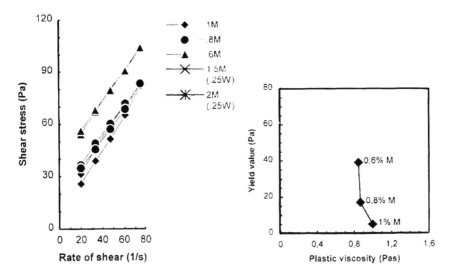

Fig. 1 Effect of Mighty 150

Table 3 Rheological results from test 2, different dosage of Mighty

	1% Mighty	0.8% Mighty	0.6% Mighty	1.5% Mighty w/c: 0.25	2% Mighty w/c: 0.25
Plastic viscosity (Pas)	1.0	0.87	0.85		
Yield value (Pa)	5	17	39	too stiff	too stiff
Correlation coeff.	0.999	0.999	1.000		
Thixotrophy (kPa/s)	<1.0	<1.0	<1.0		

cosity of the water will lead to a significant increase in the viscosity of the cement based composite.

5.2 Test 1 & test 4: Mighty 150 compared with FM 420

To evaluate the reproducibility of the mixes (and the viscometric test) 8 mixes of cement paste were tested. Four specimens of the reference mix and four with 1% Woermann FM 420 instead of 1% Mighty. Both are naphthalene based additives.

The results supported the following conclusions:

- Shear stress: 1% of Mighty gives significantly lower shear stress values than 1% of FM 420;
- Yield value: Mighty has lower yield shear stress values than FM 420 (2-11 compared to 30-52 Pa);
- Plastic viscosity: the plastic viscosity of FM 420 is lower or equal to Mighty.

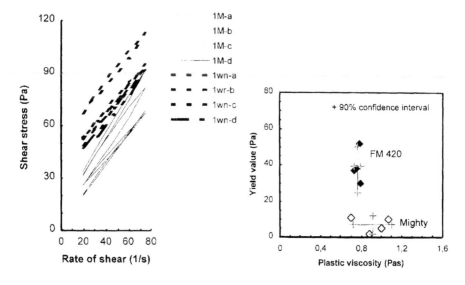

Fig. 2 Reproducibility of 8 mixes, containing 1% of Mighty and FM 420 respectively

Table 4 Rheological results from tests 1 and 4, reproducibility

	Mighty 150 (1%)				Woermann FM 420 (1%)			
	a	b	c	d	a	b	c	d
Plastic viscosity (Pas)	0.7	1.0	0.88	1.07	0.73	0.78	0.75	0.79
Yield value (Pa)	11	5	2	10	37	52	38	30
Correlation coeff.	0.999	0.999	1.000	0.999	0.996	0.996	0.999	0.994
Thixotropy (KPa/s)	<1.0	<1.0	<1.0	<1.0	<1.0	<1.0	<1.0	<1.0

5.3 Test 3: different types of cement

To get an indication of how much influence the cement has on the results, Rapid Portland cements from Denmark and Norway were also tested. First we added 1% of
Mighty to both cement type mixes and according to these results we added 0.8% of
Mighty to the Danish cement. Compared to the Icelandic cement, the Danish cement
gave lower shear stress values, a lower plastic viscosity and a very low yield value (-3
is approximately zero*). The use of less Mighty (0.8%) gave similar or a little lower
shear resistance as when 1% Mighty used with Icelandic Portland cement, however the
yield value is approximately zero and the plastic viscosity is higher. While the Danish
cement gave a lower flow resistance than the Icelandic cement, the Norwegian one
gave a higher resistance. The Norwegian RPC gives a result equivalent to Icelandic
OPC with 0.7% of Mighty. One should bear in mind that the tests with cement were
limited and no general conclusions should be made.

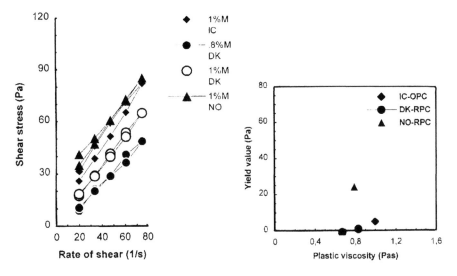

Fig. 3 Result of testing Icelandic (IC-OPC), Danish (DK-RPC) and Norwegian (NO-RPC) cements

Table 5 Rheological results from test 3, different types of cements

	IC-OPC 1% Mighty	DK-RPC 1% Mighty	DK-RPC 0.8% Mighty	NO-RPC 1% Mighty
Plastic viscosity (Pa)	1.0	0.67	0.83	0.79
Yield value (Pa)	5	-3*	1	24
Correlation coeff.	0.999	0.998	0.999	0.998
Thixotropy (kPa/s)	<1.0	<1.0	<1.0	<1.0

* It is theoretically impossible to get a negative yield value (a negative force needed to make it flow!). it can be considered as zero Pa as the value is less than 2% of the capacity of the viscometer.

5.4 Test 5: Modified lignosulphonate combined with Mighty

Modified lignosulphonate NA which is a highly refined product was also tested in a combination with Mighty. The total percentage of placticizer was kept at 1%, and the ratios between NA and Mighty were changed as follows:

    I.   0.8% Mighty - 0.2% NA
    II. 0.5% Mighty - 0.5% NA
    III. 0.2% Mighty - 0.8% NA
    IV.   1.0% NA

The results indicated that Mighty combined with NA has a lower shear resistance than Mighty alone, particulary the plastic viscosity. This difference is not significant. NA by itself produced significantly higher shear stress resulting in both higher yield value and plastic viscosity. Moreover, the use of NA without Mighty in such high dosages led to a significant thixotropic effect, namely the up-and downward curves were far from being parallel.

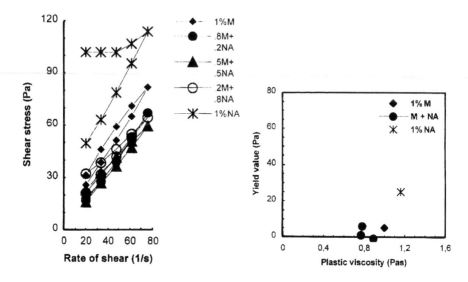

Fig. 4 Result of testing modified lignosulphonate, NA

Table 6. Rheological results from test 5, modified lignosulphonate

	1% Mighty (reference)	0.8% M 0.2% NA	0.5% M 0.5% NA	0.2% M 0.8% NA	1% NA
Plastic viscosity (Pas)	1.0	0.89	0.77	0.78	1.16
Yield value (Pa)	5	-2	1	6	25
Correlation coeff.	0.999	0.997	0.999	1.000	0.999
Thixotropy (kPa/s)	< 1.0	< 1.0	< 1.0		

5.5 Test 6: FM 420 combined with lignosulphonate

Two tests were made to see if the combination of lignosulphonate and FM 420 also gave better results. Besides NA, Woermann BV13 was also tested (BV13 is a conventional sodium lignosulphonate based admixture). Both times the total percentage of plasticizer was kept at 1%, and the dosages of FM 420 were 0.8%. The results were compared with the results of 1% Mighty and of 1% FM 420. The measured values are shown on the page after the results of test 5. Compared to 1% FM 420, the combination of both FM 420 - NA and FM 420 - BV13 gave higher shear stress values and considerably higher yield values. This indicates that, unlike the results of combinations with Mighty, the combination of FM 420 and a lignosulphonate does not improve the results. There is little difference between the results with NA or with BV 13. They both show higher values in the downwards curve than in the upwards curve (opposed to thixotropic behaviour). This indicates that the paste was already stiffening during the measurements.

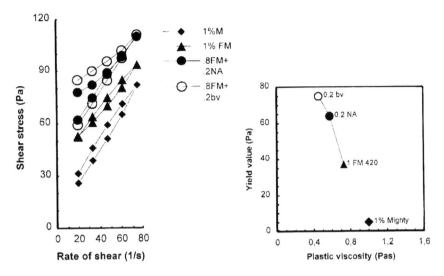

Fig. 5 Result of testing FM 420 combined with lignosulphonate

Table 7: Rheological results from test 6, FM 420 combined with lignosulphonate

	1% Mighty (reference)	1% FM 420	0.8% FM 420 0.2% NA	0.8% FM 420 0.8% bv
Plastic viscosity (Pas)	1.0	0.73	0.58	0.46
Yield value (Pa)	5	37	64	75
Correlation coeff.	0.999	0.996	0.985	0.994
Thixotropy (kPa/s)	<1.0	<1.0		

5.6 Tests 7, 8 and 9: different dosage of melamine

Three different melamine based additives were tested, Woermann FM 27, Peramine F and Melment L10 (Sikament P25 is most likely also melamine based plasticizer, incorporating a thickening agent). All these additives at dosages of 1% were too stiff to be tested. One has to bear in mind before further evaluations that the Icelandic cement is unique as it is very fine grained, (made of shell sand instead of lime stone and rhyolite instead of clay) and incorporates 7% silica fume. At lower dosages therefore and with different cement (no silica fume) the dispersing effect may be different.

0.8% of FM 27 was also tested together with 0.2% of the modified lignosulphonate NA and at double dose or 2% FM 27, respectively. In both cases the mixes were too stiff to be tested.

Peramin F was also tested at enlarged dosages of 1.5, 1.75 and 2% respectively. The viscometer could only measure shear resistance at certain rates of shear with mixes containing 1.5% and 1.75% Peramin F. It is possible to estimate the plastic viscosity for the upward curve and in both cases it was very high, or 1.6 and 1.4 Pas respectively. At the double dose (2%) the shear resistance was within the limit of the viscometer and a test could be completed. The shear resistance was higher than for the

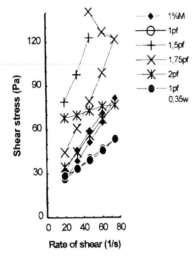
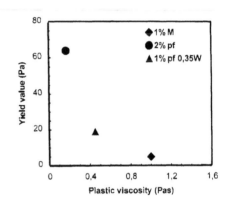

Fig. 6 Result of testing Peramin F

Table 8 Rheological results from test 8, Peramin F

	1% Mighty (reference)	1% pf	1.5% pf	1.75% pf	2% pf	2% pf
Plastic viscosity (Pas)	1.0				0.17	1.16
Yield value (Pa)	5	too stiff	too stiff	too stiff	65	25
Correlation coeff.	0.999				0.986	0.996

paste containing 1% Mighty. Peramin F led to Rheoplexic behaviour (anti-thixotropy: downward curve is higher than upward curve) which was not likely due to colloidal forces, but to chemical reactions or physical changes at the surface of the particles during testing. Finally, the Peramin F was tested at a dosage of 1% and a higher w/s ratio of 0.35 (there is a big difference in flowability of paste with w/c = 0.3 and 0.35). This mix had a lower shear resistance than the mixes with 1% Mighty (w/c = 0.3) and gave lower plastic viscosity but higher yield value.

5.7 Mapeifluid X

Three additives, namely Melment L10, Sikament P25 and Mapeifluid X, were tested in test series 9. The first two, which produced mixes too stiff for the viscometer, have already been discussed in part 5.6. Mapeifluid X is a new generation plasticizer. The main difference between the new and conventional naphthalene and melamine based plasticizer is that it contains carboxyl (COO) instead of a sulfonic (SO_3) anionic group and that the molecular ratio of the anionic group is much lower per organic monomer. The dispersing effect of the Mapeifluid is more based on steric hindrance than electro-static stabilisation which is the dominant mechanism in conventional plasticizers. The new plasticizer is designed to have lesser workability loss than the conventional plasti-cizers (not investigated here).

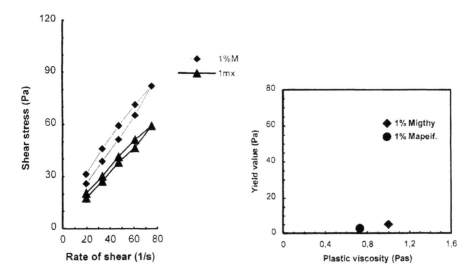

Fig. 7 Result of testing Mapeifluid X

Table 9. Rheological results, Mapeifluid X

	1% Mighty (reference)	1% Mapeifluid	2% Mapeifluid w/c: 0.25
Plastic viscosity (Pas)	1.0	0.73	
Yield value (Pa)	5	3	too stiff
Correlation coeff.	0.999	0.999	

The mix which contained 1% Mapeifluid had a little lower shear stress than the reference mix containing 1% Mighty, which was reflected in a lower plastic viscosity. The plastic viscosity was outside the 90% confidence interval for Mighty (see part 5.2), but so close that the difference could not be considered significant. An additional mix was made with 2% Mapeifluid with a lower w/c-ratio of 0.25, but the mix was too stiff to be tested in the viscometer.

6 Conclusions

Rheological measurements proved to be successful in evaluating the plasticizing effect of several admixtures. One should bear in mind that the dosages of admixtures used were relatively high and the w/c-ratios were low (w/c-ratio: 0.3); also that the cement used included 7% silica fume and had a relatively high Blaine value.

However, the results indicated that the reference admixture, Mighty 150, showed in this study to have greater plasticizing effect than other conventional plasticizing admixtures and that the melamine admixtures were not suitable under these special conditions. In addition, specially refined lignosulphates could be added on a substitute basis together with naphthalene based admixtures and thereby an additional plasticizing effect was gained. These results agree with earlier research work [1]. The most effective plasticizer in these experiments was a new type of dispersing agent named Mapeifluid X.

References

1. Wallevik, O., Iversen, K. (1995) Rheological approach in mix design of very high strength concrete. Paper presented at the ConChem Conference, Brussels, p. 11.
2. Ramachandran, V.S., et al (1984) *Concrete admixtures handbook,* Noyes Publications, New Jersey, p. 626.
3. Wallevik, O., Gjörv, O.E. (1990) Development of a coaxial cylinder for fresh concrete, *Properties of fresh concrete,* Chapman and Hall, London, pp. 213-224.
4. Collepardi, M., et al (1994) Acrylic based superplasticizer, Fourth CANMET/ACI international conference on superplasticizers and other chemical admixtures in concrete, supplementary papers, Montreal, pp. 1-9.

27 VIBRATION AND THE RHEOLOGY OF FRESH CONCRETE – A FURTHER LOOK

P.F.G. BANFILL
Department of Building Engineering and Surveying, Heriot-Watt University, Edinburgh, Scotland, UK

Abstract
Application of vibration, of appropriate amplitude and frequency, to fresh concrete reduces its yield value to zero and permits it to flow under its own weight to pass between reinforcement, fill formwork and release air bubbles. Previous work suggested that the peak velocity is the most important characteristic of the vibration and a re-examination of results obtained in a vertical pipe apparatus confirms that there is a linear relationship between fluidity of the vibrated concrete and the peak velocity (given by amplitude x frequency) of the vibration. The proportionality constant is termed the vibrational susceptibility and is characteristic of the material, being influenced by the workability of the unvibrated concrete, but decreases as frequency increases. The greatest fluidity and hence the most rapid placement of concrete is therefore achieved at low frequencies (16-30 Hz) rather than at the frequencies commonly employed in industrial vibrators for concrete (50-200 Hz).
Keywords: Fresh concrete, rheology, vibration

1 Introduction

Vibration has long been recognised as necessary for effective compaction of concrete and arising from much early work done on the effect of vibration on the properties of the hardened concrete recommendations for current practice exist in many countries. The general consensus seems to be that the higher the acceleration at moderate frequencies the more effective the vibration. However, it is much more fundamentally sound to base recommendations on the performance of fresh concrete and advances in the understanding of the rheology of fresh concrete have permitted this. This paper discusses the relationship between rheology and the behaviour under vibration together with some important implications.

Production Methods and Workability of Concrete. Edited by P.J.M. Bartos, D.L. Marrs and D.J. Cleland.
Published in 1996 by E & FN Spon, 2–6 Boundary Row, London SE1 8HN. ISBN 0 419 22070 4.

2 Effect of vibration on hardened properties

The importance of vibration as a means of compacting concrete has been recognised for a very long time and much progress was made as a result of early studies of the effect of vibration on hardened properties [1, 2, 3], which are embodied in current practice recommendations e.g. [4]. The application of vibration, of an appropriate frequency and amplitude, fluidifies the concrete enabling air pockets to be filled and air to bubble to the surface. An initial gross increase in bulk density is observed, after which prolonged vibration permits slow reorientation and packing of the particles in the concrete. The main consequence of the increased density is higher strength. The air content of the uncompacted freshly mixed concrete may exceed 10% by volume and removal of this amount of air from a test cube by vibration will double the measured compressive strength. Associated with the strength increase is the achievement of the other desirable hardened properties.

3 Rheology of fresh concrete

With advances in the understanding of its rheology came the possibility of studying the effect of vibration upon fresh concrete. It is now well established that fresh concrete, over the range of shear rates important in practice, conforms to the Bingham model and that the yield value and plastic viscosity can be measured conveniently in the two-point workability apparatus [5]. In the test the torque required to turn an impeller at several speeds in a bowl of fresh concrete is measured. From the resulting data a graph of torque T against speed N shows a straight line relationship

$$T = g + hN \tag{1}$$

where g and h can be shown theoretically to be proportional to yield value and plastic viscosity, respectively, the Bingham constants of the material. In practical terms, the fact that fresh concrete has a yield value explains why it can support its own weight, as for example in the slump test. Under vibration concrete loses that ability to support itself and flows freely under its own weight to pass between reinforcing bars and fill the corners of a mould. This suggests that the vibration either applies sufficient shear stress to exceed the yield value or causes a change in the nature of the material such that the yield value is reduced to a low level. Extensive work summarised by Tattersall [6] confirmed that the latter was the case.

4 Effect of vibration on rheology of fresh concrete

Tattersall and Baker [7] mounted the bowl of the two-point workability apparatus on an electromagnetic vibrator and constructed flow curves with and without vibration over a range of frequencies and amplitudes. They found that the flow curve was instantaneously and reversibly altered by the application of vibration and that the yield value decreased to zero (Figure 1), i.e. the constructed flow curve fits a power law model passing through the origin. This enables the concrete to flow under its own weight. Consideration of the region near the origin in Figure 1 suggests that, under vibration, and at low shear rates, fresh concrete behaves as a Newtonian liquid. This means that the viscosity of the vibrated concrete or its inverse, the fluidity, will be the key parameter in determining how it will flow.

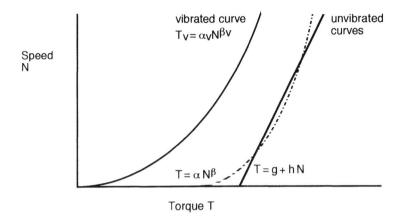

Figure 1 Flow curves for unvibrated and vibrated fresh concrete.

5 Influence of vibrational parameters

Assessing the influence of vibrational parameters on the rheology is impossibly complex when the unvibrated and vibrated curves conform to different models (Bingham and power law), so Tattersall and Baker [7] found it necessary to approximate the unvibrated curve to a power law of the form

$$T = \alpha \, N^\beta \tag{2}$$

It was then possible to study the effects of vibration by considering changes in the ratio α_v/α and the difference $(\beta_v - \beta)$, where the subscript denotes the value determined under vibration. They found that, subject to exceeding a small threshold frequency, the most important characteristic of the vibration is its peak velocity.

Constructing a flow curve from individual readings of torque and speed is cumbersome and the data analysis is complex, so Tattersall and Baker concentrated on the region near the origin of the flow curve where the slope of the curve gives a measure of the fluidity. This suggested that a simpler apparatus could be successful and in a comprehensive series of tests using a much more convenient vertical pipe apparatus they determined the fluidity of the concrete from the rate of efflux out of the bottom of the pipe into the vibrating receptacle [8]. Under vibration, the height H of the column of concrete in the pipe decreases with time as concrete flows from the bottom according to

$$dH/dt = - b \, H \tag{3}$$

where b is a constant proportional to the fluidity. Solving this equation shows that a graph of ln H against t is a straight line of slope -b and this is used to determine b.

Based upon measurements on four concretes carried out at 11 frequencies between 16 and 200 Hz and 8 accelerations between 0.85g and 8.90g, they were able to model the effect of vibrational parameters - frequency f and amplitude A - on fluidity by the equation

$$b = {}^1/_c \ln (1 - {}^f/_F) \, . \, [\, A - A_o] \tag{4}$$

where F is an upper limit of frequency (between 250 and 500 Hz), A_O is a threshold amplitude, which in turn depends on frequency, and c is a constant [8]. This equation shows that there is a threshold amplitude below which vibration has no effect, i.e. yield value is not reduced sufficiently for flow to occur, and that the effect of vibration decreases markedly as frequency approaches some upper value F. If $A >> A_O$ and $f << F$ this equation reduces to

$$b = KAf = Kv \tag{5}$$

since peak velocity v is the product of amplitude and frequency.

5.1 Vibrational susceptibility
The question arises as to whether K is a material parameter. It may be termed the vibrational susceptibility and could be understood as follows. The larger the value of K, the larger the value of b and the faster the concrete flows out of the pipe when it is subjected to vibration of peak velocity v, as would be expected of concrete with a higher vibrational susceptibility. It is reasonable to expect that it would be affected by the workability of the concrete.

5.2 Effect of workability on vibrational susceptibility
The validity of this relationship can be tested by replotting Tattersall and Baker's original data for mix A in the form of -b against v as shown by Figure 2 which shows the relationship for all four mixes and only the points for mix A. Despite the fact that the experiment was not designed to optimise the graph of -b against v and the points are bunched there is clearly a strong correlation confirming the validity of equation 5.

Tattersall and Baker's experimental procedure required repeat batches of nominally identical concretes to be produced and tested to build up the vibrational susceptibility data. Each batch of concrete was tested in the two-point workability apparatus (LM version with planetary impeller) and while the target slump was 20-25mm, Table 1 shows that the values of g and h varied over such a wide range that it is impossible to conclude that the mixes differ significantly in g or h, with the exception of mix C which has a significantly higher h. Similarly the scatter of points about the lines in Figure 2 implies that the slopes of those four lines are also not significantly different. Thus no inferences can be drawn about the effect of workability, measured by g and h, on K.

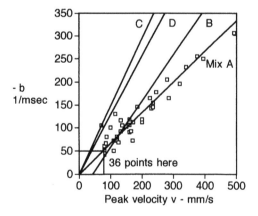

Figure 2 Replot of data showing the relationship between -b and v [8]. Points are shown for mix A only. All mixes 20-25 mm slump.

Table 1. Summary of two-point workability results for Tattersall and Baker's mixes (LM apparatus)

Mix	g range	mean	h range	mean
A	12.4-16.9	14.5	0.6-3.5	2.3
B	13.2-18.0	15.6	3.0-6.9	4.5
C	6.8-16.2	11.4	7.8-15.3	10.4
D	13.8-20.3	16.0	3.7-6.0	4.7

However, in subsequent experiments, Wood and Millar [unpublished data] worked with the same apparatus but with more workable concrete of 100mm slump. In their experiments the vibrational parameters were set to ensure a spread of values of peak velocity up to 500 mm/s so that the vibrational susceptibility could be more precisely estimated. The workability of their four concretes was determined in the two-point apparatus (MH version with uniaxial impeller) and Table 2 summarises their data. Figure 3 shows the relationship between -b and v for their results. It is clear that, instead of the values of K in the range 0.75-1.5 obtained by Tattersall and Baker, the more workable concrete of Wood and Millar gave values of K in the range 1.6-2.1. This shows that the fluidity under vibration was higher, i.e. K had changed in the sense expected and suggests that the vibrational susceptibility really is a material parameter. It is influenced by the general level of workability of the concrete, although much more work will be needed before the nature of that relationship can be established. For example, it might be expected that h, the plastic viscosity, would have a major effect on the vibrational susceptibility through its influence on the rate of flow of fresh concrete at shear stresses greater than the yield value. The limited data reported here cover an insufficiently wide range of mixes to support this assertion but some initial work to explore the effect of different combinations of g and h on K is in progress. Circumstantial evidence in support of this is also provided by high strength high workability concretes, prepared at low water/cement ratio with superplasticisers and silica fume. These are self-levelling but often very slow to flow, implying a low g and high h. It is common practice to vibrate these mixes out of delivery skips into formwork and this implies that vibration reduces the value of h.

Table 2. Summary of two-point workability results for Wood's and Millar's mixes (MH apparatus)

Mix	g range	mean	h range	mean
1	4.1-5.0	4.4	0.3-1.0	0.7
2	3.7-4.1	4.0	0.9-1.3	1.1
3	3.5-4.2	3.9	0.9-1.2	1.0
4	3.8-4.9	4.4	1.0-1.4	1.2

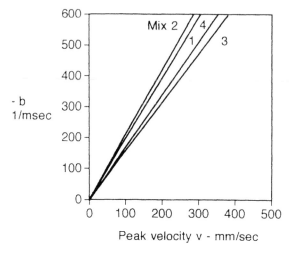

Figure 3 Relationship between -b and v for Wood's and Millar's mixes. All mixes 100 mm slump.

5.3 Effect of frequency on vibrational susceptibility

For vibrational susceptibility to be a true material parameter its value should be independent of the vibration conditions. Since Tattersall and Baker tested their concretes at up to seven levels of acceleration for each frequency, graphs of -b against v plotted from their original data at each frequency have up to seven points - sufficient to define the value of K. Figure 4 shows the effect of frequency on vibrational susceptibility resulting from such treatment. Despite the method of data treatment causing the values of K to cover a wide range for the four mixes the broad trend is quite obvious: vibrational

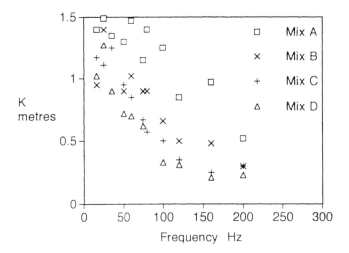

Figure 4 Relationship between vibrational susceptibility K and frequency.

susceptibility decreases with increasing frequency. In addition there is a suggestion that the relationship is becoming flatter at the high frequency end. Thus vibrational susceptibility is not independent of frequency and there are some significant practical implications of this.

6 Practical implications

Frequencies between 50 and 200 Hz are in common use for vibrators [4]. Table and surface vibrators operate towards the lower end, formwork and internal vibrators towards the top end. In contrast, Figure 4 shows that the vibrational susceptibility at 16-30 Hz is about four times that at 200 Hz. Thus, in principle, for a given mix and flow geometry the fluidity and hence the rate of flow at low frequency could be four times that at high frequency, provided the same peak velocity can be achieved. This could mean a fourfold increase in the throughput of concrete in the vibratory compaction stage of concrete placement. In practice, because peak velocity is the product of amplitude and frequency it is only possible to achieve high velocity at low frequencies if the amplitude is very high and therefore amplitude is likely to be the limiting factor in any equipment. The significant effect on site productivity implied by the greater throughput could be realised if concrete technologists and equipment manufacturers work together to carry out vibratory compaction at low frequencies in equipment with maximum amplitude.

7 Conclusions

Application of vibration to fresh concrete reduces its yield value to a level so low that it can flow under its own weight. The fluidity of the vibrated concrete is directly proportional to the peak velocity of the vibration where the constant of proportionality, the vibrational susceptibility, is a characteristic of the material which decreases as frequency increases. Vibration equipment should therefore be operated at the combination of lowest frequency and greatest amplitude possible in order to maximise the fluidity of the vibrated concrete.

Recently started research will give information on the effect of g and h on the vibration performance of normal and high strength concretes.

8 Acknowledgements

I am grateful to Dr G H Tattersall for granting free access to his original results and to M H Gharooni for assistance with processing the data.

9 References

1. Kirkham, R.H.H. (1951) The compaction of concrete slabs by surface vibration: first series of experiments. *Magazine of Concrete Research*, Vol. 3, pp.79-91.
2. Venkatramaiah, S. (1951) Measurement of work done in compacting a known weight of concrete by vibration. *Magazine of Concrete Research*, Vol. 3, pp.89-96.
3. Cusens, A.R. (1958) The influence of amplitude and frequency in the compaction of concrete by vibration. *Magazine of Concrete Research*, Aug 1958, pp.79-86.
4. ACI Committee 309 (1981) Behaviour of fresh concrete during vibration. Journal of the American Concrete Institute, Proceedings, Vol. 78, No. 1, pp.36-53.
5. Tattersall, G.H., and Banfill, P.F.G. (1983) The rheology of fresh concrete. Pitman.

6. Tattersall, G.H. (1991) Effect of vibration on the rheological properties of fresh cement pastes and concretes. Rheology of fresh cement and concrete (ed. P.F.G. Banfill) E & FN Spon, London, pp.323-337.
7. Tattersall, G.H. and Baker, P.H. (1988) The effect of vibration on the rheological properties of fresh concrete. *Magazine of Concrete Research*, Vol. 40, pp.79-89.
8. Tattersall, G.H. and Baker, P.H. (1989) An investigation into the effect of vibration on the workability of fresh concrete using a vertical pipe apparatus. *Magazine of Concrete Research*, Vol. 41, pp.3-9.

28 THE EFFICIENCY OF SNF-TYPE SUPERPLASTICISER IN PORTLAND CEMENT PASTES

R.MANNONEN and V.PENTTALA
Helsinki University of Technology, Espoo, Finland

Abstract
The efficiency of superplasticizer was determined by measuring the viscosity of cement pastes and the amount of free superplasticizer in cement pastes. Three phases of superplasticizer in cement pastes were distinguished; free, adsorbed on cement particles and absorbed into hydration products. The viscosity of cement paste correlated with the amount of the superplasticizer adsorbed on cement. The superplasticizer absorbed into hydration products was found only in the pastes where superplasticizer was added simultaneously with water. The absorbed superplasticizer did not have any effect on the viscosity of cement paste. The amount of the absorbed superplasticizer depended on the dosage of the admixture and also on the specific surface area of tricalcium aluminate in cement. When superplasticizer was added simultaneously with water white foam was formed on the surface of the cement paste. This foam was identified mainly as ettringite. Also, a synthetic ettringite was prepared in the presence of superplasticizer. There was a clear limit in the content of the superplasticizer bound by the ettringite. Below the mole ratio one between sulphonate and ettringite all superplasticizer was practically bound into ettringite but it appears to be impossible to bind more than one mole of sulphonate to a mole of ettringite.
Keywords: Absorption, addition time, adsorption, cement paste, superplasticizer, viscosity,

1 Introduction

It is widely agreed that the water reduction effect of the superplasticizer is caused by improved dispersion of the cement grains in the mixing water. The flocculation of

Production Methods and Workability of Concrete. Edited by P.J.M. Bartos, D.L. Marrs and D.J. Cleland.
Published in 1996 by E & FN Spon, 2–6 Boundary Row, London SE1 8HN. ISBN 0 419 22070 4.

cement particles is decreased or prevented, and the water otherwise immobilized within the flocs is added to that in which the particles can move [1].

The adsorption of the admixture on the hydrating cement grains could decrease flocculation at least in three ways. The first is an increase in the magnitude of the Zeta-potential; if all the particles carry a surface charge of the same sign and sufficient magnitude, they will repel each other. Secondly, the superplasticizer causes an increase in solid-liquid affinity; if the particles are more strongly attracted to the liquid than to each other, they will tend to disperse. The third is a steric hindrance; the oriented adsorption of a non-ionic polymer can weaken the attraction between solid particles [2].

It is commonly known that the organic admixtures act on the surface of solid particles in concrete. Based on this, many researchers have studied the adsorption of admixtures on cement, on clinker minerals and on the hydratation products. Sulphonated naphthalene formaldehyde condensate (SNF) superplasticizer has been found to adsorb on the calcium silicate hydrate gel in monolayer fashion and on the ettringite in multilayer fashion [3]. The adsoption of superplasticizers was high on the monosulphate and very small on anhydrous C_3A [4]. The aluminate-gypsum-water system has been noticed to adsorb large amounts of sulphonated melamine formaldehyde condensate (SMF) or SNF type superplasticizers in water while in a non-aqueous medium only small amounts will be adsorbed [5].

The factors affecting the admixture adsoption have been widely studied. The degree of polymerization of naphthalene sulphonated formaldehyde condensates is found to have influence on the fluidity of cement paste. Polymers of high molecular weight have been observed to be more effective in increasing the fluidity of cement pastes than those of low molecular weight. The phenomenon seems to be related to an increase in both the Zeta-potential and the polymer adsorption which causes steric hindrance [6]. The adsorption of different types of superplasticizer such as sulfonated naphthalene, melamine and styrene polymers has also been studied. The naphthalene based superplasticizer which had the lowest molecular weight had the highest adsorption ability on cement. Furthermore its Zeta-potential was the weakest, e.g. the dispersing capability was lower than that of the other two. The length of carbon chains was suggested to be the reason for the two different modes of adsorption [7].

Empirically there has been found a negative correlation between the specific surface area of tricalcium aluminate and the efficiency of the naphthalene based superplasticizer. An increase in the fineness and the C_3A and C_4AF contents of the cement resulted in a decrease in the fluidity of the grout [8].

Only a few studies have been conducted on attempts to understand the importance of the timing of the addition of superplasticizer in concrete mix. In the middle of the 80's, the effect of the addition time of the admixture on the workability of cement paste and on adsorption of superplasticizer was investigated. Naphthalene and melamine sulphonate based superplasticizers were used with Portland cement. The workability was found to increase when a delayded addition of superplasticizer was used, and the maximum workability was obtained when superplasticizer was added 2 minutes after water. This was called the optimum addition time [9]. Similar results have been obtained later in concrete tests. The effect of the delayed addition of superplasticizer on the production of high strength concrete has also been studied. The time of superplasticizer addition was one minute after water. When extra rapid hardening Portland cement was used the delayed addition of superplasticizer had a remarkable effect on the consistency of the mix [10].

The adsorption of the superplasticizer in the delayed addition has also been investigated [9]. The adsorption was found to decrease profoundly as time of delay increased from 0 to 2 minutes. No further decrease in the adsorption was observed when the time of delay was longer than 2 minutes. In fact, the results showed that when a naphthalene based superplasticizer was used, no significant decrease in the adsorption was observed if delay was more than 1 minute. Similarly, in another study, the viscosity of cement paste was reduced when using delayed addition. The reduced efficiency in the simultaneous addition was assumed to be caused by the adsorption of superplasticizer on the C_3A-gypsum mixture leading to a reduced amount of admixture for the promotion of dispersion of the C-S-H phases. Ultimately, a reduction in the viscosity of the system is expected [11].

In the present study, the aim was to determine the connection between the properties of cement, viscosity of cement paste and the adsorption of superplasticizer. The main interest was focused on the effect of the addition time.

2 Experimental

2.1 Materials
Three Finnish and one German Portland cements were used. The oxide and mineral compositions as well as the specific surface areas are presented in Table 1. Mighty 150 (sodium salt of a sulphonated naphthalene formaldehyde condensate, SNF) was used as superplasticizer.

Table 1.   Chemical and physical composition of the cements

	PZ55 (German extra rapid)	P40/3 (Finnish extra rapid)	OPC (Finnish ordinary)	LH (Finnish low heat)
Chemical composition				
CaO	64.5%	61.4%	62.6%	62.4%
SiO_2	20.5%	19.7%	20.5%	23.1%
Al_2O_3	5.3%	4.7%	4.9%	2.8%
Fe_2O_3	2.0%	2.5%	2.7%	4.1%
MgO	1.3%	3.0%	2.7%	2.4%
Na_2O	0.3%	0.8%	0.9%	0.4%
K_2O	0.9%	0.9%	1.0%	0.5%
SO_3	3.8%	4.0%	2.6%	1.9%
Mineralogical composition				
C_3S	54.6%	55.6%	54.8%	48.3%
C_2S	17.7%	14.6%	17.5%	36.4%
C_3A	10.7%	8.2%	8.4%	0.5%
C_4AF	6.1%	7.6%	8.2%	12.5%
Specific surface area (Blaine)	520 m^2/kg	613 m^2/kg	377 m^2/kg	440 m^2/kg

2.2 Methods

The viscosity of cement pastes was determinated by the use of Haake Viscometer RV 100. The rotation speed and the rotor was selected to give a shear rate of 200 s^{-1}. The paste was homogenized by a dissolver with a blade of 4 cm in diameter and with a rotation speed of 4000 min^{-1}. The water-cement ratio was chosen to give easily measurable viscosities of plain cement pastes (200...1000 mPas). The maximum dosage of the superplasticizer was chosen to give a viscosity of not significantly less than 10 mPas. The composition of the cement pastes has been presented in Table 2.

Table 2.   Composition of the cement pastes.

Cement	W/C	Maximun dosage of superplasticizer (%)	
		simultaneous addition	delayed addition
PZ55	0.40	3.0	3.0
P40/3	0.40	2.5	2.0
OPC	0.35	1.5	1.0
LH SR PC	0.35	1.0	1.0

All samples were tested 10 minutes after the addition of water. Superplasticizer was added either simultaneously with water or 1 minute after the mixing water. The delay was chosen as 1 minute according to the results of the previous studies [9] [10]. Also, the preliminary tests showed that no further decrease in the adsorption of admixture is expected after delay of 1 minute.

2.3 Analysis of SNF superplasticizer by UV-spectroscopy

The total amount of bound superplasticizer and the amount of absorbed superplasticizer was determinated using Hitachi 2000 UV-spectrophotometer. Absorption at 227.8 nm corresponds to the amount of free admixture in the water phase. The amount of SNF superplasticizer in the water phase was read from the standard curve which was made with a dilution series of SNF in deionized water. The difference between the amounts of added and measured superplasticizer gives the bound proportion.

For UV spectroscopical analysis, the cement paste was prepared by mixing cement and water in 1:1 ratio. The superplasticizer was added either simultaneously or one minute after water. The higher water-cement ratio superplasticizer (1.0) was chosen in order to separate the solid particles from the liquid. The solid particles were separated by centrifugation at 2000 g (20 000 m/s^2) for 5 minutes. The solutions were diluted to 1:1000 with deionized water before the determination.

3 Results and discussion

3.1 Adsorption of SNF superplasticizer

The adsorption of the SNF superplasticizer was calculated as the difference of the added superplasticizer and the measured free superplasticizer. The calculated amounts of adsorbed superplasticizer on four different Portland cements are shown in Figs. 1 and 2. The adsorption in the simultaneous addition with the extra rapid cements was much higher than that in OPC or in low heat cement, especially with high admixture dosages. In the delayed addition, the adsorption between different cements was quite similar .

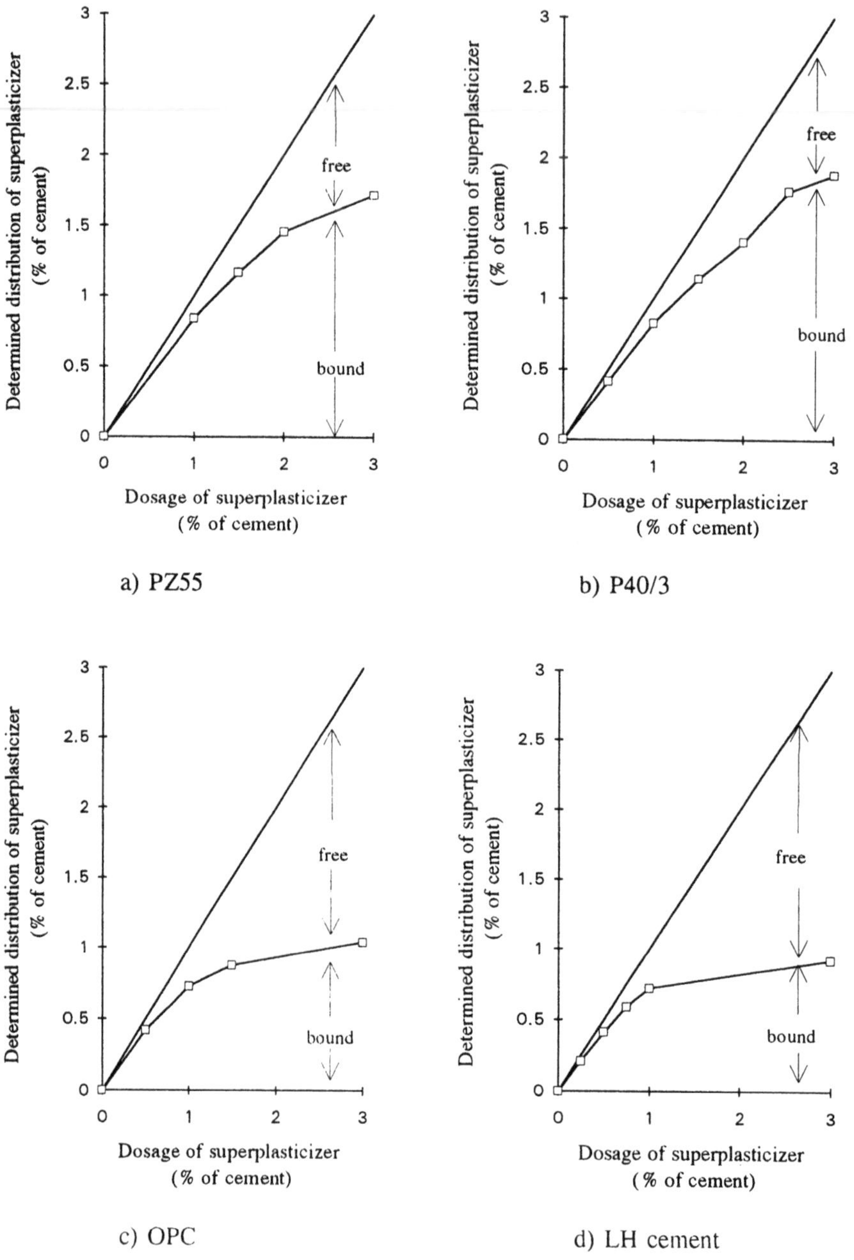

a) PZ55

b) P40/3

c) OPC

d) LH cement

Fig. 1.   Distribution of the SNF-based superplasticizer in different cement pastes when superplasticizer was added simultaneously with water. All amounts are presented as 42% solution of sulphonated naphthalene formaldehyde condensate.

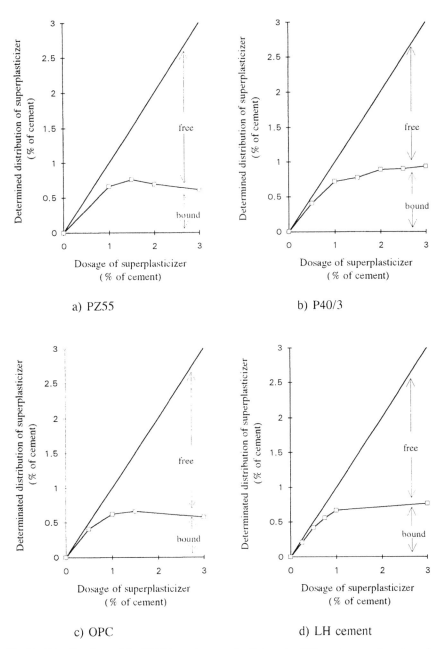

a) PZ55

b) P40/3

c) OPC

d) LH cement

Fig.2. Distribution of the SNF-based superplasticizer in different cement pastes when superplasticizer was added one minute after water. All amounts are presented as 42% solution of sulphonated naphthalene formaldehyde condensate.

Before dilution with deionized water the pH of the samples varied between 13 and 14 and after dilution pH decreased to 10-11. The pH of standards were 7. However, pH does not interfere with the results, because the UV adsorption of SNF is not dependent on the alkalinity in used pH range [12].

3.2 Viscosity of cement paste

The efficiency of the superplasticizer was determined by measuring the viscosity of cement pastes. Cement paste is responsible for the major part of the total specific surface area of concrete, and the reological properties of cement paste are much easier to study than those of concrete. A water-cement ratio of 1.0 was used in the adsorption tests and the ratio of 0.35 or 0.40 in the viscosity tests.

The effects of the dosage and the addition time of superpasticizer are presented in Fig. 3. These figures show that the viscosity decreased faster in the delayed addition compared with the simultaneous addition, especially when rapid hardening cements were used. Furthermore, the amount of the bound superplasticizer by the cement was less in the delayed addition. These results indicate that the viscosity correlates with the amount of free superplasticizer in the mix.

In Fig. 4 the viscosity of the cement paste is presented as a function of the free superplasticizer amount. The differences between the viscosity curves of the simultaneous and the delayed addition were significantly smaller when the viscosity is presented as a function of the amount of the free superplasticizer as in Fig. 4 compared to the situation where the viscosity is presented as a function of the amount of the total superplasticizer as in Fig. 3. From these results it can be concluded that the free superplasticizer in liquid phase can decrease the viscosity of the cement paste. It is widely accepted that the decrease in the viscosity of the cement paste depends on the ability of the admixture to be adsorbed on the surface of cement. For that reason, a more probable explanation for the connection between viscosity and the amount of free superplasticizer is that the free superplasticizer and the superplasticizer adsorbed on cement are in an equilibrium. A higher amount of free superplasticizer causes a higher adsorption on the cement particles and, therefore, this decreases the viscosity of the cement paste.

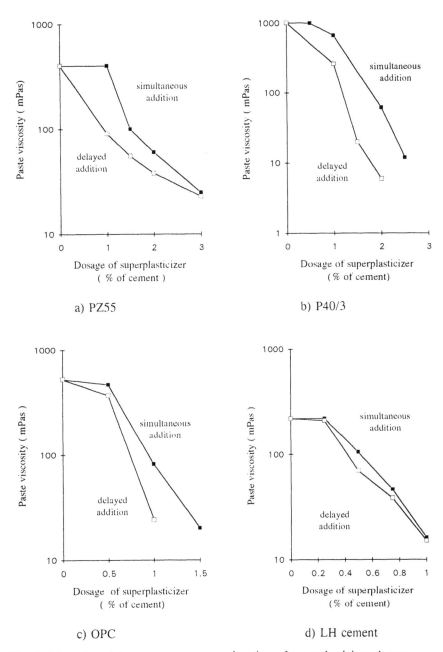

Fig. 3. Viscosity of the cement paste as a function of superplasticizer dosage.

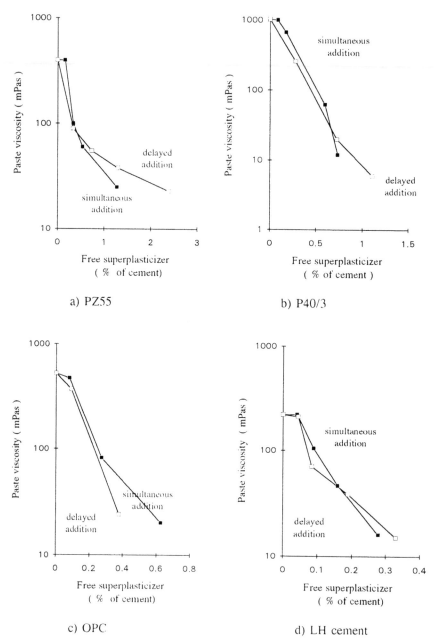

a) PZ55

b) P40/3

c) OPC

d) LH cement

Fig. 4. Viscosity of the cement paste as a function of the free superplasticizer amount.

3.3 Distribution of admixture in cement paste

In the cement paste, superplasticizer appears in three phases: as free admixture in water, as admixture adsorbed on cement particles and as admixture absorbed into hydration products. In the delayed addition only the two first phases are present and the phase can be directly determined from the UV absorption results.

In the case of simultaneous addition, all three phases of admixtures are present. The amount of the adsorbed superplasticizer can be calculated from the determined free superplasticizer amounts. The superplasticizer amount in liquid phase is supposed to cause a proportional adsorption on cement grains. In other words, the amount of the free superplasticizer in water phase governs the admixture amount which is adsorbed on the binder particles. Fig. 5 shows the distribution of superplasticizer in the simultaneous addition.

3.4 Adsorption of superplasticizer onto hydration products

An explanation for the different adsorption properties of the cements can be seen from Fig. 5. The adsorption on the cement particles in the delayed addition of superplasticizer is similar for all cement types. However, the distribution of the phases in the simultaneous addition is different. The properties of cement and the adsorption rates of admixture for different cement types are presented in Table 3.

Table 3. Adsorption rate of admixture for different cement types.

Cement	Specific surface area	C_3A-content in cement	Rate of adsorption
PZ55	high	very high	high
P40/3	very high	high	high
OPC	low	high	moderate
LH cement	moderate	very low	low

The absorption of the admixture into hydration products of the different cement types is high for both extra rapid cements (PZ55 and P40/3), significantly lower for OPC and almost nonexistent for low heat cement. The difference between P40/3 and OPC is the fineness and the essential difference between OPC and LH SR cement is the tricalcium aluminate (C_3A) content. A mathematical connection between the absorption of superplasticizer into hydration products and the specific surface area of C_3A was calculated. The specific surface area of C_3A was calculated by multiplying the specific surface area of the cement by the amount of C_3A. The connection is presented in Fig. 6 for the admixture dosages of 1% and 3%. The linear regression lines for both series are also presented.

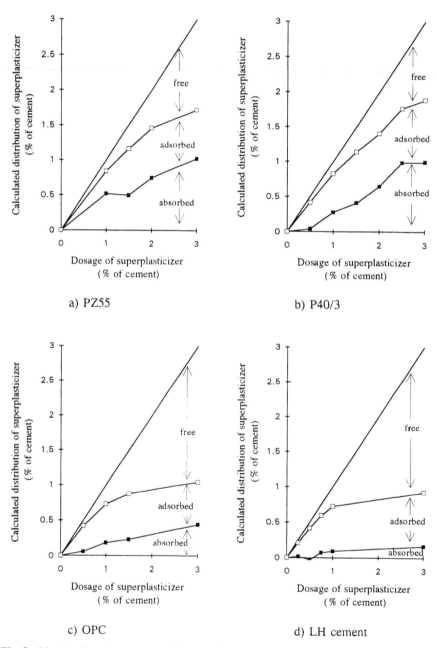

a) PZ55

b) P40/3

c) OPC

d) LH cement

Fig.5. Distribution of superplasticizer in simultaneous addition.

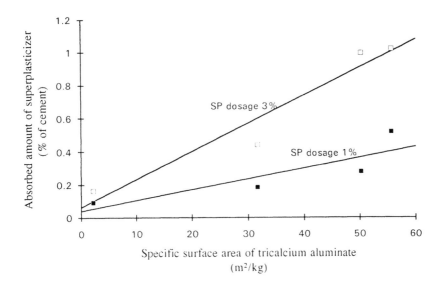

Fig. 6. Connection between the absorbed amount of superplasticizer and the specific surface area of tricalcium aluminate in the different cement types.

The admixture absorption phenomenon has also been reported earlier. It was noticed that the C_3A-gypsum mixture in contact with a solution of superplasticizer binds large amounts of admixture, whereas the mixture prehydrated for a few minutes absorbs substantially less [5]. Also the hydration process of tricalcium aluminate and gypsum with and without superplasticizer has been examined [13]. SMF-based superplasticizer, as an ionic compound, was noticed to react strongly with calcium aluminate hydrates and related substances. Although the molecule is too large to be included in the crystalline phases it can be bound into the gel. This gel was called ettringite precursor. In the present work, a mathematical connection between the amount of the absorbed admixture in the hydratation products and the specific surface area of C_3A was found. Also, two different ways of binding, absorption in the hydration products and adsorption onto cement, were separated.

3.5 Absorption into ettringite

During the determination of free superplasticizer amount in the water phase, a layer of white foam was observed on the samples where superplasticizer was added simultaneously with water. In the samples where the admixture was added one minute after water no foam was observed. The foam was collected and further analysed using X-ray diffractiometer (XRD) and environmental scanning electron microscopy (ESEM). The XRD pattern was typically that of ettringite. The ESEM and elemental analysis results confirmed that foam was ettringite. The total amount of foam was collected from cement paste made with superplasticizer by using simultaneous addition. It was shown that only a very small fraction of the superplasticizer is adsorbed tightly enough to cause the flotation of ettringite particles. Only 0.1% of the total adsorbed superplasticizer was in the foam. However, the result indicates that SNF-type superplasticizer participates in the reaction between tricalcium aluminate and gypsum.

A synthetic ettringite was prepared by mixing $Al_2(SO_4)_3$ and CaO in water phase at the stoichiometric ratio in the presence of the sugar and variable amount of naphthalene based superplasticizer. The precipitation was washed by large amount of water. There was a clear limit in the content of the superplasticizer bound by the ettringite. Below the mole ratio one between sulphonate and ettringite all superplasticizer was practically bound into ettringite but it appears to be impossible to bind more than one mole of sulphonate to a mole of ettringite.

In addition to the groups of naphthalene sulphonate there are also enough sulphate ions to form ettringite with calcium and aluminium. Therefore, this test simulates more the situation when new ettingite is forming during the first day. This test does not simulate the situation during first seconds after water addition and superplasticizer in simultaneous addition because at that moment the water phase does not yet contain sulphates from gypsum.

4 Conclusion

In the present study, two kind of binding types of SNF superplasticizer in cement pastes were distinguished; admixture adsorbed on cement particles and as admixture absorbed into hydration products. Only the adsorbed admixture causes a reduction of viscosity of the cement paste.

In the simultaneous addition both types of binding are present. Part of the superplasticizer is absorbed into the hydration products and part is adsorbed on cement particles. The absorption appears to be dependent on the specific surface area of tricalcium aluminate. This absorption had no effect on the viscosity of cement paste and this portion of the superplasticizer is lost from performing its task. A small amount of the superplasticizer was bound tightly into the hydration products and caused a flotation phenomenon.

The advantage of the delayed addition of superplasticizer was noticed to be bigger with the cements having a high tricalcium aluminate content compared to those having a low tricalcium aluminate content.

5 References

1. Taylor,H.F.W. (1990) *Cement Chemistry*, Academic Press, London, pp. 352-357
2. Daimon, M. and Roy, D.M. (1978) Reological properties of cement mixes: I. methods, preliminary experiments, and adsorption studies. *Cement and Concrete Research*, 8, pp. 753-764
3. Fukaya, Y. and Kato, K. (1986) Adsorption of superplasticizers on CSH (I) and ettringite, *Proc. 8th int. congr. chem. cem.*, Rio de Janeiro, Vol.3, pp. 142-147
4. Massazza,F., Costa, U. and Barilla, A. (1981) Adsorption of Superplasticizers on Calcium Aluminate Monosulphate Hydrate, In: *Development in the use of superplasticizer*, ACI, SP-68, ed. Malhotra,V.M., pp. 499-514
5. Ramachandran, V.S. (1983) Adsorption and hydratation behavior of tricalcium aluminate-water and tricalcium aluminate-gypsum-water system in the presence of superplasticizer. *J. am. Concr. Inst.*, 80, pp. 235-241
6. Collepardi, M., Corradi, M. and Valente, M. (1981) Influence of polymerization of sulfonated naphthalene condensate and its interaction with cement. In: *Development in the use of superplasticizer*, ACI, SP-68, ed. Malhotra,V.M., pp. 485-498
7. Andersen, P.J., Roy, D.M. and Gaidis, J.M 1987. The effects of adsorption of superpasticizer on the surface of cement. *Cement and Concrete Research*, 17, 5, pp. 805-813
8. Hanna, E., Luke, K., Perraton, D. and Aitcin, P.-C. (1989) Rheological Behavior of Portland Cement In the Precence of a Superplasticizer, *Superplasticizer and Other Chemical Admixtures in Concrete: 3rd Intrnational Conference*, ACI SP-119, pp. 171-188
9. Chiocchio,G. and Paolini, A.E. 1985 Optimum time for adding superplasticizer to portland cement pastes. *Cement and Concrete Research*, 15, 5, pp. 901-908
10. Penttala, V. (1990) Possibilities of increasing the workability time of high strength concretes. Proceedings of the RILEM Colloquium on *Properties of Fresh Concrete*, October 3-5, 1990, Hannover, West Germany. ed. Wierig, H.-J., pp. 92-100
11. Masood, I. and Agarwal, S.K. (1994) Effect of various superplasticizer on reological properties of cement paste and mortars. *Cement and Concrete Research*, 24,2, pp. 291-302
12. Yilmaz, V.T., Kindness, A. and Glasser, F.P. 1992 Determination of sulphonated naphtalene formaldehyde superplasticizer in cement: a new spectrofluorometric method and assessment of the UV method. *Cement and Concrete Research*, 22, 4, pp. 663-670
13. Yilmaz, V.T. and Glasser, F.P. 1991 Early hydratation of tricalcium aluminate-gypsum mixtures in the presence of sulphonated melamine formaldehyde superplasticizer. *Cement and Concrete Research*, 21, 5, pp. 765-776

29 RHEOLOGICAL BEHAVIOURS OF MORTARS AND CONCRETES: EXPERIMENTAL APPROACH

C. LANOS, M. LAQUERBE and C. CASANDJIAN
Laboratoire GTMa, INSA de Rennes, Rennes, France

Abstract
The sifting of a concrete gives some indications about the cement paste bonding with aggregates. Results of this sifting test lead to the formulation of a method to carry out concrete composition. The use of a plastometer test accompanied with a suitable data processing method gives a global identification of the rheological behaviours of fresh mixes. According to the beginning of this study, some specific mortars with or without additives are tested with a plastometer. These tests show the apparition of a cement paste migration through a coated aggregates matrix. The proposed method of formulation gives mortars presenting essentially a plastic behaviour with a Coulomb's criterion. More supple mortars present a viscoplastic behaviour. But evolution of the plastic yield is subordinated to the possible consolidation of the material.
Keywords: behaviours of fresh mortars and concretes, composition method, additives, plastometer, plasticity, consolidation.

1 Introduction

The paste has to be smooth enough to be poured in its fresh state, whereas an optimum compacity with a limited water content has to be reached to obtain a high mechanical strength. The improvement in the particle packing obtained by addition of fine aggregates can be neutralised by required water content to obtain a suitable paste workability. In this case admixture of additives, as plasticizers or superplasticizers, has to be used to fit the concrete formulation to the set up (moulding, pumping, vibration). It is necessary to control the additives' influence on the rheological fresh concrete behaviour.

An original analysis of concrete sifting results clearly shows the interaction between

Production Methods and Workability of Concrete. Edited by P.J.M. Bartos, D.L. Marrs and D.J. Cleland. Published in 1996 by E & FN Spon, 2–6 Boundary Row, London SE1 8HN. ISBN 0 419 22070 4.

aggregate areas and quantity of paste in the mix. The distinction between a linked paste to the aggregates and a free paste permits the study of the rheological behaviour of a mix with a new approach. It is then possible to propose a new method of mortar formulation taking into account rheological parameters.

The new formulation concept is supported by an analysis of global rheological behaviours. Identification of these rheological behaviours is given by plastometer tests. These tests are carried out on cement pastes, mortars and mortars with additives. The tested mixes are relatively firm and with a mainly plastic behaviour. The obtained results, under different testing conditions, allow an understanding of the influence of the water content, of the aggregates, and of the additives, on the mix behaviour.

The obtained results for these materials show a plastic behaviour which depends on the pressure, sometimes attached to a kind of consolidation resulting from the fluid migration inside the "granular" matrix.

2 Mortars and concretes rheology

2.1 Mortars and concretes
The size of the biggest aggregate in a concrete prohibits the use of traditional rheometers such as capillary viscometer, cylindrical viscometer. Because of the difficulty in identifying the true rheological behaviour of the concretes, some comparative tests are used. Some of them can be standardised (Lessage's plasticimeter, flow-test...). These tests are easy to implement and inexpensive. They give roughly the static or dynamic behaviour of the paste through global values (setting, spreading, flowing duration) which are hardly linked to the specific values of the tested material.

If the behaviour of mortars and concretes can be considered as a viscoplastic one, an affine relationship between the mixing torque and the rotation speed of the mixer is proposed by Tattersall [1] and Banfill [2]. The value of a shearing torque threshold and of a "plastic viscosity" is proposed. This method always uses global parameters. It is well adapted to follow up production. For instance, the measurement of the mixing torque is usual practice in a concrete mixer. In the same way, Wallevik and Gjørv [3] use results of tests with coaxial cylinders viscometer. De Larrard, Szitkar and Hu [4] propose to identify the concrete behaviour (considered as a Bingham fluid) with a torque viscometer. Recently, by using a cylinder-viscometer, the thixotropic behaviour of mortars, in which the solid volume part is 0.48, is shown by Costil and Chappuis [5]. The viscoplastic character of concretes and mortars is confirmed by all these different results. Another approach is proposed by Sedran, De Larrard and Angot [6]. The Mooney's model is adapted (viscosity of grains in a newtonian fluid mix) and this allows an equivalent viscosity of the mix, which is considered as a multimodal solid suspension, to be proposed. This concept is, however, limited to the optimisation of the compacity in a dry aggregates mix.

2.2 Pastes and cement grouts
The rheology of such fluids has been more often studied. Their behaviour is a viscoplastic one, with shear-thinning or shear-thickening, depending on the water content (Legrand [7]). With a **W/C** ratio in weight between 0.32 to 0.40, Papo [8] shows, by

using a coaxial cylinder viscometer, that these fluids have a Bingham or a Herchel-Bulkley behaviour. Such an identification cannot be used with $W/C<0.32$. For $W/C = 0.22$ and using a plastometer, Kendall [9] suggests that plastic behaviour of the paste, depending on the pressure, respects a Coulomb's criterion.

3 Experimental approach

3.1 Tests with plastometer

Because of the main plastic character of the tested mix, a plastometer is used for the rheological behaviour identification. The plastometer is a compression apparatus with reduced slenderness. The squeezing of a sample between two parallel plates creates a radial flow. A non-viscometric flow is induced. However, the recording of the compression load **F** versus the height between the two plates **h**, their radius **R**, and the compression speed **c**, allows the identification of parameters linked to an average behaviour relation. The way of test analysis is related in detail by Lanos and Doustens [10] [11]. For such a test, the obtained solutions (Kendall [9]) for a Von Mises' plastic fluid (plastic yield **K**) and for a viscous newtonian fluid (viscosity μ) are:

$$\text{viscous fluid:} \quad F = \frac{3\mu\pi\, cR^4}{2h^3} \tag{1}$$

$$\text{plastic fluid:} \quad F = \pi R^2 K\left[1.5 + \frac{R}{h}\right] \tag{2}$$

All these tests are carried out with scratched plates. Such a surface condition is supposed to avoid slipping of the fluid along the wall. The flow created by compression is analysed, for plastic fluids, as if it was a succession of limit equilibrium states. The **F(h,R,c)** records are shown in the reference system **(P,K)**. **K** is the plastic yield value (by Von Mises) calculated with the relationship (2) and **P** is the average compression stress. Otherwise, [10] and [11] show that, for small **h/R** ratio, the average induced compression strain rate is proportional to $a = (R.c)/h^2$.

3.2 Concrete sifting

The cement paste is often considered as a colloid with a real capacity for coating the aggregates. This concept is used in many composition methods. It seems interesting to make a distinction between the strongly linked paste on the surface of the aggregate and the free paste acting as a plasticizer in the mix.

A test of concrete sifting with vibration gives the following results:

- The sieve refuse is composed of a mix of aggregates and cement paste. The granularity of aggregates that constitute the mortar passing through the sieve, and the granularity of dry concrete aggregates passing through the next sieve, approximately one size minus, are the same.
- The granularity of aggregates that constitute the mortar passing through the sieve allows an estimation of the surface of the aggregates in this mortar. These aggre-

gates are supposed spherical. A relationship can be found between this surface and the cement paste content in the mortar passing through the sieve. This relationship is linear. Fig. 1 gives the result of the sifting of a French standard mortar.

A distinction can be made between a linked paste on the aggregate and a free one. The composition of these pastes is not studied. A diminution of the content of the linked paste, per square metre of aggregate surface, can be observed, induced by a plasticizer. A too fluid free paste or an excess of it favours consolidation and segregation in concrete.

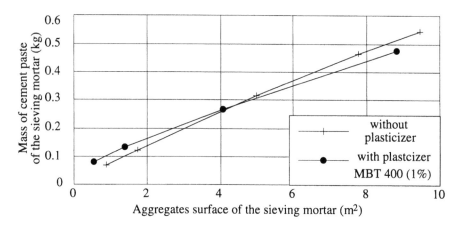

Fig. 1 Results of the sifting of a standard mortar $W/C = 0.5$ and $C/S = 1/3$

Then concrete can be compared with a multimodal suspension of aggregates coated with a linked paste in a fluid constituted by the free paste. The rheological behaviour of the concrete could be separated into two components: a viscous one concerning the fluid and a plastic one for the aggregates coated with the linked paste. This plastic component does not allow an extrapolation of an equivalent viscosity of the material, according to the Mooney's model.

A concrete mixture design can be imagined with the following criteria: the free paste content has to be minimum, the granularity of the coated grains has to be optimum. The linked paste quantity is calculated with respect to the aggregate surface. The rheological behaviour of this concrete is essentially plastic, whose characteristics are mainly depending on the behaviour of the linked paste and on its binding with the aggregates.

This formulation has been applied to realise three different mortars submitted to tests with the plastometer. The sands in these three mortars present different granularities and the paste content is calculated such as the whole paste is linked (no free paste).

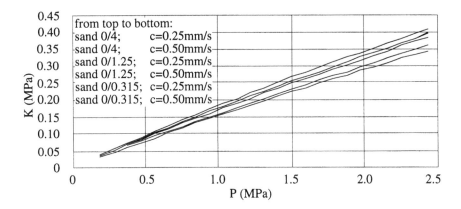

Fig. 2 Plastometer tests on different mortars composed with the proposed method.

The results (Fig. 2) show that the plastic yield **K** is related to the average compression stress **P**. The variations are very closed for the three mortars. The influence of compression speed is slight. Thus the behaviour of these mortars can be considered as a plastic behaviour influenced by **P**.

The formulation of a cement paste with a given rheological behaviour is possible if the behaviours of pure paste or paste with fillers or additives are well known.

4 Identification of rheological behaviours

4.1 Cement pastes
The materials used are: cement CPJ-CEM II/A 32.5, synthetic amorphous silica fume CAB-O-SIL M5, French normal sand (norm NF P 15-403, 0.08/2 mm) and three additives of the SIKA commercial range: superplasticizer A1 (SIKAMENT FF 86), plasticizer A2 (SIKAFLUID), plasticizer water reducer A3 (PLASTIMENT BV 40).

The first tests are carried out on a pure cement paste. Different water contents are used (**W/C** near by 0.23). For these ratios, the behaviour seems to be essentially plastic. The treatment of our experimental results, as Kendall [8], considers a plastic behaviour. The results for **F(Rh,c)** are expressed in (**K,P**) system.

The curves obtained for weight ratios **W/C**: 0.20, 0.23 and 0.26 are drawn in Fig. 3. The plastic behaviour (in Coulomb's criterion meaning) is confirmed by the linear features of these curves. An increase of **W/C** value for these pastes, reduces the apparent coefficient of internal friction (see the curve slope). No conclusion concerning the cohesion is possible.

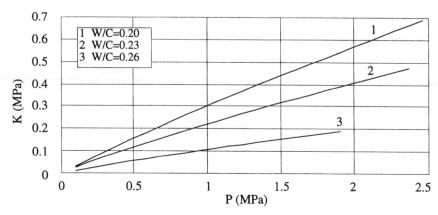

Fig. 3 Tests on cement paste **c** = 0.15 mm/s

4.2 Cement pastes and silica fume

Fig. 4 shows the results obtained with pastes in which silica fume is added (weight **FS**). The quasi linear feature of these curves shows, once more, that pastes present a Coulomb's criterion. Addition of silica fume into the paste increase slope of the curves.

The water is absorbed by the silica fume constituted with grains finer then the cement grains. A gel appears and presents the same function concerning the cement as the function of the linked paste concerning the aggregates in a mortar. An increase of the content of silica fume decreases the free water and leads to a greater apparent plastic yield.

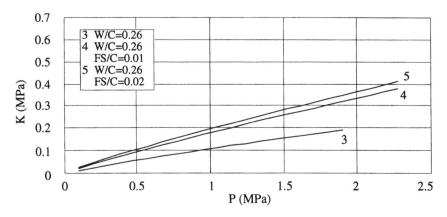

Fig. 4 Tests on cement pastes with silica fume: **c** = 0.15 mm/s

Tests with different compression speeds are necessary to detect a viscous component. The results of such tests, carried out on water-cement-silica fume mix, are shown in Fig. 5. The compression stress decreases as the speed increases. This phenomenon cannot be explained by the viscosity of the tested fluid alone. The plastic criterion can be questioned, or a kind of hardening in relation to migration of the fluid inside the granular matrix can be presumed. After the compression, a radial heterogeneity in the sample appears. Its centre (high pressure) is more drained than the periphery. As clay-sanded soil, a rapid test of the mix limits the fluid migration and an "undrained" shear stress appears, which is less than the drained one.

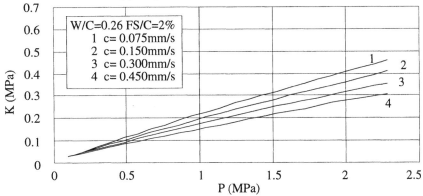

Fig. 5 Tests on cement pastes with silica fume, compression speed influence

4.3 Mortars with additives

A pure mortar, the same mortar with silica fume and this last one with a superplasticizer additive (A1), are compared. The results are shown in Fig. 6. The addition of silica fume induces a slope increase of the curve. On the opposite side, a speed increase incudes a diminution of the same slope (exactly as cement). If a superplasticizer is added to silica fume, the mortar becomes supple without influence of the speed modifications.

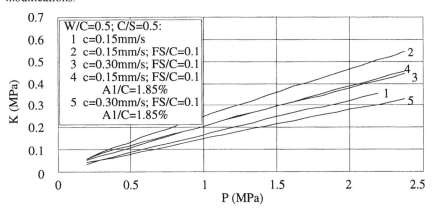

Fig. 6 Tests on mortars with silica fume and superplasticizer A1

Use of the plasticizer (A2) in the mortar **W/C** = 0.30, gives results shown in Fig. 7. The speed has a small influence upon these results. With little additive contents, appears the plasticizer effect. An additive content increase does not involve an important increase in fluidity.

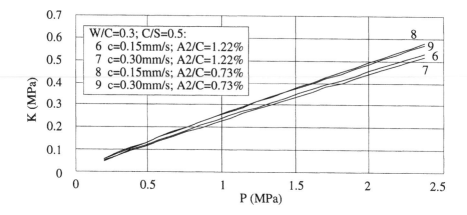

Fig. 7 Tests on a mortar with plasticizer A2

The results in Fig. 8 are obtained with a plasticizer water reducer (A3) in very little content inside the mortar: **W/C** = 0.32. The effect of the speed is quasi ineffective.

The plastic yield of a mortar with silica fume decreases by using the superplasticizer A1 but the kinetic of consolidation does not change.

For the two other additives (A2 and A3) with little water content mortars, the compression velocity is also ineffective, as mortars composed with the proposed method of composition (Fig. 2).

It seems that the used additives fluidify the mortars and decrease the plastic yield. They give better coating of the aggregates (lubrification effect) and at the same time decrease the average thickness of the coating.

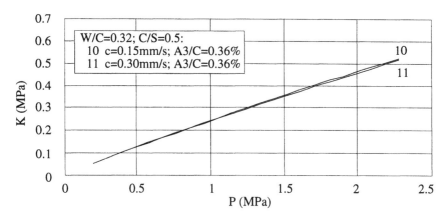

Fig. 8 Tests on a mortar with plasticizer water reducer A3.

Then the composition method can be amended. In a mortar with not enough water content the addition of additive (weight **A**) must be enough to allow the cement paste **(W+C+A)** to coat correctly the aggregates without an excess of free paste.

4.4 Approach of the consolidation phenomenon

The consoldiation phenomenon appears during tests on cement pastes with silica fume. The three tests with different compression speeds are presented in the (**a,K,P**) system where **a** is proportional to the average strain rate. The (**a,K**) system gives the global evolution of the behaviour connected to the pressure.

Fig. 9 shows the curves obtained in (**a,K**) system for three values of **P**. The average shear stress **K** decreases appreciably for an increase of the **a** value. So the mortar behaviour presents a shear-thinning. This variation of the shear stress **K**, given for a **P** value, recalls the results obtained by Costil and Chappuis [5]. They record a decrease of the shear stress associated with a thixotropy of the mix. Similarly, Ducerf and Piau [12] show that some thick suspensions present a plastic shear stress yield and a shear stress decrease for small strain rate. The existence of a minimal shear stress value in the flow-curve is also indicated by Coussot and Piau [13] for a suspension of polystyrene balls in a water and clay mix.

The curves obtained in Fig. 9 allow the drawing of curves in the (**P,K**) system for some **a** values (see Fig. 10). These curves are linear. If their slopes are influenced by **a**, cohesions (value of **K** for **P** null) are not.

The curve obtained for **a** null is not disturbed by the compression speed. This curve is the representation of the plastic criterion of the tested material, expressed in the (**P,K**) system. Then the mix has a Coulomb's plastic behaviour.

The behaviours of mortars present an analogy with plastic hardening soils behaviours. The water role, in a soil, is the same as that of the cement paste around the aggregates. The curve expressed in the (**P,K**) system, for **a** null, is identical to a normally consolidated test answer. For **a** positive, material has an unconsolidated answer.

A complete identification of the mortars behaviour necessitates a relationship between the solid volume part and **P**.

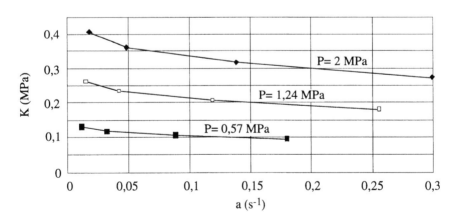

Fig. 9 Tests on cement pastes with silica fume, results for various **P** values

With a hypothesis concerning pressure distribution (Terzaghi principle) it should be possible to build the relationship between the solid volume part and the fluid pressure.

So the kinetic consolidation and permeability parameters should be deduced from this relationship.

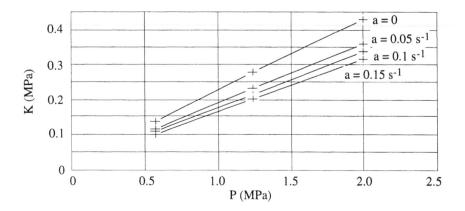

Fig. 10. Tests on cement pastes with silica fume, results for values of **a**

5 Conclusions

These rheological studies lay upon global parameters. The following conclusions appear:

- The concrete looks like a suspension of coated aggregates, with linked cement paste, in a free cement paste. It is shown by a simple sifting. The difference between linked paste and free paste depends essentially on the sifting energy.
- Many just coated aggregates create a cohesive material directly visible in sieves. The free paste constitutes the fluid part of the suspension.
- The identification of the cement paste quantity, just sufficient to coat one square metre of aggregate surface, gives a composition method. This method provides an economic and firm concrete. To increase fluidity requires the use of additives or a vigorous setting up. Then the free paste part increases. As in a concentrated suspension, the fluid part increase lays down a decrease of the apparent solid volume part and, consequently, the mix viscosity decreases.
- For mortars composed with the proposed method, the independence between the results, obtained with the plastometer, and the compression speed shows an essentially plastic behaviour.
- For mortars with free past, a migration of paste appears through the grain matrix. The apparent thixotropy of those mortars, or the decrease of the apparent internal friction coefficient, are two interpretations of the phenomenon of paste migration. This phenomenon is governed by the consolidation theory.
- The mortar without free paste presents an essentially plastic behaviour with a Coulomb's criterion. The knowledge of the relationship between solid volume part of aggregates and pressure remains to be built.

- The mortar with free paste presents a viscoplastic behaviour. But evolution of the plastic yield is subordinated to the possible consolidation of the material.

6. References

1.   Tattersall, G.H. (1990) Application of rheological measurements to practical control of concrete. *Rheology of fresh cement and concrete*, Liverpool, pp. 270-280.
2.   Banfill, P.F.G. (1991) The rheology of fresh mortar, *Magazine of concrete research*, Vol. 43, No. 154, pp.13-21.
3.   Wallevik, O.H., Gjørv, O.E. (1990) Development of a coaxial viscometer for fresh concrete, *Properties of fresh concrete*, RILEM, Hanover.
4.   De Larrard, F., Szitkar, J.C., Hu, C., (1993) Conception d'un rhéomètre pour bétons fluides, *Bulletin*, LPC, No. 186, pp. 55-59.
5.   Costil, V., Chappuis, J., (1994) Etude des phénomènes de thixotropie dans les suspensions concentrées de particules minérales, *Les cahiers de rhéologie*, Vol. XIII, No. 1-2, pp.61-70.
6.   Sedrant, T., De Larrard, F., Angot, D. (1994) Prévision de la compacité des mélanges granulaires par le modèle de suspension solide, LPC No. 194, pp. 59-93.
7.   Legrand, C. (1982) La structure des suspensions de ciment et le comportement des suspensions de ciment, *Le béton hydraulique*, Presses ENPC, Paris.
8.   Papo, A. (1988) Rheological models for cement paste, *Matériaux et Constructions*, Vol. 21, pp. 41-46.
9.   Kendall, K. (1987) Interparticle friction in slurries, *Tribology in particulate technology*, edited by Briscoe and Adam, pp. 91-103/
10.   Lanos, C., Doustens, A. (1993) Exploitation du test au plastomètre, *11ème congé AUM*, Vol. 3, pp.297-300.
11.   Lanos, C., Doustens, A. (1994) Rhéométrie des écoulements entre plateaux parallèles: réflexions, *European Journal Mech. Eng.*, Vol. 39, No. 2, pp. 77-89.
12.   Ducerf, S., Piau, J.M. (1994) Influence des forces inter-particulaires sur la structure et les propriétés rhéologiques de suspensions denses micronisées, *Les Cahiers de Rhéologie*, Vol. XIII, No. 1-2, pp. 120-129.
13.   Coussot, P., Piau, J.M. (1994) Rhéologie des suspensions très concentrées en grosses particules, *Les Cahiers de Rhéologie*, Vol. XIII, No. 1-2, pp. 266-277.

PART SEVEN
TEST METHODS

30 EQUIVALENT SLUMP

K.W. DAY
Concrete Advice Pty Ltd, Melbourne, Australia

Abstract
The paper advocates the concept of a slump value adjusted for the effects of time and temperature so as to more faithfully represent water content variations. The concept employs Arrhenius Equivalent Age and Popovics i value for rate of slump loss. It is shown to be simple to derive the necessary constants, and to obtain the proposed Equivalent Slump, using a computer spreadsheet
Keywords: slump, temperature, time, computer, water requirement.

1 Introduction

The slump test is at the same time the most used and the most vilified test for the workability of concrete. It can give substantially different answers on the same sample of concrete when performed by different operators, and even when repeated by the same operator. Even more seriously, the true value of their slump does not necessarily rank different concrete mixes in the correct order of their workability or compactability.

On the positive side, the test is very sensitive to small variations of water content. A difference of 10 mm in slump will be caused by a difference of somewhere between 1 and 4 litres of water per cubic metre (or say 0.5 ins caused by 2 to 8 lbs/cu yd), depending on the nature of the concrete and especially on the average slump. This may be equivalent to a strength change of about 0.5 to 2.0 MPa (say 100 to 300 psi). Such a strength change is of the same order as the average difference to be expected between two compression test specimens from the same sample of concrete.

Furthermore, if the results of slump tests are graphed as cusums (cumulative sums) alongside cumulative sums of strength for a given mix, it is virtually certain that a

Production Methods and Workability of Concrete. Edited by P.J.M. Bartos, D.L. Marrs and D.J. Cleland. Published in 1996 by E & FN Spon, 2–6 Boundary Row, London SE1 8HN. ISBN 0 419 22070 4.

distinct and sustained change in average slump of as little as 10 mm will be mirrored by a distinct change in strength. It does not work the other way around because other factors produce strength changes without affecting slump (eg cement content or quality changes). Due to the latter fact, it is not necessarily to be anticipated that a good relationship between strength and slump will be obtained over a sustained period of time. The slump test works best on cohesive concrete of between 50 and 150 mm slump. If concrete falls apart or gives widely different values on the same sample when slump tested by a competent person, it is probably not cohesive concrete. This is worth knowing. It does not necessarily mean the concrete is unsuitable for the purpose for which it was supplied, but it does mean it is likely to segregate in some uses. Slump is only part of the story, but it is an important part of the story and should not be discarded because it is not the whole story.

The upshot of this is that slump is not such a bad test if its limitations are understood. We are going to have to live with the slump test for a good number of years yet. Even if a better replacement were to be devised tomorrow, it would take a decade to be generally accepted by the highly conservative concrete industry. It is therefore desirable that we do our best to understand as many of the foibles of this test as possible.

2 The problem

Everyone realises that the slump of concrete reduces with time after batching (assuming no water or chemical is added). Everyone except Bryant Mather[1] realises that it takes more water to produce a given slump on a hot day. Strictly, Bryant is the one who is correct in that his contention is that **initial** slump (at essentially zero time, so early that one cannot normally arrange to measure it, ie usually while still mixing) is not affected by temperature, but slump loss is faster on hotter days, which makes it **appear** to require more water for a given slump. Everyone else is also correct because this means that it **does** require more water to obtain a given slump *at a given time after the cement meets the water* on a hot day.

3 Purpose of slump test

The main purpose of a slump test is usually to indicate whether the correct amount of water has been used. There can be other reasons, eg to indicate when the risks of segregation, excessive bleeding or incomplete compaction have increased, but these do not negate the contention that control of water content is the most usual reason. If this is the reason, then it makes no sense to come to a different conclusion on concrete with the same water content according to when it is tested or what its temperature is. It is exactly the same as compression testing a concrete cylinder without considering how old it is or at what temperature it has been stored. If the situation is the same, then it will not be surprising if the solution is the same.

4 Solution to the problem

Of course it is not possible to require that slump tests take place at a fixed time after batching. It is even less practical to require that the concrete be kept at a standard temperature for such a period. However a great deal of work has been done on the interpretation of early age compression testing of concrete and it is this work which is applicable to the slump problem. The compression work is aimed at transforming an early age strength into an **equivalent strength** after 28 days of standard curing. One way of doing this is by ascertaining the **equivalent age** of the test and using Arrhenius equation to effect the transformation.

What is proposed is that "equivalent slump" be defined as the slump a concrete would have had if it had been kept 30 minutes at 20°C prior to testing (from the time of the water first meeting the cement).

The problem now becomes to translate a slump value taken after some other period of time at some other temperature to the standard equivalent slump. This needs mathematical consideration to understand what is happening, but it must reduce to a simple, if necessary approximate, method if it is to gain acceptance.

There are two parts to the problem. One is to establish the equivalent age (equivalent time may be more appropriate, since minutes rather than days will be the units). The other is to relate the rate of slump loss to the elapsed time.

For equivalent time, the Arrhenius relationship is proposed:
Equivalent time = actual time x exponential($-Q(1/t_1-1/t_2)$)
Where: Q = a rate constant
$\quad t_1$ = actual temperature (°K = °C + 273)
$\quad t_2$ = standard temperature (20 + 273 = 293°K)

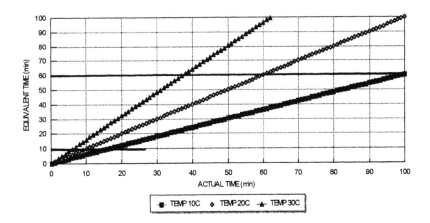

Fig 1. Equivalent age

For rate of slump loss Popovics' assumption[2] of a straight line relationship between the logarithm of slump and the logarithm of time is proposed.

The Q value of a concrete will be affected by its chemical composition, fineness of grind, use of admixtures etc. For some concretes its value may have been determined from strength testing (it is assumed the same value would apply). It may be that an inaccurate assumption of the value of Q will only result in a different value being obtained for the slope (Popovics i value[2]) of the log(slump)/log(equivalent time) relationship. If so it may be sufficient to assume a Q value of say 4200, or perhaps any value between say 4000 and 5000.

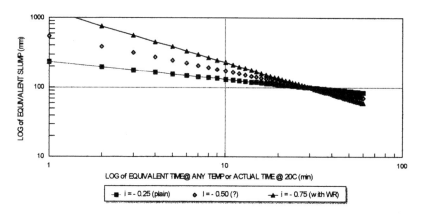

Fig 2. Log slump v log time at various i values

It is very simple to set up (eg on a Lotus spreadsheet) an equation to automatically translate actual values of slump, concrete temperature, and time after batching, into equivalent slump.

The equation is:
Equivalent slump = $\log^{-1}(i \times \log 0.5 + \log S)/\log^{-1}(i \times \log A)$
Where: S = measured slump

A = Arrhenius equivalent age

i = Popovics' constant, for which he suggests [2] values of -0.25 for plain concrete and - 0.75 for concrete with a water- reducing admixture

It is equally simple to set up a spreadsheet area which analyses two readings of slump taken 20 or 30 minutes apart on the same concrete to give the value of the constant i.

The expression is: $i = (\log S_1 - \log S_2)/(\log A_1 - \log A_2)$
Where: S_1 and S_2 are the two slumps

A_1 and A_2 are the two equivalent ages

Fig 3 shows a spreadsheet set up both to generate an average i value and to adjust actual slumps to equivalent slumps using that average value. A Q value and the desired reference temperature are entered in cells B2 and B3. Normal single slumps are

	A	B	C	D	E	F	G	H	I	J	K	L
1			EQUIVALENT SLUMP $PREADSHEET									
2	Q VALUE USE	4200	< set to appropriate value for your cement					i VALUE	<criterion range for database extraction			
3	REF TEMP >	20	< set to desired value					1	<entered as - H7 > 0			
4		MEASURED VALUES			MEASURED VALUES							
5	READING No	SLUMP No 1			SLIUMP No 2		AVG i VAL>	-0.27633	EQUIVALENT SLUMPS		EQUIVALENT AGES	
6		SLUMP	TEMP	TIME	SLUMP	TEMP	TIME	i VALUE	SLUMP1	SLUMP2	SLUMP1	SLUMP2
7		96	21	12	73	23	35	-0.2448	76	78	0.210	0.643
8		87	24	18				0.0000	80	ERR	0.364	0.000
9		80	28	10	48	33	65	-0.2571	66	68	0.244	1.779
10		60	20	22				0.0000	55	ERR	0.367	0.000
11		97	15	15				0.0000	75	ERR	0.195	0.000
12		55	25	21	45	27	37	-0.3271	53	52	0.445	0.822
13								0.0000	ERR	ERR	0.000	0.000
14								0.0000	ERR	ERR	0.000	0.000

Fig. 3. Lotus spreadsheet for "equivalent slump"

entered in column B with their associated temperature and delay since mixing in columns C and D. A calculated equivalent slump appears automatically in column I. This calculation uses the value of i in cell H5, which is the average of all the values in column H. From time to time a second slump is taken and entered in column E, with its associated temperature and time in columns F and G. This entry automatically causes an i value to be calculated in column H. Columns K and L could be hidden, they are only an intermediate stage in the calculations. Of course there is no necessity to use a spreadsheet in a compiled computer program. The same values would be entered, and the same results output, but the workings would remain unseen. However there is some value in using a spreadsheet in that the variation in obtained i value, and the agreement or otherwise of the two calculations of the same equivalent slump, are readily observed.

5 Application to field testing

The above is easy to apply to values of slump being entered into a control/analysis system and will hopefully give distinctly better relationships between strength and slump via water content. It is a little harder to apply the proposal to field acceptance testing of concrete. One solution is for the testing officer to carry a notebook computer or programmable calculator. Another is to use an off-site computer to generate transformation tables, graphs, or conversion factors applicable to the particular concrete being tested. Even if none of these were done, it would be a distinct step forward if the situation were recognised and at least some rule of thumb latitude or adjustment allowed or imposed in appropriate cases. At least we could avoid the situation of rejecting concrete of high slump on a cold morning and accepting concrete of higher water content but lower slump on a hot afternoon.

6 Graphical relationships

It is interesting to see the outworking of the above assumptions in practical terms. Fig 4 shows the effect of i value in rather more comprehensible terms than Fig 2. Obviously the effect of changing i value is very substantial and, contrary to the probable situation with Q value, it is not safe to simply assume a value for it without prior checking (although continual checking of the same mix may be unnecessary).

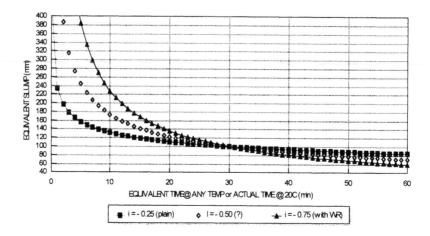

Fig 4 Variation of slump with time for plain and admixture concrete

Values of time below 10 minutes and slumps in excess of 300 mm are obviously purely theoretical but are shown in Figs 4 & 5 to explain the otherwise surprising Fig 6, which shows concrete losing slump more rapidly at 10°C than at 30°C. The explanation is that the concrete at 30°C **has already lost a substantial amount of slump before the earliest practicable slump measurement at 10 mins**. and is then on a less steeply sloping part of the slump loss curve. In Fig 5 the water content is held constant and the three concretes all start at the same (theoretically infinite!) initial slump. It can be seen why those familiar only with the data illustrated in Fig 7 would assume that water requirement varies with temperature and slump loss does not, whereas this correctly observed effect is fully explained by the different rate of slump loss in the first 10 mins.

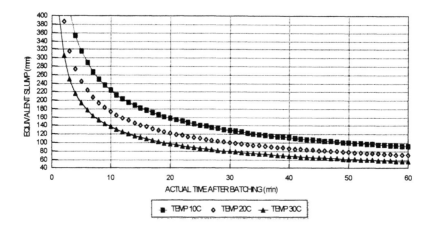

Fig 5. Variation of slump with actual time and temperature (i value of - 0.50)

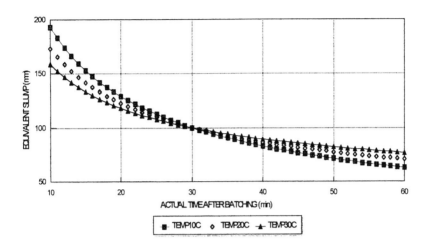

Fig 6 Variation of slump with time and temperature at an **i** value of - 0.50 note that the curves at different temperatures are **not** at the same water content.

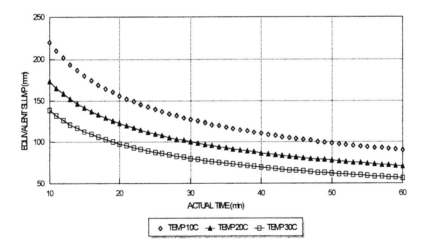

Fig 7 shows the variation of slump with time and temperature *at a given water content*, again assuming i = - 0.50. (this is the same graph as Fig 5 but with the initial 10 mins deleted)

Field testing officers could be provided with a set of graphs similar to Fig 7 and with **i** values of say - 0.25, - 0.50, and - 0.75. They could then be advised which i value was appropriate to the concrete they were testing and could obtain its temperature and age for themselves (assuming the delivery docket provided a batching time, as it certainly should).

8 Conclusion

It has been demonstrated that it is quite simple to adjust measured slumps to an **"Equivalent Slump"** allowing for age and temperature.

Some have expressed surprise, and even doubt, at Bryant Mather's contention that the higher water of hotter concrete could be fully explained by faster hydration for promptly taken slumps. The current examination shows that this is indeed to be expected. In fact the hotter concrete has lost so much slump in the first 10 minutes (compared to the colder concrete) that it thereafter loses slump more slowly than the colder concrete.

The proposal is a starting point rather than a definitive conclusion. It is possible, and even likely, that more will be learned about the factors affecting slump loss if and when we start to use the proposed "double slump giving i value" technique. It may be that investigations will be carried out to establish or modify the proposed mathematical basis and to provide relationships between the i constant and such factors as w/c ratio, type of cement and admixture etc. It may also be that the concept presented does not adequately cover the process of actual setting as opposed to mere slump loss, although the equivalent age concept certainly applies to the setting process.

The immediate need is an initial acceptance of the principle that "slumps ain't slumps" unless they are adjusted to a standardised time and temperature.

9 References

1. Mather, Bryant The warmer the concrete the faster the cement hydrates, *Concrete International*, August 1987
2. Popovics, S *Fundamentals of Portland Cement Concrete Vol 1 Fresh Concrete* Wiley-Interscience. (1982)

31 CONTINUOUS CONTROL OF FRESH CONCRETE USING THE FCT 101 TESTER

T. STEINER
Resarch and Consulting Institute of the Swiss Cement Industry,
Wildegg, Switzerland

Abstract
The new test method provides results combining those of the traditional testing methods of fresh concrete. Advantages of the new method are the speed of the test procedure, the possibility to test concrete in situ without removing any material, ability to carry out a number of workability tests in a rapid succession on the same sample, to store the results and print them out with a portable PC and provide complete practical test documentation.
Keywords: Fresh concrete tests, FCT-tester, workability, slump, water-cement ratio, continuous recording, computer processing, concrete temperature.

1 Introduction

Tests such as the slump, flow, compaction, Vebe and other test methods are usually used to determine the variability of the workability of a fresh concrete mix.
Such tests are described in appropriate standards and guidelines exist for use of these tests in practical production quality control. The standard tests tend to provide reproducible results and the basic tests are indispensable tools in modern concrete construction practice.

All the standard workability tests are relatively labour intensive. A considerable time usually elapses between the extraction of a test sample and the final results being available. In practice this means that the control of the workability of concrete is generally only made at intervals or at random. A systematic, virtually gap-less, continuous control of workability dependent on the process of stiffening of the cement paste and the water loss is practically impossible with the traditional methods.

Production Methods and Workability of Concrete. Edited by P.J.M. Bartos, D.L. Marrs and D.J. Cleland.
Published in 1996 by E & FN Spon, 2–6 Boundary Row, London SE1 8HN. ISBN 0 419 22070 4.

Fig. 1 . Assessment of workability of a fresh mix using the FCT 101 tester.

Furthermore, it has to be considered that in each test in which a sample of fresh concrete is removed there is a probability of the sample not representing the in-situ concrete on the building site.

Practical concrete construction requires a method which allows to judge the workability of a mix within max. 2 minutes if a fresh concrete is to be placed in a shuttering or it is to be rejected because of its low workability. Not only the consideration of easy pouring and compaction but also workability required for all finishing work which has to be undertaken must be taken in order to determine the correct workability. Concrete mixes of good workability are particularly necessary in the case of air entrained concrete, concrete with silica-fume or for concretes used for building conservation.

2 The FCT tester

The operating principle of the test is based on a determination of a torque of a rotating probe. The tester gives values for concrete with coarse aggregate up to 32 mm and a slump between 35 and 180 mm with an accuracy of $\pm$ 3%. The equipment is hand-held (2.4kg) and battery operated. The tests can be carried out in any location provided there is at least a volume of 50 litres of fresh concrete available. The calibration of the tester is very simple. The required FCT values can be determined at the same time as the fresh concrete is tested in the usual conventional manner. Depending of the requirements and problems which have to be solved several test-series may be necessary. Selection of integrated working characteristics allows the required quality parameters to be called upon in the shortest of time (slump, flow, compaction...) under the condition that the mix design has not changed. Within one minute the workability can be measured and with the established correlation factors other parameters like water-cement-ratio can be determined.

A significant advantage is that the tests can be repeated using the same batch of fresh concrete. It is therefore possible to observe the development of the stiffness of the mix. The measured values are stored in the tester and can be recalled and printed out via a PC. The quality-control can be thus documented continuously.

3 Application

High performance reinforced concrete construction elements require a continuous quality control. In such cases a single non conforming mix can cause the failure of the whole construction. Bridge decks, waterproof concrete containers, pipes, slim supporting piers and similar structures have to guarantee the required safety demands even in their weakest parts. In such circumstances it is necessary to test not just some of the concrete, selected batches only, but much more of the concrete, virtually all the mix prior to being poured into the formwork must be tested. The continuous testing should also eliminate external effects such as long transportation times, weather etc. The continuous and inexpensive monitoring proposed will always make it cheaper to reject doubtful concrete rather than accept the potential cost of repair of the failed part of a construction.

32 A PRESENTATION OF THE BML VISCOMETER

J.H. MORK
Euroc Resaerch AB, Slite, Sweden

Abstract
This paper describes the advantages and disadvantages of using a coaxial cylindrical viscometer for concrete for investigating the rheological characteristics of fresh concrete. The BML-Viscometer was developed at the Norwegian Institute of Technology, NTH, in the eighties. It was based on a theoretical approach as well as a review of other possible viscometers. The BML-Viscometer is controlled by a computer with MS-Windows based software. The viscometer assures accurate and reliable measurements of the rheological properties of the concrete with necessary parameters to describe the relationship between shearing rate and force.
Keywords: Coaxial cylinders viscometer, concrete, rheology, testing.

1 Introduction

During the years, many attempts have been made to study the rheological properties of concrete. These are important in defining the workability and fluidity of concrete when certain requirements are laid down. The rheological properties of concrete are, however, not easy to study. Empirical methods such as slump do not provide the correct parameters necessary to describe the rheological properties of fresh concrete. Techniques for measuring the rheological properties of concrete, often lack any theoretical basis, they may be too operator dependent and even sample separation may occur. The slump provides one parameter necessary to describe the rheological properties, but it is not capable of, for example, describing the relationship between shear rate and force in a concrete mix or giving rheological parameters as the initial flow resistance, the viscosity or quantifying the thixotropic effects.

The need for accurate measurement of rheological properties of fresh concrete,

Production Methods and Workability of Concrete. Edited by P.J.M. Bartos, D.L. Marrs and D.J. Cleland. Published in 1996 by E & FN Spon, 2–6 Boundary Row, London SE1 8HN. ISBN 0 419 22070 4.

have increased as new materials for mix design have been introduced (plasticizers, silica fume), and the requirements of fresh concrete have become more specific.

A project was initiated at the Department of Building Material at the Norwegian Institute of Technology in the early eighties. The aim was to develop an instrument with a solid theoretical basis, which could measure the rheological parameters in concrete more easily and provide accurate and reliable results.

2 Description of the BML-Viscometer

The BML-Viscometer type <u>WO-3</u>, which is the latest model in the BML-Viscometer family, is a constant rate coaxial cylindrical viscometer working according to the principle of a Couette viscometer [1]. It consists of a measuring unit and a computer that controls the instrument (Figure 1).

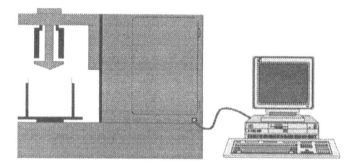

Fig. 1. The BML-Viscometer.

Figure 2 shows a cross section of the inner and outer cylinders in contact with the sample. The radius of the inner cylinder is 10 cm, the outer 20 cm and the height of the cylinders are 20 cm. Figure 3 shows the outer and inner cylinder from above with the ribs to prevent slippage and to have the interface concrete-concrete in the shearing zone. The software that controls the BML-Viscometer is based on MS-Windows, with possibilities to define the measuring procedures and also to study measured rheological variables.

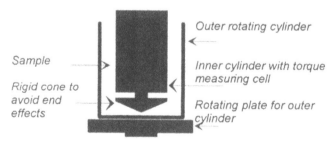

Fig. 2. Cross section of the viscometer cylinders.

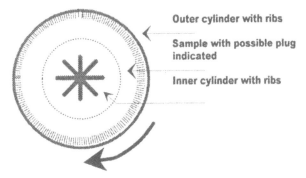

Fig. 3. The outer and inner cylinder from above.

There exist several possibilities for graphical presentations and modelling using rheological models or simply raw data as measured torque and the speed of the outer cylinder. Transferring data to other software such as Excel is also possible, and in a new version of the software to come, it will also be possible to cut and paste graphics, define measuring procedures with both increasing and decreasing shear rate on the same sample besides some other improved features.

The construction ensures the minimum of end effects affecting the measured values. It is possible to have control with separation under measurement, since the applied force is constant all over the shearing zone.

The speed range of the outer cylinder is 0.001-0.9 r.p.s., and the torque applied is maximum 100 Nm, though the normal maximum torque is set to 66 Nm to obtain the highest accuracy [2]. These ranges are suitable for different concrete mixes when a test is to be undertaken. Normally concrete exhibits a torque in the range 0.5-25 Nm during measuring when speed is set between 0.1-0.5 r.p.s.. It is even possible to undertake rheological tests on zero-slump concrete, since all parts of the viscometer are intentionally oversized, however accurate results from such concrete are not possible.

Sometimes it seems to be possible to obtain reliable measurements with a slump above 20 mm, but often a slump of 40-50 mm is necessary depending on the mix.

3 Calibration and accuracy during measurement

The variation in calibration constants for speed and torque are determined over a wide range of different speeds and moments. The difference in values measured under calibration, are in percentage within two decimal places, well below the variation observed for concrete. The reproducibility of the instrument is therefore very good. The calibration is easily done with standard weights, however, calibration fluids (oils) can also be used. The precision in measuring also seems to be satisfactory.

A test of 7 equal concrete mixes, showed a maximum variation of the measured torque of ± 0.20 Nm, which shows a satisfactory reproducibility as the variation between the concrete mixes is included.

4 Running the BML-Viscometer

When testing, the preferred procedure (which influences the results for concrete, as well as for any particle systems) is set up using the computer. The outer cylinder is filled with approximately 12 litres of concrete and testing can begin. Normally it will take a minute to finish a test, dependent on the set-up. The outer cylinder with the concrete is easily transported between the mixer and the viscometer by a small hand handled electric truck.

5 Are the results reliable, and what parameters are adequate ?

One of the problems with viscometric measurements on concrete, is the concentrated and dense particle system. Figure 4, tables 1 and 2 shows a division of suspensions with different particle concentration into regions and indicates the most adequate measuring techniques in these regions [3]. Concrete is in an area where viscometric tests could be acceptable, but also in an area where other techniques could be preferable. Suspensions with low to medium concentration may cause a flow profile in the sample, but concrete has a too high particle concentration for this to occur completely.

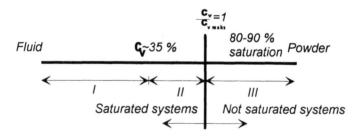

Fig. 4. Systems of different particle concentration, C_V = concentration.

Table 1. Characterisation of particle systems: Influence of concentration.

Region I	Suspensions with low to medium concentration. May be treated as pseudo homogeneous.
Region II	Poor reproducibility in viscometric tests. The flow behaviour is regulated by packing effects. Suspensions can show a granule-viscous behaviour.
Region III	Not saturated, compressible 3-phase materials. Viscometers and rheometers cannot in most cases be used.

Table 2. Techniques for the different regions.

	Rheology	Soil mechanics
Region	I & II	III (II)
Technique used	Viscometers Rheometers	(1) Shear box (2) Shear cell (3) Triaxial apparatus (4) IC-Tester [4]

One of the problems arising then, is dilation in the shearing zone, causing the values measured to be lower than they should be. This dilatation effect is greatly dependent on the shear rate, the higher shear rate causing increased separation [5]. Studies undertaken in the shearing zone of the BML-Viscometer, show that the dilatation effects are only pronounced, when the sample is sheared for the first time. The paste and liquid are then sucked into the shearing zone, but during subsequent shearing this does not occur as the zone has already separated [6]. The dilatation effects in concrete, and the permeability of the fresh concrete, have also been studied with techniques used in soil mechanics (triaxial tests) [6].

The nature of concrete also makes it impossible to obtain shear or velocity profiles, in opposite to many other materials, such as oils or suspensions at lower solid concentrations [7]. The viscometer will with concrete behave like an "advanced shear box", with the shearing zone just outside the inner cylinder. The remainder of the concrete, between the inner and outer cylinder, will act as a plug.

All this indicates that it may be inadequate to speak about stresses in fresh concrete, since there are so many special factors influencing the results. It is more or less impossible to determine velocity profiles or accurate stress profiles. Rheological models such as the Rheiner-Rivlin equation are therefore not adequate for concrete. Besides, it is impossible to have stress control in a composite material such as concrete: is it the stresses in the paste, around the particles, between the particles, or just some average stress in the material that is measured ?

The most natural choice of parameters used for presentations are therefore the two parameters necessary to describe the relationship between the shear speed and the measured torque, which for concrete happens to be linear as in the Bingham-model. The two parameters have been given the name g [Nm] (initial flow resistance) and h [Nms] (a plastic viscous parameter), as for the two-point tester [4]. The g- and h-parameters are illustrated in figure 5.

There exists a non-linear relationship between the slump and g-value, but not between the h-value and the slump or between the g- and h-values, as expected, as the latter two describes two different properties of the concrete. This is shown in figure 6 for some mixes, both with and without admixture (a superplasticizer). The slump, therefore does not indicate anything about the viscous properties of concrete, or other important rheological properties as for example thixotropic effects. These properties are of special importance in both high performance concrete and self levelling concrete (vibration free concrete).

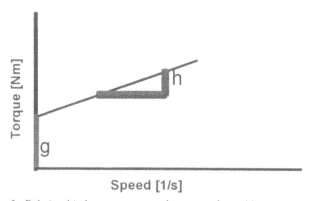

Fig. 5. Relationship between measured torque and speed in concrete.

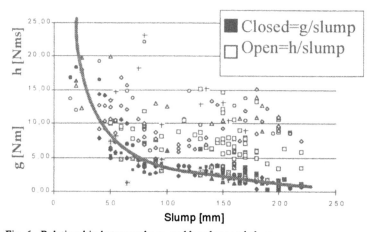

Fig. 6. Relationship between the g- and h-values and slump.

6 How can the results be implemented ?

The use of the BML-Viscometer at Euroc Research AB is R&D in general, mix design optimising and quality control of both cements and admixtures in concrete. A couple of examples of the use in R&D are shown in figures 7 and 8. Figure 7 shows the changes in the rheological properties when different amounts of silica are replacing the cement on a substitution basis in some concrete mixes with different cement content from a reference mix containing no silica. Figure 8 shows how different amounts of some admixtures added to a admixture free reference concrete can affect the rheological properties in terms of the g- and h-parameters.

The BML-Viscometer has shown the ability to present very useful results. One of the reasons, is the geometry that ensures control with separation (the effect of

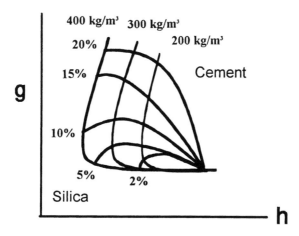

Fig. 7. The effect of silica in concrete [6].

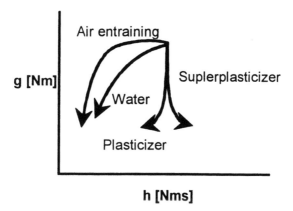

Fig. 8. The effect of admixture in concrete [8].

separation is the same all over the shear zone), coaxial forces in the concrete (material lines the same as shear lines), control of the end effects, and that it is oversized to avoid differences in results depending on the concrete mix.

At Euroc Research work is also progressing in explaining the different rheological behaviour of different concrete mixes by studying: changes in pore water, consumption of the different components in the cement and admixtures, and the formation of different hydration products, both qualitative and quantitative.   This is being undertaken using techniques such as: XRD (X-ray diffraction), ICP (Inductively coupled plasma), UV (Ultra-violet scanning spectrophotometer) and DSC (Differential scanning calorimeter).

A lot of work has also been carried out to study the rheological properties of the binder by means of elastoplastic parameters and structure development with rheometers such as the Bohlin CS rheometer [6].

7   Conclusions

The BML-Viscometer has the ability to provide very useful results, despite the fact that it is not an easy matter to perform rheological tests on concrete. Experience has shown that the instrument has the ability to study the rheological behaviour of different mix designs (parameter studies) in R&D, to be a useful tool in optimising mix designs and in quality control of cements and admixtures, which claims high demands on the accuracy and capacity of the instrument.

8   References

1.   Wallevik, O.H. and Gjørv O.E. (1990) *Development of a coaxial cylinders viscometer for fresh concrete*. Proceedings of the RILEM Colloquium on Properties of Fresh Concrete, H.J. Wierig, Ed., Chapman & Hall.

2.   Wallevik, O. H. (1995). *Calibration of the BML-Viscometer for EUROC Research AB* with documentation (Preliminary), Reykjavik, (in Norwegian).

3.   *Fluid Rheology* (1987). A course arranged by Skandinavisk Teknikförmedling International AB, Uppsala.

4.   Tattersall, G.H. (1991) *Workability and Quality Control of Concrete*, E & FN Spon, London.

5.   Cheng, D.C-H. and Richmond R.A. (1978) *Some observations on the rheological behaviour of dense suspensions*. Rheol Acta 17, pp 446-453, Dr. Dietrich Steinkopff Verlag, Darmstadt.

6.   Mork, J.H. *Effect of gypsum-hemihydrate ratio in cement on the rheological properties of fresh concrete*. The Norwegian Institute of Technology-NTH, Dr.Ing. dissertation 1994:4, Trondheim, 286p, (in Norwegian).

7.   Cheng, D.C-H. (1984) *Further observations on the rheological behaviour of dense suspensions*. Powder Technology 37, pp 255-273, Elsevier.

8.   Wallevik, O.H. *The rheology of the fresh concrete and applications for concrete with and without silica fume*, The Norwegian Institute of Technology-NTH, Dr.Ing. dissertation 1990:45, Trondheim, 185p, (in Norwegian).

33 EVOLUTION OF THE WORKABILITY OF SUPERPLASTICISED CONCRETES: ASSESSMENT WITH THE BTRHEOM RHEOMETER

F. de LARRARD, T. SEDRAN, C. HU*, J.C. SZITKAR, M. JOLY and F. DERKX
Laboratoire Central des Ponts et Chaussées (LCPC), Paris, France
* currently employed by Lafarge, L'Isle d'Abeau, France

Abstract
Superplasticizers provide spectacular gains in fresh concrete workability. However, the effectiveness of these products sometimes does not last more than some 10-20 minutes. It is therefore essential, when using such admixtures, to assess not only the initial workability, but also its evolution. BTRHEOM, the new rheometer developed at LCPC, is especially suitable for doing this job with only one 7-litre specimen. The quantification of the friction of the seal between container and rotating part is first presented. A model is built which allows to deduce from bulk measurements the contribution of concrete to the apparent rheological properties.

Three phenomena may influence the early evolution of workability: water absorption of aggregate, chemical activity of cement, and segregation. A guide is given to deduce the nature of the phenomena from the rheological assessment, which allows the user to find remedies and design a stable concrete. Finally, it is shown that the fact of using the same sample for successive characterisations does not affect the measurements significantly.
Keywords: Bingham's mode, fresh concrete, plastic viscosity, rheology, rheometer, superplasticizer, workability, yield stress.

1 Introduction

Any concrete designed for field utilisation or the precast industry must have a suitable workability at the very moment when it is cast, say several minutes (sometimes several hours) after mixing. Various phenomena make the workability evolve once the cement is in contact with water. This evolution may be very rapid in the presence of superplasticizer if there is incompatibility between the cement and the admixture.

Production Methods and Workability of Concrete. Edited by P.J.M. Bartos, D.L. Marrs and D.J. Cleland. Published in 1996 by E & FN Spon, 2–6 Boundary Row, London SE1 8HN. ISBN 0 419 22070 4.

It is therefore important to provide tests that can be used to assess the workability of fresh concrete for a certain time. Up to now, the only way was to produce a batch of at least some 50 litres, then sample it from time to time and perform successive tests. This process requires a significant amount of materials, time and manpower. There are always discussions on whether the concrete should be continuously mixed or not, and the results can be significantly affected by evaporation (the free surface of the batch being relatively large compared to its volume). Finally, it is difficult to state what is the cause of loss of workability, as the consistency measurement is generally a crude test, giving only one value.

With the new test presented below, the assessment of workability vs. time is greatly simplified. Only one small sample has to be prepared, which is placed in the rheometer for successive tests under shear. The apparatus will be presented first. The question of extracting the contribution of concrete from the bulk measurements (which are partially influenced by the seal friction) will then be discussed. Finally, some typical behaviours (rheological measurements vs. time) will be highlighted. It will be shown that the rheometer allows the user not only to detect losses of workability but also to know the origin of the phenomenon (water absorption of aggregate, segregation and/or chemical activity of the binder/superplasticizer system).

2 The BTRHEOM rheometer for soft-to-fluid fresh concrete

The apparatus, which is a plane-to-plane rheometer, has already been presented in various publications [1-8]. The principle is to place a hollow cylinder-shaped concrete specimen in a container, then shearing it between the top section (in rotation around a vertical axis) and the bottom one, which is fixed. The container is placed on an integrated vibrating plate which may be activated if the aim is to assess the behaviour under vibration. The apparatus is compact and computer-controlled. It can be used on site as well as in the laboratory, for mixes having a soft-to-fluid consistency (say, a slump higher than 10 cm). A dozen have been produced to date. They are used both in industrial practice and for research.

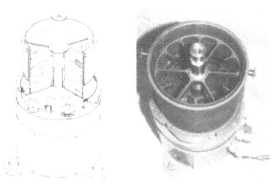

BTRHEOM provides several physical parameters that characterise the rheological behaviour of fresh concrete (see Fig. 2):
- the shear yield stress τ_0 (in Pa) and the plastic viscosity μ (in Pa.s); they are often called Bingham's parameters, as a Bingham fluid is a body, the rheological

behaviour of which may be represented by a straight line in a diagram of shear stress plotted against shear rate;
- the same parameters under vibration (τ_{0v}, in Pa and μ_v, in Pa.s);
- the shear yield stress at rest τ_{0r} (in Pa), which illustrates the thixotropy of fresh concrete; and
- the dilatancy, in percent (volume change of concrete between the sheared state and the consolidated state).

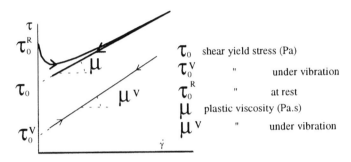

Fig. 2. Some rheological parameters of fresh concrete measured with BTRHEOM.

Actually, the aim is not to measure as many parameters as possible for themselves, but to assess physical quantities that can be scientifically related to various stages of concrete utilization. For instance:
- the shear yield stress describes the minimum stress that should be applied to deform fresh concrete just after shearing (*reworked* concrete). The angle of equilibrium of the material may be calculated from the yield stress [3, 4]. Also, τ_0 is directly correlated with the Abrams cone slump [3, 4, 7];
- the plastic viscosity expresses the increment of stress necessary to provide a certain flow rate. The rate at which the concrete slumps in the Abrams cone test is a direct illustration of plastic viscosity. In field practice, plastic viscosity could play an important role: it controls the pumping rate, and the ease of finishing the concrete surfaces.

The experience gained to date with the other parameters is less. However, τ_{0r} and τ_{0v} have obvious links with the ease of re-working fresh concrete and placing it under vibration, respectively. Also, the dilatancy is expected to influence the pumpability of concrete.

3 Friction created by the seal between the container and the rotating blades

The shear motion is imposed in the test by a rotating part, made of an external skirt, linked to the axis by a system of blades. In order to avoid the penetration of grout and fine materials between the container and the rotating part, a circular seal is placed on the external face of the skirt. During the early development of BTRHEOM, a *felt* seal was used. Investigations had been carried out in order to evaluate the part of the torque due to this seal [3]. However, it had an imperfect tightness, which could undermine the significance of measurements with ultra-fluid mixtures as self-levelling concretes.

Therefore, it was decided to adopt a new seal, made up of three layers of different materials.

3.1 Experiments

In the practical use of BTRHEOM, the friction of the seal is assessed by performing a test without concrete in the rheometer. During this preliminary test, the container is filled up with water. However, the friction changes when the fine part of the concrete (i.e. grout or fine mortar) is in contact with the seal. Systematic measurements of the relationship between torque and speed of rotation have been undertaken, with different fine cementitious materials, with and without vibration. These data were converted to equivalent stress and strain rate values. It was found that, in all cases, these relationships were straight lines, so that equivalent yield stress and plastic viscosity could be affected to each material/seal couple tested (at a certain time), while the contribution of the material itself (i.e. its shear resistance) was negligible.

The materials tested were designed to cover the range of fine mixture that may penetrate into the seal during a concrete test (i.e. very fluid grouts and fine mortars, with water/binder ratios ranging from 0.33 to 0.5, some of which containing silica fume or limestone filler). With each seal, two successive water test were carried out. Then, the normal test (i.e. determination of Bingham characteristics, without or with vibration, see ref. [2, 3, 8]) was performed every 15 minutes. A system of dead loads was fixed on the container to simulate the mass of concrete that is normally placed in the apparatus.

3.2 Modelling the effect of seal friction in a concrete test

The response of the seal in the presence of water was found essentially horizontal (i.e. the torque did not depend on the speed of rotation). Therefore, a given seal was characterized by a mean (equivalent) shear stress $\langle\tau_w\rangle$. Then, the aim was to predict from this datum the values of the seal yield stress and plastic viscosity when water was replaced by cementitious material.

The values of plastic viscosity, with or without vibration, did not appear to depend on $\langle\tau_w\rangle$, unlike the ones of yield stress which are linearly related to $\langle\tau_w\rangle$. From the data collected, simple models were adopted, which are given in Tab. 1. To calibrate these equations, the bulk measurements (in a stress vs. strain rate diagram) were fitted with straight lines of constant slope (representing the plastic viscosity). An experimental yield stress was obtained. At a second step, the yield stress values were plotted against the $\langle\tau_w\rangle$ ones, and linear equations were generated by the least squares method (see Fig. 3 & 4).

Tab.1. Relationships between the equivalent Bingham's parameters of the seal in the presence of cementious materials and the equivalent yield stress in the presence of water.

	Equivalent yield stress (Pa)	Mean error (Pa)	Equivalent plastic viscosity (Pa.s)	Mean error (Pa.s)
without vibration	$\tau_0^s = 0.76\langle\tau_w\rangle + 131$	60	$\mu^s = 20$	6
under vibration	$\tau_0^s = 0.75\langle\tau_w\rangle + 47$	45	$\mu^s = 40$	10

To summarize, when a concrete test is performed with BTRHEOM, bulk values of Bingham's parameters are obtained. They must be corrected by substracting the seal values given in Tab. 1. The correction on plastic viscosity is minor, and the error on this correction is negligible. The correction on yield stress is more significant, and leads to a certain inaccuracy ($\pm$ 100 Pa, equivalent to $\pm$ 1 cm in the slump test). To limit this error, it is wise to try to limit the value of $<\tau_w>$ below 400 Pa. Then, the error on assessing the seal yield stress falls (mean error value: 44 and 30 Pa, without and with vibration, respectively).

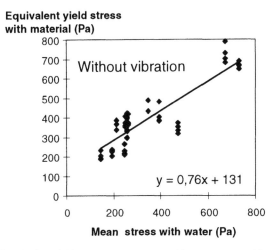

Fig. 3. seal equivalent yield stress vs. mean stress with water $<\tau_w>$, without vibration.

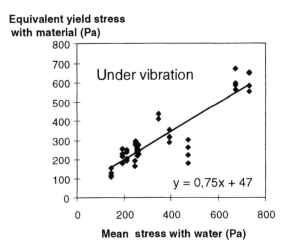

Fig. 4. Seal equivalent yield stress vs. mean stress with water $<\tau_w>$, with vibration.

4 Assessment of losses of workability

4.1 Early changes occurring in a concrete sample after mixing - Effects on the rheological behaviour

If one considers fresh concrete as a dense suspension of particles, two effects may influence the rheological behaviour.

First, the aggregate may be porous, and concrete may be prepared with unsaturated aggregate. As soon as the aggregate is in contact with the cement paste, it tends to absorb water to reach the saturated state. This phenomenon is especially noticeable in lightweight aggregate concrete, but is also encountered with normal-weight aggregate. The effect on rheology will be the same as the one observed when a series of batches are produced with a decreasing water dosage: both yield stress and plastic viscosity will increase [9, 10].

Second, the mixture which is homogenous after an efficient mixing may segregate for various reasons:
- *at rest,* gravity may create a settlement of coarse aggregate, if the aggregate skeleton is not well proportioned, especially when it is gap-graded; this settlement may be considered as a contraction of the granular phase;
- *under vibration,* the phenomenon is amplified;
- *under horizontal shear,* if the coarse aggregate dosage is high (close to the packing value), the aggregate phase tends to expand. This expansion creates an accumulation of aggregate in the dead zones (not submitted to shear), if any. As the aggregate movement is limited by the bottom of the container, aggregates tend to rise out of the concrete.

Experience has shown that there is a balance between the two kinds of segregation in the BTRHEOM apparatus when concrete is well-proportioned. Otherwise, coarse aggregate may rise to the top of the specimen, and the sheared zone becomes richer (as far as the mortar content is concerned), creating an apparent decrease of Bingham's parameters.

Third, there can be 'abnormal' chemical activity in fresh concrete. Cement in contact with water always develops some hydrates, and even during the so-called 'dormant' period, some chemical phenomena occur (if not, the mixture would never set). Moreover, in the presence of superplasticizer, the gypsum added in the Portland cement has sometimes difficulties in controlling the early hydration of C_3A (calcium aluminate). Consumption of the adsorbed superplasticizer occurs. The effect on the rheological behaviour is the same as the one observed when the initial amount of admixture decreases in a fresh mix: the yield stress increases, while the plastic viscosity is hardly affected. All these effects are summarized in Fig. 5 & 6.

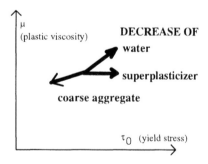

Fig. 5. Effects of changes in initial dosages on the rheological behaviour of fresh concrete (inspired from [10]).

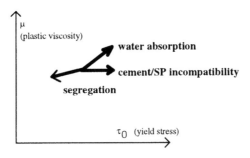

Fig. 6. Effects of different processes on the evolution of apparent rheological parameters with the BTRHEOM rheometer.

4.2 Typical behaviours and interpretation

From the considerations presented above, a guide for interpreting the changes in rheological behaviour may be deduced, given in Tab. 2.

Tab. 2. Evolution of rheological parameters during the period of use of fresh concrete. Interpretation and remedies.

Case	Yield stress	Plastic viscosity	Slump	Top surface of sample	Interpretation	Remedy
I	$\rightarrow$	$\rightarrow$	$\rightarrow$	-	Stable mix	-
II	$\uparrow$	$\rightarrow$	$\downarrow$	-	Chemical activity	Add a retarder or change of cement/SP system
III	$\uparrow$	$\uparrow$	$\downarrow$	-	Water absorption	Pre-saturate aggregates
IV	$\downarrow$	$\rightarrow/\downarrow$	$\rightarrow/\downarrow$	coarse agg. rising	Segregation between mortar and coarse aggregate	Change the grading of aggregate or add a viscosity agent

Let us now examine some typical examples of the behaviours presented in Tab. 2. Most examples presented hereafter are high-performance concretes (i.e. fluid mixtures having low water/binder ratios).

In Fig. 7, the case of a stable concrete (case I) is presented. This mixture did not exhibit any significant rheological evolution during the first hour. This was the result of a thorough mix-design study, related elsewhere [11].

In the same study, a case of incompatibility between cement and superplasticizer was found (see Fig. 8). Later, the increase of yield stress was diminished with the help of a retarding agent, but a stable behaviour could not be reached, even with high dosages of retarder.

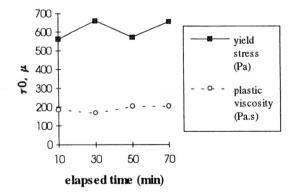

Fig. 7. Evolution of Bingham's parameters, for a mixture corresponding to case # I in Tab. 2 (stable behaviour).

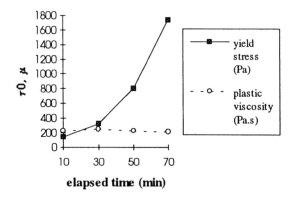

Fig. 8. Evolution of Bingham's parameters, for a mixture corresponding to case # II in Tab. 2 (cement/SP incompatibility).

Before a suitable mixture was found (Fig. 7), some preliminary batches were produced with dry aggregates. A crushed limestone with a 1-1.5 % water absorption was used. Therefore, some 10-30 litres of water could migrate from the paste into the aggregate during the first hour. The effect may be noted in Fig. 9. Later, the mixing process was changed: aggregates with water were first introduced in the mixer, pre-mixed together and let rest for 30 minutes; then, other materials were introduced and the mixing was completed. No significant increase of plastic viscosity was found anymore.

In another study, normal-strength fluid concretes with a water/cement ratio of 0.55/0.60 were to be produced with a fluid consistency. However, a gap-graded aggregate was used, which made the mixtures prone to segregate. While this proneness was not apparent in the mixer, it was possible to detect this hazard with BTRHEOM, not only from the aspect of top concrete surface, but also from the trend observed on Fig. 10.

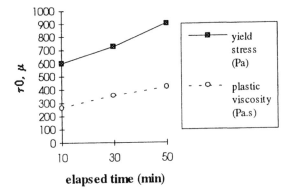

Fig. 9. Evolution of Bingham's parameters, for a mixture corresponding to case # III in Tab. 2 (water absorption of aggregate).

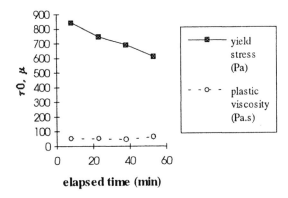

Fig. 10. Evolution of Bingham's parameters, for a mixture corresponding to case # IV in Tab. 10 (segregation between mortar and coarse aggregate).

4.3 Does the shear motion during testing influence the rheological behaviour of fresh concrete?

A question often arises about the way of assessing rheological evolution with BTRHEOM: could the fact of shearing concrete from time to time significantly affect the material ageing, compared with what would happen in a sample remaining at rest?

A first argument is to refer to the continuous agitation any concrete suffers in a truck during transportation. To this extent, it is more representative of the industrial process to shear the sample than to keep it a rest. Another way of facing this question is to compare the evolution of yield stress measured by BTRHEOM with the trend observed on successive slump tests (performed on different samples). Such an experiment has been recently performed at LCPC, on a fluid mixture having a water/cement ratio of 0.55 (see Fig. 11). In Fig. 12, it can be seen that the two quantities are closely related. The linear correlation found is not far from the theoretical relationship established by Hu [3, 7]. If the shear process had changed the thickening phenomenon in BTRHEOM, the increase of the yield stress would have been less.

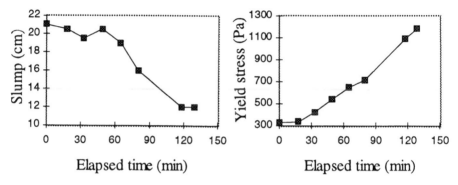

Fig. 11. Evolutions of slump and yield stress during 2 hours, for a fluid mixture.

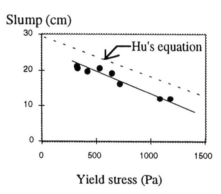

Fig. 12. Correlation between yield stress and slump for the same mixture.

An even more direct verification has been made in the following experiment, dealing with a fluid high-performance concrete [3]. Here, in addition to comparing the evolution of slump and yield stress, another batch was prepared, and successive samples were tested with BTRHEOM. The yield stress values are not consistently different, as can be seen in Fig. 13.

5 Conclusion

The question addressed in the present paper was to propose a methodology for an easy and thorough assessment of the losses of workability of fresh superplasticized concrete after mixing. It has been shown that

- the loss of workability may originate in two main phenomena, having different rheological signatures: water absorption by the aggregate, or chemical activity of the cement in combination with the superplasticizer;
- segregation between mortar and coarse aggregate may counteract the effect of the above-mentioned phenomena;

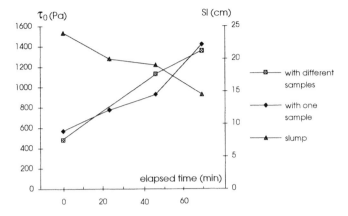

Fig. 13. Evolution of slump and yield stress, for measurements performed on several samples or one single sample.

- the BTRHEOM rheometer is suitable for assessing the evolution of rheological parameters, the cause and the remedies of which most often can be deduced;
- experiments have been related dealing with the contribution of the seal to the bulk measurements of the apparatus, from which a model has been calibrated. This model will be implemented in the test control software, so that neat values of the rheological parameters will be directly displayed for the user;
- the fact of monitoring the rheological evolution with only one sample does not influence the measurement significantly.

Finally, rheometers like BTRHEOM are precious instruments not only for research, but also for practical studies and quality-control measurements. They give much more information than conventional empirical tests, lowering the material and personnel expenses. Also, the information provided is more objective, as the test is fully automated and computer-controlled.

6 References

1. de Larrard F., Szitkar J.C., Hu C., Joly M. (1994), "Design of a rheometer for fluid concretes", Proceedings of the International RILEM workshop *Special Concretes: Workability and Mixing*, P.J.M. Bartos ed., E & FN SPON, pp. 201-208.
2. de Larrard F., Hu C., Sitzkar J.C., Joly M., Claux F., Sedran T. (1994), "Le nouveau rhéomètre LCPC pour bétons très plastiques à fluides" (The new LCPC rheometer for soft-to-fluid concretes), *Annales de l'Institut Technique du Bâtiment et des Travaux Publics*, N° 527, October, pp. 17-47 (in French).
3. Hu C.(1995), "*Rhéologie des bétons fluides*" (The Rheology of Fluid Concretes), Doctoral Thesis of Ecole Nationale des Ponts et Chaussées, January. Etudes et Recherches des LPC, OA 16, LCPC, Paris (in French).
4. de Larrard F., Hu C., Sedran T. (1995), "Best Packing and Specified Rheology: Two Key Concepts in High-Performance Concrete Mixture-Proportioning", *CANMET Adam Neville Symposium*, Las Vegas, June.

5. Hu C., de Larrard F. (1996), "The Rheology of Fresh High-Performance Concrete", *Cement and Concrete Research*, Vol. 26, No. 2.
6. Hu C., de Larrard F., Sedran T. (1996), "A new rheometer for High-Performance Concrete", BHP 96, *4th International Symposium on the Utilization of High-Strength/High-Performance Concrete*, Paris, May,.
7. Hu C., de Larrard F., Sedran T., Boulay C., Bosc F., Deflorenne F.(1996), "Validation of BTRHEOM, the new rheometer for soft-to-fluid concrete", to appear in *Materials and Structures*, RILEM, Vol. 29, No. 192, October.
8. de Larrard F., Hu C.,Sedran T., Sitzkar J.C., Joly M., Claux F., Derkx F., "A New Rheometer for Soft-to-Fluid Fresh Concrete", accepted for publication in *ACI Materials Journal*.
9. Tattersall G.H. (1991), *Workability and Quality Control of Concrete*, E & FN Spon, London, 262 p.
10. Gjørv O.E. (1992), "High-Strength Concrete", *Proceeding of the ACI/CANMET International Conference on Advances in Concrete Technology*, pp. 19-82, Athens.
11. de Larrard F., Gillet G., Canitrot B. (1996), "Preliminary HPC mix-design study for the 'Grand Viaduc de Millau': an example of LCPC's approach", BHP 96, *4th International Symposium on the Utilization of High-Strength/High-Performance Concrete*, Paris, May.

34 A NEW WORKABILITY TEST ON CONSOLIDATION-FREE FLOWING (CFF) CONCRETE

S. KAKUTA
Department of Civil Engineering, Akashi College of Technology, Akashi, Japan
T. KOKADO
Technical Development Bureau, Nippon Steel Co., Shintomi, Japan

Abstract
This study was conducted to offer a new rheological test method for consolidation-free flowing concrete (CFF concrete). This concrete is characterized as high fluidity, non-segregation and self-levelling in the fresh state. Rheological properties of paste and mortar composed in CFF concrete were tested by using a viscometer. Furthermore, the rheology of the concrete was measured by using a coaxial cylindrical viscometer. In the next step, a spiral-flow twin-shaft mixer fitted with a rotating speed controller and the electric power of the mixing motor, connected to a measuring system, was used to obtain the relationship between mixing power and rotating shaft speed. The results of mixer tests indicated that the rheological quantities could be obtained. Test results indicated that the CFF concrete has a small yield value and higher viscosity than normal concrete. The measurement of rheology using the proposed mixer test is a useful method of controlling the workability of CFF concrete.
Keywords: Rheology, coaxial cylinder viscometer, consolidation-free, fresh concrete, superplasticizer.

1 Introduction

The consolidation-free flowing (CFF) concrete was developed as a labour cost saving concrete in Japan[1]. This concrete is characterized as high fluidity, non-segregating and with an ability to self-level when fresh. This concrete is more effective when used with heavy reinforcement or in composite structures of steel and reinforced concrete.

The application of this type of concrete to complicated sections of concrete structures is increasing. The high fluidity, which is one of the properties, is produced by use of a superplasticizer at more than about two percent by weight of cement. Lack of

Production Methods and Workability of Concrete. Edited by P.J.M. Bartos, D.L. Marrs and D.J. Cleland. Published in 1996 by E & FN Spon, 2–6 Boundary Row, London SE1 8HN. ISBN 0 419 22070 4.

segregation comes from sticky paste that is produced by a viscosity increasing agent of aqueous polymer or a very fine powder such as ground granulated blast-furnace slag.

In Japan, at present, this type of fresh concrete is controlled by the slump flow value at field. It is the average diameter of the spread of concrete after the slump test. Results from the slump or the slump-flow test, however, cannot properly evaluate the flowing behaviour of this concrete. Workability of this type of concrete should be evaluated by rheology. The first stage of this study deals with rheology tests on pastes and mortars using a B-type viscometer. The rheology of concrete was measured by a coaxial-cylinder viscometer for concrete. In the second stage, the relationship between electric load of a mixer and the rotating speed of the mixer shaft was investigated. A spiral flow type twin shaft mixer was installed with both a rotating speed controller and an electric motor power measuring system. This was used to obtain the relationship between mixing electric power and mixing speed. The results of these tests were closely related to the results of the rheology tests obtained using a viscometer.

Test results indicated that CFF concrete has a very small yield value and higher viscosity than normal concrete. It was found that the mixer, with a rotating speed controller and an electric power measuring system, is suitable for use as a workability control for this concrete.

2 Experimental

2.1 Test apparatus
A Brookfield type (B-type) viscometer was used to obtain the flow curves of CFF paste and CFF mortars.

A coaxial-cylinder viscometer for concrete[2] was used to measure the flow curve of CFF concrete. The inner cylinder rotating coaxial viscometer is not a satisfactory rheometer for use with concrete because segregation of aggregate will occur both in the vertical and centrifugal directions. In addition, slip will be induced by a weak mortar layer at the interface of the surface of inner cylinder and concrete. For these reasons, the rheology factors estimated for ordinary concrete will be smaller than the true values Since CFF concrete does not segregate, the coaxial-cylinder viscometer is useful to obtain flow curves and rheology factors.

Provided the CFF concrete behaves as a Bingham fluid, rheology factors are given by equation (1).

$$\frac{d\omega}{dM} = \frac{1}{4\pi R^2 h \mu} - \frac{\tau_y}{(2\mu M)} \tag{1}$$

where ω is angular velocity,
M is torque of inner cylinder,
R is radius of inner cylinder,
h is depth of concrete,
μ is plastic viscosity,
τ_y is yield value

Details of the coaxial-cylinder viscometer for concrete are shown in Fig.1. The inner cylinder has a diameter of 200 mm and a depth of 24 mm. The outer container has a diameter of 300 mm and is 320 mm deep.

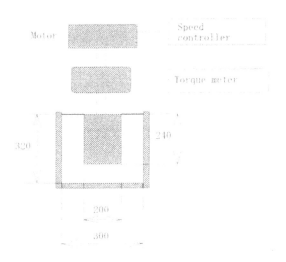

Fig. 1. Set up of a coaxial-cylinder viscometer for concrete.

2.2 Twin shaft mixer
A spiral flow twin shaft mixer with a capacity of sixty litres was used and was remodelled for the experiment. It was fitted with a rotating speed controller for the twin-shaft and an electric power meter link to motor. The rotating speed of the twin shaft relates to the shear velocity, and the electric power relates to shear stress during concrete mixing. The rotating speed and the electric power of the motor must agree with the result of a concrete rheology test by a coaxial-cylinder viscometer. The set-up of the twin shaft mixer is shown in Fig. 2.

3 Materials and mix proportions

3.1 Cement, blast-furnace slag and aggregates
Ordinary portland cement, a blast-furnace slag (BFS), a river sand and a crushed stone were used in the experiment. Details of materials are given in Table 1. The blast-furnace slag was used to increase fluidity and reduce segregation. The reduced segregation arises from the use of the very fine granulated powder. In this study, a 6000 Blaine of BFS powder was used. The physical and chemical properties of the BFS are given in Table 2.

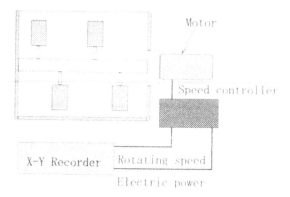

Fig. 2 Set up of the remodelled twin-shaft spiral-flow mixer

Table 1. Details of materials

Material	Type	Details
Cement	Ordinary portland cement	Specific gravity = 3.16
Fine aggregate	River sand	Absorption = 2.3%, FM. = 2.81 Specific gravity = 2.53,
Coarse aggregate	Crushed stone	M.S. = 20 mm, Specific gravity = 2.61, Absorption = 0.9%, F.M. = 7.13
Blast-furnace slag	6000B	Specific gravity = 2.88, Specific surface = 6190 cm^2/g

Table 2. Chemical compositions of blast-furnace slag (%)

SiO$_2$	Al$_2$O$_3$	FeO	CaO	MgO	SO$_3$
31.6	12.8	0.3	43.4	6.3	2.1

3.2 Admixtures
The high fluidity of the CFF concrete comes from the use of a large dosage of super-plasticizer. Three types of superplasticizers were used to compare their effect on CFF concrete. The admixtures used are listed in Table 3.

Table 3. Admixtures

SP	Type
A	Aromatic aminosulfonic acid based AE superplasticizer
B	Naphthalene based superplasticizer
C	Acrylic polycarboxylate based AE superplasticizer

3.3 Concrete mixes

Mortars for the rheology tests to determine the effect of the sand volume in paste were mixed by varying the sand/binder ratio from zero to 1.5; the OPC/BFS ratio was kept constant at 0.43.

Mix proportions of the CFF concrete and the normal mixture concrete are given in Table 4[3].

Table 4. Mix proportions of consolidation-free flowing concrete

Concrete Mixes	W/(C+B) (%)	W	Unit Contents (kg/m^3)				
			OPC	BFS	S	G	AESP
Normal	55.0	201	364		904	890	
CFF	31.6	179	179	387	807	794	9.6

4 Test procedures

4.1 Mortar test

Mortars were mixed in accordance with JIS R 5201 Strength Test of Cement. The flow curves of 1000 ml of mortars were measured by the B-type viscometer. The rotating speeds were 2, 4, 10 and 20 r.p.m.. The rheology factors were calculated by using Bingham's model equation.

4.2 Concrete test

The flow test was done before the rheology tests. To obtain the flow curve of CFF concrete, a coaxial-cylinder viscometer for concrete was used. The measured rotating speeds of the inner cylinder were 20, 40, 60 and 80 r.p.m. respectively. The speed of the inner cylinder was changed from zero to a maximum of 80 r.p.m. and then down to zero. Torque was measured at each rotating speed. The torque and rotating speed were recorded by an X-t recorder.

The mixer test was carried out at the same time as the rheology test. Mixing speeds were 16.2, 24.0, 33.2, 41.5, 50.1, 58.4 and 67.0 r.p.m. The electric power of the motor was measured by a power meter and was recorded by an X-Y recorder.

The specimens for compressive strength test of CFF concrete were made in the following manner. The CFF concrete was poured into a ϕ 10 × 20 cm mould without any consolidation. The specimens were moist after two days in a curing room at temperature of 20 ± 3° C and at 95% RH. The compressive strength tests were carried out at 28 days and the unit weights were determined at 2 days and 28 days.

5 Results and discussions

5.1 Mortar test

The flow curves of CFF mortars were obtained from the rheology test. Fig. 3 shows that the sand to paste ratio significantly affects the flow curves. This figure shows that large viscosities of CFF mortar can be obtained with a sand/binder ratio exceeding 1.25. The yield values and plastic viscosities of CFF mortars are given in Table 5.

Table 5 Rheology factors of CFF mortars

Sand/binder ratio	0	0.50	0.75	1.00	1.25	1.50
	(0.000)*	(0.227)	(0.306)	(0.368)	(0.424)	(0.469)
Plastic viscosity (Pa · sec)	0.085	0.231	0.328	0.598	0.680	0.731
Yield value (Pa)	0.050	0.133	0.520	1.508	2.450	5.470
Relative viscosity	1.00	2.72	3.86	7.04	8.02	8.60

* Numerical values in parentheses mean volume concentration of cement.

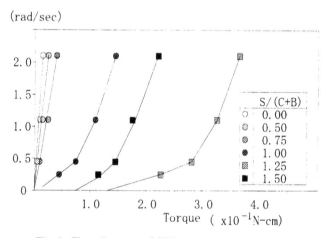

Fig. 3. Flow diagrams of CFF mortars.

5.2 Concrete test

The rheology of CFF concrete was measured for a given set of mix proportions with different kinds of superplasticizers. Table 6 gives flow values and rheology factors of CFF concrete. Type A concrete contains an aromatic aminosulfonic acid based AE superplasticizer. Type B contains a naphthalene based superplasticizer. Type C contains an acrylic polycarboxylate based AE superplasticizer. Results from the experiment indicate that the plastic viscosity of the CFF concrete is very large and the yield values are very small in comparison with normal concrete. These data indicate that a small yield value leads to a self-levelling and flowing concrete, and a large viscosity leads to non-segregation of the CFF concrete. These results indicate that different superplasticizers in CFF concrete react differently. The flow values do not correctly indicate this viscous flow property. To understand the workability of CFF concrete, it is necessary to use an adequate test such as a rheology test.

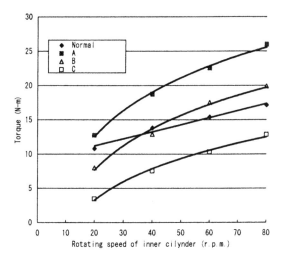

Fig. 4 Flow curve of CFF concrete

Table 6 Rheology factors of CFF concrete

Type of superplasticizer	Normal	A	B	C
Slump flow (cm)	29.6	78.0	80.7	66.0
Plastic viscosity (Pa·s)	0.060	0.346	0.258	0.158
Yield value (Pa)	0.501	0.331	0.090	0.058

5.3 Mixer test

As described above, some practical simplified rheology test method is needed for controlling this concrete in the fresh state. The electric power of the motor and rotating speed of the twin shaft mixer relationshp is shown in Fig. 3. These curves are similar to the flow curves of Bingham fluid in the rheology test for concrete. The workability of CFF concrete must be controlled by the value of a curve crossing the electric power axis, called yield electric power Ey, the related yield value and a gradient of (electric power)/(rotating speed), called velocity gradient D, and the related viscosity. In comparison with normal concrete and CFF concrete, the CFF concrete curves indicate same tendency based on the rheology test results. The relationship between mixer test and rheology test is shown in Fig. 4. These relations can be roughly estimated. Provided these relations can be measured before concreting, workability quality control of CFF concrete becomes easier in the field.

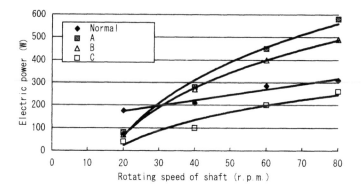

Fig. 5. Relationship between electric power and rotating speed of shaft.

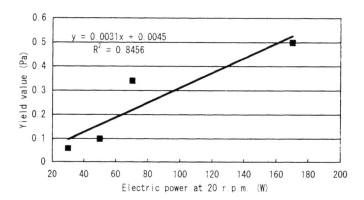

Fig. 6. Relationship between yield value and electric power at 20 r.p.m.

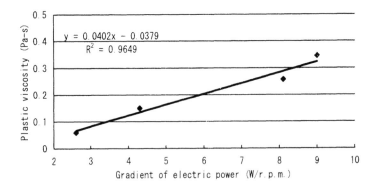

Fig. 7. Relationship between gradient of electric power and plastic viscosity.

5.4 Hardened concrete

Unit weights and compressive strengths of CFF concrete at 14 and 28 days are cured shown in Table 7. Specimens produced without consolidation performed almost normal concrete. Even without consolidation, this concrete has good quality.

Table 7. Compressive strength and unit weight of CFF concrete

Concrete mixes	Compressive strength (MPa)		Unit weight (kg/m^3)	
	Age (days)			
	14	28	14	28
Normal concrete	365	406	2360	2360
Type A CFF	340	444	2380	2380
Type B CFF	384	462	2350	2350
Typce C CFF	322	436	2360	2360

6 Conclusions

1. Consolidation-free flowing concrete was produced by mixing about 68% BFS and 32% cement with 1.7% by weight of cement and AE-superplasticizer.
2. The workability of consolidation-free flowing concrete can be determined in rheology test by a coaxial viscometer for concrete. Results from cylindrical rheology test indicated that fluidity of CFF concrete is affected by superplasticizer.
3. A twin shaft spiral flow mixer fitted with a rotating speed controller and an indicator of electric power of motor can control the workability of a CFF concrete by utilizing an empirically developed relationship between rheological properties and mixer test results.
4. Even without consolidation, a hardened CFF concrete has almost the same performance as normal concrete, insofar as strength and unit weight are concerned.

7 References

1. 27 papers are published in *Proceedings of Japan Concrete Institute Symposium on Super-Flowable Concrete*, (1993) JCI Tokyo, Japan.
2. Murata, J. and Kikukawa H. (1973) *Studies on Rheological Analysis of Fresh Concrete; Important Properties and their Measurement*, Proceedings of a RILEM Seminar, Vol. 1, pp. 1.2-1 to 1.2-33.
3. Kokado, T. and Miyake, M. (1991) *Basic Properties of a Slag Based Consolidation-free Concrete and its Filling Test*, Proceedings of the Japan Concrete Institute, No. 13-1, pp. 875-880.

35 ASSESSMENT OF WASHOUT RESISTANCE OF A FRESH CONCRETE BY THE MC-1 TEST

M. CEZA and P.J.M. BARTOS
Advanced Concrete Technology Group, Department of Civil, Structural and Environmental Engineering, University of Paisley, Paisley, Scotland, UK

Abstract
Underwater concrete which does not disperse significantly when placed in water was tried first about 10 years ago. This type of underwater concrete is usually known as a " non-dispersable concrete" or an " antiwashout underwater concrete ". The admixture used to produce this special concrete is often marketed as a "nondispersable underwater concrete admixture" or an "antiwashout admixture ".

Several test methods and equipment have been developed previously to asses the required properties of the fresh underwater mixes which include devices to evaluate their washout resistance such as the Plunge test However one of the problems facing both producers and users of underwater (UW) concrete admixtures up to now has been an absence of a standardised and realistic test method for the assessment of performance of such products in a practical concrete construction. This paper shows results of a research project in which an efficient new test for the practical evaluation of the washout resistance / non-dispersability of such fresh concrete mixes has been developed. Other properties of the underwater concrete such as workability, setting time and compressive strength have been also investigated.

The new MC-1 test designed at the Advanced Concrete Technology Group at the University of Paisley by Ceza is introduced. The principle of the new test, the apparatus required and the test procedure are described. Results of tests on practical underwater mixes are shown and effects of UW admixtures are discussed.

Keywords : non-dispersable concrete mix, antiwashout underwater mix, underwater concrete,nondispersable underwater concrete admixture, antiwashout admixture, Plunge test

Production Methods and Workability of Concrete. Edited by P.J.M. Bartos, D.L. Marrs and D.J. Cleland. Published in 1996 by E & FN Spon, 2–6 Boundary Row, London SE1 8HN. ISBN 0 419 22070 4.

1 Introduction

Admixtures for concrete placed underwater have been developed to a stage where they are now widely used. The use of this type of product allows concrete to be placed underwater with far less risk than has been previously possible. Such concrete should conform to appropriate requirements when fresh, eg. workability, washout resistance etc., as well as when hardened, eg. strength, durability etc. Several test methods and devices to evaluate the required properties of the fresh mixes have been proposed before. These include the washout resistance of the concrete mix, a very important requirement for underwater placing. One of the problems facing both producers and users of underwater (UW) concrete admixtures up to now has been an absence of a standardised and realistic test method for the assessment of the washout resistance of the concrete mix. This research project aims to develop efficient new tests for the practical evaluation of the washout resistance / non-dispersability of fresh concrete mixes placed underwater.

2 Previous tests

The tests previously proposed for the assessment of the washout resistance of fresh concrete mix can be separated into categories according to the basic principles adopted and the form of the results, namely :

A a. Direct tests : The concrete sample is in contact with water during the test.
   b. Indirect tests: The assessment is based on a determination of parallel properties of the mixes, such as viscosity etc.

B a. Tests which produce quantitative results in SI or other numerical units of measurement.
   b. Tests which evaluate the mixes by a visual qualitative classification only.

Adopting the classification mentioned above the existing test method fall into the following categories :

Direct tests - Stream test, Drop test, pH factor test, Plunge test
Indirect tests - Orimet test, rheometers (generally)

The most widely used test method at present to measure the performance of the UW admixtures commercially available is the "Plunge test".

The Plunge test together with the pH factor test and the new proposed MC-1 test are the only existing direct tests which provide a numerical value of the test result. The Plunge test is in a category Aa and Ba/b.

The basic test procedure consists of a concrete sample being placed in a wire mesh basket which is then allowed to drop freely several times through a selected height of water, see Fig. 1.

The exact test procedure has varied from one research centre to another. In each case a different size of the basket and a different diameter (3,5,20 mm) of the holes in the wire mesh have been used. The number of 'plunges' of the basket through the water has also varied from three to five [1][2][3].

The test result is the mass of the concrete sample lost after the required number of 'plunges' usually expressed as a percentage of the original mass.

The Plunge test has provided the performance parameters for assessments of effectiveness of most of the UW admixtures currently available.

There are fundamental problems associated with the test :

(a) The wire mesh of the basket restrains the concrete sample. The test is thus
     likely to indicate a 'washout' smaller than in a real situation [1][3].
(b) Fresh concrete placed underwater must have a very high workability.

A proportion of the test sample will therefore flow out from the basket without it being subjected to a washout.

Any reduction in the size of the holes in the mesh of the basket will only increase further the restraint of the concrete sample, as mentioned in (a) above. The assessment of the washout by this test does not relate to practice placing and its results become inherently unreliable.

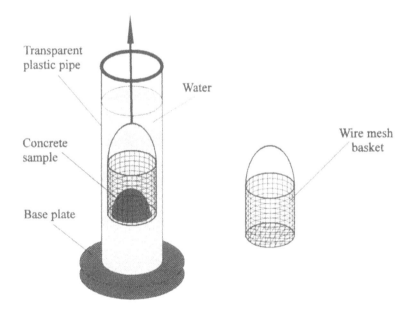

Fig. 1. The Plunge test

A previous review of the authors [4] indicates that until now no apparatus has been developed which could unambiguously and quantitatively (e.g. in SI units) measure the resistance to washout of the fresh concrete mix.

Recent research [4][5] shows that viscosity has an influence on the non-dispersability of fresh concrete mixes but this effect is not a direct one [2]. In such a case the indirect method can be useful as only a quick site test.

3 Development of the new test

The new test apparatus is based on the principle of a direct determination of the resistance to washout of a concrete mix which produces quantitative results. The test is in a category Aa,Ba.

3.1. Basic principles of the new test method
• A sample of concrete is put into the test apparatus and weighed before it is washed out by water.
• The duration of the washing-out is determined by preliminary trials. The concrete sample is weighed continuously throughout the test.
• The test result is the ratio of the mass of the sample before and at the end of the washing out.
Fig. 2 illustrates the principle on which the new "MC-1" test for washout resistance has been designed.

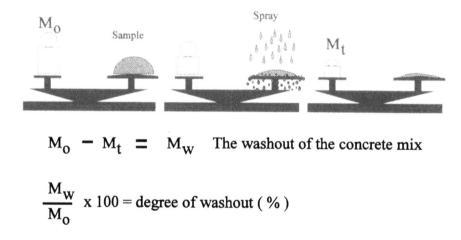

$$M_0 - M_t = M_w \quad \text{The washout of the concrete mix}$$

$$\frac{M_w}{M_0} \times 100 = \text{degree of washout (%)}$$

Fig. 2. Principle and results of the new test method.

3.2 Prototype of the test apparatus
The test apparatus consists of a container containing approx. 30 l of water with a pipe connected at the bottom and an overflow at the top. The pipe leads water into the spray-head above the test sample. The test sample is placed on a frame which is suspended from an electronic balance. The balance itself is supported on a bench.

For the purpose of assessing the performance of different test arrangements the prototype of the test device has been designed to permit setting out of different horizontal and vertical distances (angle) between the spray head and the sample using different water tanks, water supply pressures and volumes, and different designs of the spray head. A diagram on Fig. 3 shows the main parts of the MC-1 test apparatus.

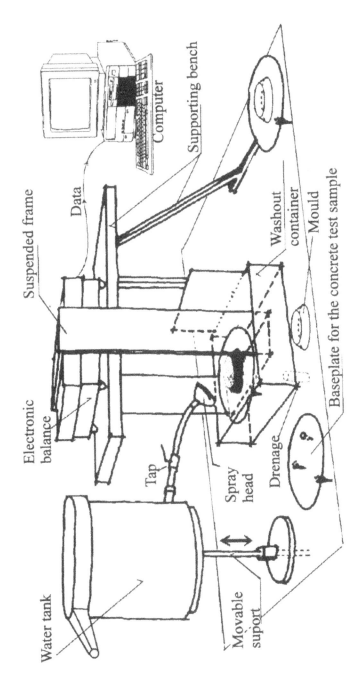

Fig. 3. Prototype of the new test apparatus.

3.3 Test procedure

The test procedure is simple. A mould is placed onto an impervious baseplate and filled with a sample of a concrete mix, see Fig. 4. The mould is removed and the baseplate with the sample is placed on the frame suspended from the balance. The computer is connected to the balance and it is set into a mode in which an increase of the pressure on the baseplate by the water spray will automatically switch on the recording of the washout process. The tap on the pipe connecting the water tank with the spray-head is turned on and thus the recording begins automatically. Water from the spray head is normally allowed to wash out the sample for 4 minutes with a constant head of water being maintained in the container, see Fig. 5. The duration of the standard test procedure was established following trial test series on different types of concrete.

The measurement recorded directly from the balance in every two seconds is a mass of the sample resting on the 'plate' and the mass and pressure of the water which is poured onto the sample. A net amount of the lost material is then obtained as the direct measurement received by the balance less the effect of the pressure and mass of the poured water see Fig. 6. The pressure and mass of the poured water have been previously established in a 'calibration' procedure in which a hardened concrete mock-up of the test sample was used instead of the fresh concrete.

The final results are diagrams showing the loss of mass due to the washout during the test. The 'washout ' is expressed as a percentage of the original mass, see Fig. 2.

Fig. 4. Baseplate and a concrete sample before the test procedure

Fig. 5. The ' washout ' in progress

Table 1. Proportions of concrete mixes

Composition :				R	RU	RUS
Free-water content	w/c - 0.5		195 kg			
Cement content			390 kg			
Fine aggregate	45%		794 kg			
Coarse aggregate	55%		971 kg			
Total aggregate content			1765 kg			
Admixtures - 1% of cement			3.9 kg			
Superpl. 1l / 50kg cem			7.8 l			

3.4 Typical test results

A comparison of washout resistance of three types of a concrete mix is shown on Figs 7a,b. The proportions of the mixes used are provided in Table 1. The following identification codes were adopted for the mixes tested:

" R "   - control mix, reference sample
" RU "   - mix " R " as above + an underwater admixture " S "
" RUS "   - mix as " RU " + superplasticiser

Three tests have been done carried out for each type of the mix. The results are plotted on Figs 7a,b. The diagrams show a clear a difference between washout resistance of each type of the concrete mix. The three curves for each of the mixes used are relatively close together. The deviations between the curves inherently increase when more mass has been washed out from the original sample. However, this variation does not have a significant effect on the final evaluation of the sample. Figs 7a,b, 8 and Table 2 show a comparison between the results of a slump test and the washout test.

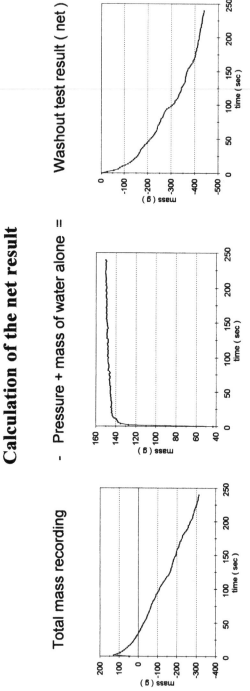

Fig. 6. Calculation of the net washout test result

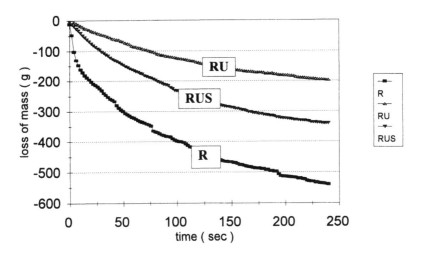

Fig. 7a. Net washout test result

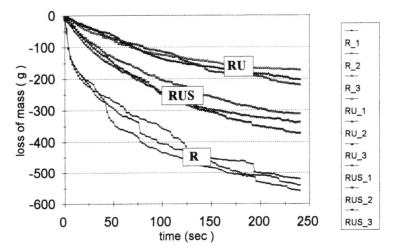

Fig. 7b. Washout test result (average of three tests each)

Table 2. Practical test results

	R Control.sample	RU UW admixture 1%	RUS UW ad. 1%+superpl
washout %	52	21	38
slump (mm)	170	(220)	(250)

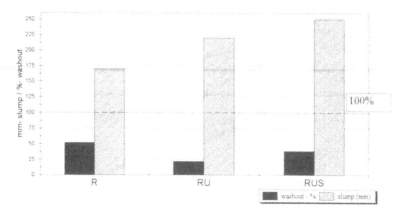

Fig. 8. A comparison between slump and washout

4 An effect of dosage of UW admixtures

Following the successful completion of the prototype testing stage the research project has moved into a large-scale testing stage aimed to evaluate the complete behaviour of the underwater mix and its relationship with other workability / rheology characteristics.

The MC-1 test has shown a good potential for its development into a standard test method for assessment of washout resistance of the non-dispersable underwater mixes which is required by the industry worldwide. An assessment of the behaviour of two UW admixtures available on the market is shown below.

A comparison of washout resistance of four types of concrete mixes is shown on Table 3 and Figs 9,10. The basic proportions of the mixes used are given in Table 1. The following identification codes were adopted for the mixes tested :

0%A   control mix sample 0 percent of "A" type UW admixture
0.5%A   mix as 0%A above + 0.5 percent of "A" type UW admixture by weight of cement
1%A   mix as 0%A above + 1.0 percent of "A" type UW admixture by weight of cement
1.5%A   mix as 0%A above + 1.5 percent of "A" type UW admixture by weight of cement
2%A   mix as 0%A above + 2.0 percent of "A" type UW admixture by weight of cement

Table 4 and Figs 11,12 show results of using a " B " type UW admixture. The proportions of the mix were identical to the previous mix. The identification codes are (0%B, 0.5%B, 1%B, 1.5%B, 2%B).

The graphs show clearly the differences between each type of the concrete mix and the admixture. The washout resistance increases in both cases with an increase of the UWA dosage. Different types of curves and different final results are obtained by the MC-1 test for admixtures " A " and " B ".

The admixture " A " in a dosage range of up to 1 % increases slump, in the range of 1- 1.5% it has an opposite effect. A further increase of the dosage increases the slump. The range of dosage from 0.5-1% has no significant effect on the final washout result. However, diagram on Fig.10 shows that the mix with a dosage of 1%

is more washout resistant at the beginning of the test compared with the mix of 0.5% dosage of admixture " A ".

The admixture " B " does not have a significant influence on the slump test, only the recommended dosage of 1% and an overdosage of 2% produce a slightly higher slump. The washout resistance increases steadily with an increase of the dosage of the UW admixtures almost regularly.

Table 3. Numerical results of the dosage variation of an admixture " A "

Dosage (in %) by weight of cement	0 %A	0.5 %A	1 %A	1.5 %A	2 %A
washout %	52.5	24.0	21.6	6.6	0.6
slump (mm)	170	(205)	(220)	115	(195)

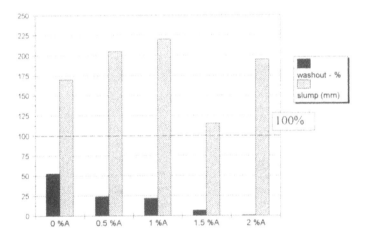

Fig. 9. A comparison between slump and MC-1 washout test result

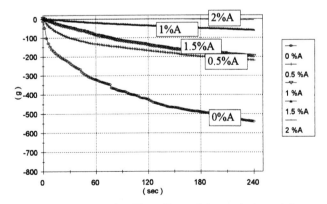

Fig. 10. MC-1 washout test results. The effect of the variation of dosage
     of the admixture " A "

Table 4. Numerical results of the dosage variation of an admixture " B "

Dosage (in %) by weight of cement	0 %B	0.5 %B	1 %B	1.5 %B	2 %B
washout %	52.5	21.8	14.4	5.0	1.5
slump (mm)	170	165	(190)	170	(200)

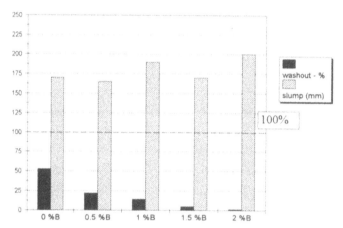

Fig. 11. A comparison between slump and MC-1 washout test result

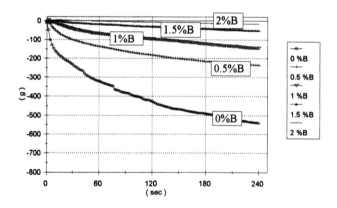

Fig. 12. MC-1 washout test results. The effect of the variation of dosage
        of the admixture " B "

5 An influence of UW admixture on the strength of a cement paste.

The dosage of both admixtures had an effect on the strength of the cement paste. The strength tests have been carried out on 50 mm cubes. The cement paste (w/c = 0.5) was mixed for 2 min, then the admixture " A " or " B " was added and mixed for another 2 minutes. The dosage varied from 0 to 1.0 percent by weight of cement. The hardened cubes were crushed after 7, 14 and 28 days. Diagrams on Fig. 13, Fig. 14 show the results. Both admixtures when their are added reduce significantly the compressive strength. The reduction of the strength is not directly related to the dosage of UW admixtures. The same conclusions have been reached for concrete by Barry [6] and Neeley [7] .

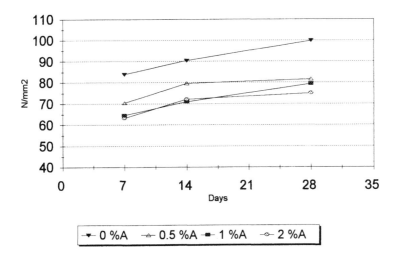

Fig. 13. Effects of a variation of dosage of admixture "A" on compressive strength of cement paste

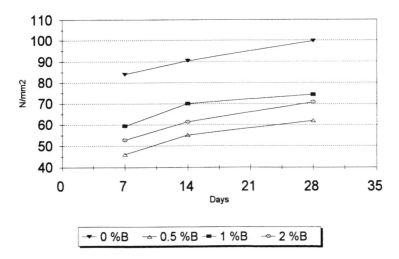

Fig. 14. Effects of a variation of dosage of admixture "B" on compressive strength of cement paste

6 Conclusions

The UW concrete is a relatively recent addition to special concretes used in construction practice. The behaviour of fresh UW concrete mixes differs considerably from that of ordinary mixes and is difficult to assess and predict. It is therefore necessary before using such a product on a large scale to ensure its required properties when it is fresh and hardened. Many of the existing tests for concrete mixes are not appropriate for such a viscous material. The MC-1 test was designed to provide means of an effective evaluation of the UW admixtures used to provide the washout resistance of fresh concrete.

Compared with the washout tests proposed previously the new "MC-1" test offers the following advantages:

- The test procedure is simple, logical and realistic.
- The mass of the test samples (1kg) is convenient.
- The final test result is provided in SI units, however, an additional simultaneous visual assessment of the washout is also available see Fig. 15.
- The test procedure is highly automated with a negligible potential operator error.

The apparatus has been primarily designed as a tool for a realistic assessment of efficacy of UW admixtures in the R&D laboratories of manufactures of admixtures and in independent testing authorities. However, the apparatus in its final form is expected to become more compact and more portable than many of the other washout tests proposed before.

Fig. 15. Visual assessment of the concrete sample after the MC-1 washout test procedure

7 References

1   Davies, B.A. " Laboratory methods of testing concrete for placement underwater " Marine Concrete '86 -International Conference on Concrete in the Marine Environment, London, 22-24 Sept 1986, pp 279- 285

2   Neeley, B. D. " Evaluation of concrete mixes for use in underwater repairs " Technical report of US Army Engineering, Waterway Experiment Station, 1988 REMR- CS-18 pp 1-103

3   Seaton, J. "Underwater Concrete", 5th Year BEng hons. project, Department of Civil Engineering at University of Paisley, Scotland, January, 1991.

4   Ceza, M. and Bartos, P. "A New Test for Washout Resistance of Underwater Concrete" Paper presented at The International RILEM Seminar at CBI in Stockholm, Sweden 7-8. May 1995

5   Bartos, P. and Tamimi, A. "Orimet Test as A Method for Site Assessment of Workability of Fresh Underwater Concrete ", Paper presented at 17th Conference on Our World in Concrete & Structures : 25-27 Aug. 1992, Singapore

6   Barry, W. and Brian, O. " The Role of Polymers in Underwater and Slurry Trench Construction " Proc. 5th International Congress on Polymers in Concrete, 22-24 Sept 1987, Brighton, England, pp 363-368

7   Neeley, B. " Antiwashout Admixtures for Underwater Concrete ", REMR Technical Note CS-MR-7.2, US Army Engineer Waterways Experiment Station, 1991

36 SETTING AND HARDENING OF CONCRETE CONTINUOSLY MONITORED BY ELASTIC WAVES

H.W. REINHARDT
Institute of Construction Materials, University of Stuttgart, and FMPA
BW Otto-Graf-Institute, Stuttgart, Germany
C.U. GROSSE
Institute of Construction Materials, University of Stuttgart, Stuttgart,
Germany

Abstract
A short pulse either generated by an ultrasonic generator or by a steel ball impact travels through a fresh concrete specimen. Three features are measured: the propagation velocity of the compressional wave, the transmitted energy, and the frequency spectrum. All are sensitive to the age of concrete and to the composition of the mix. The varying parameters are water-cement ratio, dosage of retarder and super retarder, and type of cement. The setting and hardening of the concrete can be followed accurately.
Keywords: Fresh concrete, setting, hardening, non-destructive testing, ultrasound, retarder, wave propagation.

1 Introduction

Modern concrete technology faces several challenges: there is a great demand from the design engineer for high-strength concrete, high-performance concrete, fibre concrete; from the contractor for highly workable concrete, self-levelling concrete, slip formed concrete, retarded mixes; there is less workmanship on the construction site available; there is increasing quality required for durable concrete structures in an agressive environment. The materials producers have a basket full of admixtures and additions which are deemed to affect the fresh or the hardened state of concrete. The user is sometimes inclined to combine various products in order to achieve the maximum success. However, not all mixtures lead to the expected result.

An advanced process technolgoy needs proper control by reliable and - as much as possible - objective measurements [1]. One of such methods will be presented and discussed in the following. It concerns the continuous monitoring of the setting and hardening of concrete. These properties are very relevant for the slip-forming process and also for concretes which contain long-time retarders.

Production Methods and Workability of Concrete. Edited by P.J.M. Bartos, D.L. Marrs and D.J. Cleland. Published in 1996 by E & FN Spon, 2–6 Boundary Row, London SE1 8HN. ISBN 0 419 22070 4.

2 Test method

An elastic wave travels faster through a medium the higher the elastic modulus is at constant density. As shown earlier by l'Hermite [2], van der Winden [3], Niyogi et al. [4], Reichelt et al. [5], Keating et al. [6, 7], Boutin and Arnaud [8] the wave velocity depends clearly on the state of concrete and the course of setting and hardening. But not only the velocity changes but also the amplitude and the frequency content of the signal with increasing age of the concrete a has been shown by Jonas [9] and Sayers & Dahlin [10], for mortar. The test set-up (Fig. 1) consists of a container made of 40 mm thick styrene foam (Styropor) plates.

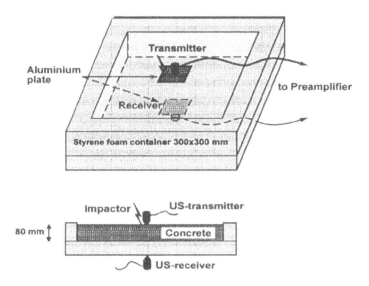

Fig. 1 Set-up with ultrasonic generator and receiver. Dimensions of the container: 300 mm x 300 mm x 80 mm

At the top and the bottom two small aluminium plates were placed in contact with the fresh concrete. US transducers (Geotron Elektronik, UPG-D and UEAE, resp.) touch the aluminium plates. A signal is generated by a USG (Geotron Elektronik) and sent to the upper transducer which triggers the data acquisition system. The transducers are broad-band transducers. The signal is preamplified, digitized and recorded by a transient recorder with a sampling rate of 100 ns and a 12 bit resolution. This test set-up has proven successfull after a precursor failed due to the coupling of the transducers to the container walls. Due to this coupling, waves travelled through the walls and interfered with the waves travelling through the concrete. This new test set-up is used for the measurement of the compressional wave propagation velocity in an automated manner up to every 6 minutes.

When the frequency was to be analysed a signal has been generated by a 4 mm steel ball which was shot from a mechanical gun to the upper aluminium plate. The receiver is a piezo transducer of the UEAE series. This method was superior to the US generator since a larger frequency range could be received and evaluated.

3 Testing materials

The composition of the reference concrete is given in Table 1.

Table 1. Concrete composition

Component		Mass (kg/m³)
Cement CEM I 32.5 R		320.0
Aggregates	0/2 mm	635.3
	2/8 mm	610.7
	8/16 mm	531.9
Water		176.0

The resulting water-cement ratio was 0.55. The concrete strength class is C20/25 (i.e. 25 MPa cube compressive strength at 28 days) and the consistence is KP (which means "plastic" with a spread table value of 350-410 mm).

The second concrete which is the same as the first one contained a commercial retarder (Pozzutec 50 G). The amount was 8, 16 and 25 mg per kilogramm of cement, resp. In a third mix a super retarder (Delvo Stop 10 G) was used with a dosage of 11, 22 and 33 ml per kilogramm of cement, resp.

The water-cement ratio has been varied between 0.40 and 0.55 while the remaining composition was kept alike. The consistence has been controlled by a small amount of plasticizer.

Finally the portland cement has been replaced by a furnace slag cement CEM III/A. The mixes tested are summarized in Table 2.

Table 2. Series of mixes tested

Series number	Important parameter
0	Reference mix acc. to Table 1
1	Commercial retarder Pozzutec 50 G
2	Super retarder Delvo Stop 10 G
3	Water-cement ratio 0.40, 0.45, 0.50
4	Blast furnace slag cement, CEM III/A 32,5 R

4 Testing results and discussion

4.1 Compressional wave velocity
The velocity of the compressional wave is shown in Fig. 2 as function of age. It is clear that the velocity increases continuously with the age of the concrete.

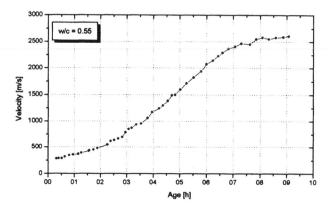

Fig. 2 Velocity of the compressional wave vs. age of concrete

The low value at the beginning is smaller than that of water (1430m/s). L'Hermite [2] has measured 150 to 300 m/s and explained this phenomena by a spring-mass model of many degrees of freedom. Biot's theory of the behaviour of masses in a fluid has also been applied and it was found that this theory can describe this phenomena [10]. After about 9 hours the velocity increases only slowly.

A plot of the results of tests with four water-cement ratios is shown in Fig. 3.

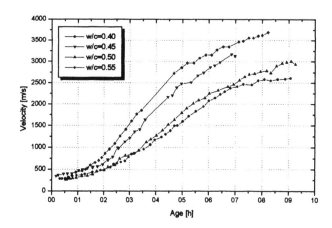

Fig. 3 Velocity of the compressional wave as function of age and w/c ratio

The lines start at about 300 m/s and develop continuously to values between 2500 and 3500 m/s after 8 hours. The lower the water-cement ratio the faster the velocity increases. This feature is well known from measurements of the compressive strength and the elastic modulus of young concrete. If a certain wave velocity is reached the concrete is not workable any more. Van der Winden [3] considered a velocity of 1000 to 1500 m/s as the end of workability. Regarding Fig. 3 the concrete with w/c = 0.40 would reach this after about 2.5 hours and the concrete with w/c = 0.55 at about 4.5 hours.

The dosage of a commercial retarder can be applied such that the duration of workability can be adjusted to the need of the construction site. Fig. 4 shows that the increase of velocity depends strongly on the dosage of retarder.

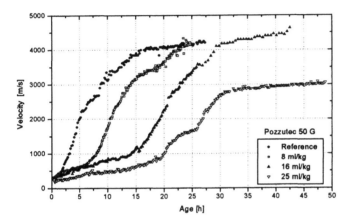

Fig. 4 Velocity of the compressional wave as function of age and dosage of retarder

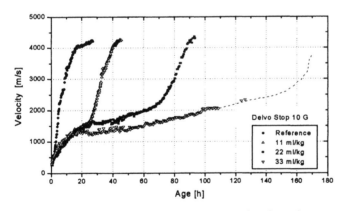

Fig. 5   Velocity of the compressional wave as function of age and dosage of super retarder

When 25 mg per kilogramm of cement are used the velocity after 30 h stays at a value which is only about two third of the value reached without or with little retarder.

The super retarder is aimed to stop hydration for many hours or even several days. When a concrete cannot be placed on a certain day it would be possible to store it to a later time and to use it when needed. Fig. 5 shows that the super retarder has a great effect already at a dosage of 11 ml per kilogramm of cement. With 33 ml, the end of workability can be delayed until about 3 days.

Finally, the effect of the cement type on the wave velocity is shown in Fig. 6.

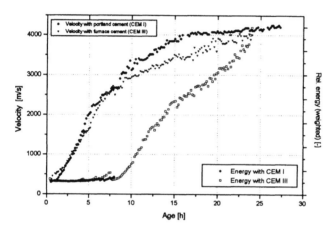

Fig. 6 Velocity of compressional wave in concrete with CEM III and CEM I as function of age

There is almost no difference until 8 h. After that time the CEM I concrete develops the wave velocity faster than CEM III does.

4.2 Energy transmission

Some of the energy of the input signal is dispersed into the concrete and not picked up by the receiver, another part is converted to heat. That part which is transported directly from the input to the output transmitter can be measured by evaluating the amplitude spectrum of all frequencies. The more elastic the material the larger the transmitted energy, the more viscous the less.

The US signals which have been used to produce the results of chapter 4.1 were not strong enough to transmit a measured energy up to about an age of 6 h for the reference concrete. However, the mix had set already and was not workable anymore. This means that the energy transmission from the US signals could not be used as a characterizing property.

Therefore, the impact by a steel ball was evaluated. This impact generates more energy and a broad frequency spectrum. Unfortunately, only some of the results have been processed yet. Fig. 7 shows the energy which has been transmitted through 80 mm concrete.

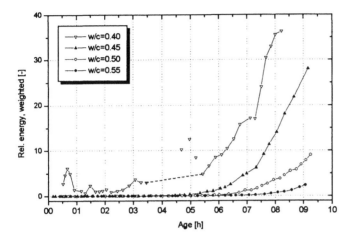

Fig. 7 Relative transmitted energy from a steel ball impact

The measured energy is given relative with respect to the maximum value measured when this maximum is adjusted to the growing stiffnes of the concrete. Fig. 7 shows that the energy increases at the age of 3 h for w/c = 0.40 and at 6.0 h for w/c = 0.55. The other concretes are between these limits. These ages compare very well with the end of workability as discussed in chapter 4.1.

This result may be pure chance and another geometry of the testing specimen may had lead to another answer. The systematic investigation on the relation between energy transmission and end of workability is still to be performed.

4.3 Frequency spectrum
When a steel ball hits a surface the resulting signal depends on mass diameter, and velocity of the steel ball, on the hardness of the impacting bodies and on the properties of the impacted material. When all parameters are constant with exeption of the impacted material this will determine the impact signal. The transmitted signal is additionally a function of the propagation distance. The Fresh concrete is more viscous than hardened concrete which makes that the signal has a low frequency in the beginning and a higher one later. Fig. 8 shows an example of the waves in the reference concrete from 22 minutes to almost 26 hours.

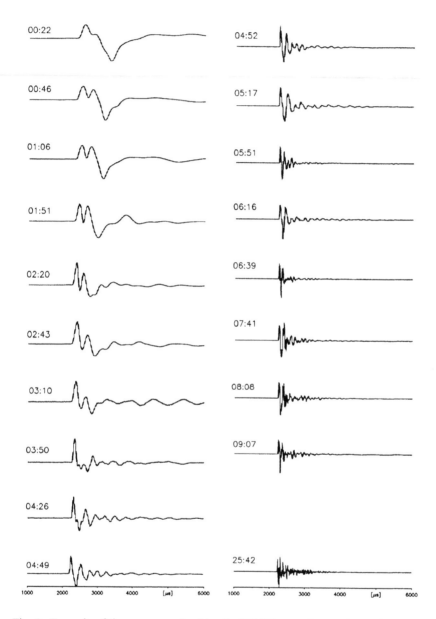

Fig. 8 Example of the waves received by the UEAE transmitter at increasing concrete age (Hours : minutes)

It can be clearly seen that the waves change considerably and that the frequencies increase with setting and hardening of the concrete. To make this phenomena more obvious the signals of Fig. 8 have been converted to the frequency domain by Fast Fourier Transform (FFT). Fig. 9 shows in the upper part the waterfall plot of the frequencies as function of age and, in the lower part, the contour plot.

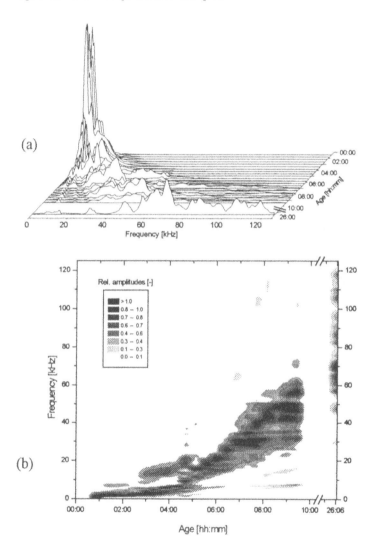

Fig. 9 Frequencies as function of age
a) Waterfall plot (Amplitudes at early ages are enlarged)
b) Contour plot

It is obvious that the low frequencies which are due to the viscous material behaviour disappear with hardening time of the concrete.

5 Future needs and development

The test method has provided a set of three measurements: the propagation velocity of the compressional wave, the transmitted energy of the pulse, and the frequency spectra of the transmitted pulse. All three are sensitive to the age of the concrete and to variations of the composition of the concrete. Deviations of the dosage of retarder, of the water-cement ratio can be detected. On the other hand, as long as the cement has not different setting properties a blast furnace slag cement cannot be distinguished from a portland cement. This is of course the general result: what is measured is essentially the dynamic behaviour of a body which changes from viscous to elastic. All effects which influence this behaviour can be detected. This means that the test has to be carried out in real time.

This is not satisfying yet. What would be needed is the prediction of setting and hardening behaviour from an accelerated test or from the measurements in the first hour after mixing. The compressional wave velocity at the age of 0.5 h is different for different water-cement ratios (Fig. 3) and dosages of retarding agents (Fig. 4). The restricted number of tests do not allow, however, to generalize this result yet, i.e. other mixes should be tested. Simultaneously, Biot's theory has to be extended to concrete and validated by testing. Both activities are on the way.

The test set-up should be analyzed again with respect to a smaller specimen, the discontinuous coupling of the receiver and the use of a single receiver in order to test several specimens subsequently with one testing device. This would be useful for a continuous production control in the ready mix plant and at the construction site as acceptance test if necessary.

6 Acknowledgement

The contribution to testing and data analysis by Mr. B. Weiler and Mr. J.C. Fischer is greatly appreciated. The financial support by the German Reinforced Concrete Committee (DAfStb, V 345) is gratefully acknowledged.

7 References

1. Bartos, P.J.M. (ed.) (1994) *Special concretes. Workability and mixing,* E & FN Spon, London, 264 pp.
2. L'Hermite, R. (1955) Idées actuelles sur la technologie du béton. *Documentation TBTP,* Paris, pp. 193-200.
3. van der Winden, N.G.B. (1990) Ultrasonic measurement for setting control of concrete. In *"Testing during concrete construction",* ed. by H.W. Reinhardt, Chapman & Hall, London, pp. 122-137.
4. Niyogi, S.K., Das Roy, P.K., Roychaudhuri, M. (1990) Acousto-ultrasonic study on hydration of portland cement. *Cer. Trans.,* Vol. 16, pp. 137-145.

5. Reichelt, U., Nickel, U., Röthig, H. (1991) Möglichkeiten für die Qualitätskontrolle von Frischbeton mit der Ultraschall-Messtechnik. *Beton- und Stahlbetonbau 86*, H. 6, pp. 147-148.

6. Keating, J., Hannant, D.J., Hibbert, A.P. (1989a) Comparison of shear modulus and pulse velocity techniques to measure the build-up structure in fresh cement pastes used in oil well cementing. *Cem. Conc. Res.*, Vol 19, pp. 554-566.

7. Keating, J., Hannant, D.J., Hibbert, A.P. (1989b) Correlation between cube strength, ultrasonic pulse velocity and volume change for oil well cement slurries. *Cem. Conc. Res.*, Vol. 19, pp. 715-726.

8. Boutin, C., Arnaud, L. (1995) Mechanical characterization of heterogeneous materials during setting. *Eur. J. Mech.*, A/Solids 14, No. 4, pp 633-656

9. Jonas, M. (1991) Einsatzmöglichkeiten einer Ultraschall-Frequenzanalyse bei der Erhärtung anorganischer, mineralischer Bindemittel. *Forschungskolloquium DAfStb, Bochum*, pp. 187-191.

10. Sayers, C.M., Dahlin, A. (1993) Propagation of ultrasound through hydrating cement parts at early times. *Advanced Cement Based Materials 1*, pp. 12-21.

11. Berryman, J.G. (1980) Confirmation of Biot's theory. *Appl. Phys. Lett.*, Vol. 37, No. 4. pp. 382-384.

PART EIGHT
MIX DESIGN AND MODELS

37 A PARTICLE–MATRIX MODEL FOR PREDICTION OF WORKABILITY OF CONCRETE

E. MØRTSELL
Aker Betong A.S., Trondheim, Norway
M. MAAGE
Selmer A.S., Trondheim, Norway
S. SMEPLASS
Norwegian University of Science and Technology, Trondheim, Norway

Abstract
Fresh concrete may be considered a two-phase material, composed of a matrix and a particle phase, or described by their properties, a **fluid material** and a **friction material**. This simple model is suitable for most common concretes. This approach has been chosen within a Dr. thesis [1] to develop a material model suitable for data based calibration and adjusting of mix design in the concrete industry.

In this model, the matrix phase is defined to consist of the cementitious materials, the water, and the aggregate fines ($< 125 \ \mu m$). The particle phase correspondingly consists of the remaining aggregates. The inputs of the model are the properties of each phase. The challenge is to describe and determine these properties in a consistent way, and to model the interference between them when combining the phases in different proportions in concrete mixes.

The matrix phase (fluid material) is described by a new testing method giving the flow resistance ratio of the matrix. The friction phase (aggregates $> 125 \ \mu m$) is described by the air voids modulus of the particles. An empirical model, based on the properties for the two phases and the interference between them, for the workability of fresh concrete, is developed. Workability is defined by the slump test or the flow table test.
Keywords: Fresh concrete, particle-matrix model, workability.

1   Scope

The main scope of work has been to develop a model for the workability of fresh concrete. Emphasis ha been put on suitability for use in computer controlled concrete mixing plants to adjust the mix design when input materials are changed.

Production Methods and Workability of Concrete. Edited by P.J.M. Bartos, D.L. Marrs and D.J. Cleland. Published in 1996 by E & FN Spon, 2–6 Boundary Row, London SE1 8HN. ISBN 0 419 22070 4.

2   Introduction to the particle-matrix model

The workability of concrete is a result of the inherent properties of the constituents, the mix proportions and the physical and chemical interference between them. The simplest possible way of modelling this complex system is to consider the concrete a two-phase material composed of a matrix and a particle phase, or described by their properties, a **fluid material** and a **friction material**. This model is considered to be suitable for most commonly used concretes, including high performance concrete [1], [2]. A similar approach is used in [6], but the matrix is not including all the fine particles.

   If the phases are to be defined according to their properties as described, a rather unusual division of the concrete constituents into these two phases has to be done:

1.   The matrix is defined to consist of the water, possible chemical additives and all fines, including cement, pozzolanes and aggregate fines, ie particles < 0.125 mm.
2.   The particle phase is defined to consist of aggregate particles larger than 0.125 mm.

The basic challenge is then to describe and determine the properties of the phases in a consistent, simple and robust way, and further to model the impact of these properties when combining the phases in different proportions in concrete mixes. The simplicity of this model is illustrated in Figure 1.

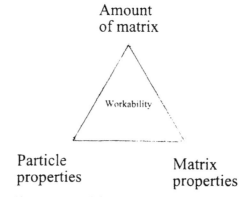

Fig. 1.   A particle-matrix model for concrete.

The present approach is based on single parameter characterization of each phase:

1.   The flow resistance ratio of the matrix
2.   The air voids modulus of the particles

3   The flow resistance ratio (λ_Q)

The flow resistance ratio (λ_Q) is calculated based on the FlowCyl test [1], [2], [3], a modification of the Marsh Cone test, originally developed to characterize oil well cements. The apparatus consists of a vertical cylindrical steel tube with a bottom outlet formed as a cone ending in a narrow nozzle, and an electronic scale connected to a data logging computer. The flow properties of the test material are characterized by the accumulated flow through the nozzle. Calibration tests performed on fluids like water, glycerine and different oils have shown that the flow is an approximately unique function of the viscosity of the material. The tests so far indicate that the test provides consistent data for cement pastes and matrix materials as defined in the model.

The flow resistance ratio (λ_Q) is a representation of the difference in accumulated flow between the test material and an ideal fluid flowing through the FlowCyl. It is defined as the ratio between the area under the loss-curve in Fig. 2 (F_t) and the area under the curve for the ideal fluid without any loss in Fig. 2 (F_i):

$$\lambda_Q = F_t / F_i \tag{1}$$

This means that λ_Q has a value between 0 (for an ideal fluid without any loss) and 1.0 (for a non flowing matrix).

For matrix materials, which not necessarily may be characterized as linear viscous, λ_Q is assumed to be a better representation of the inherent properties than viscosity.

A Microsoft Excel spreadsheet is designed for calculating λ_Q based on the data collected from the scale during the FlowCyl test. An example of a computer output for such calculations is shown in Figure 2.

When no information is available on the correlation between λ_Q and mix characteristics for the matrix, λ_Q has to be found by performing the FlowCyl test. Based on testing of 160 filler modified cement pastes (fmpastes), the following correlation is found:

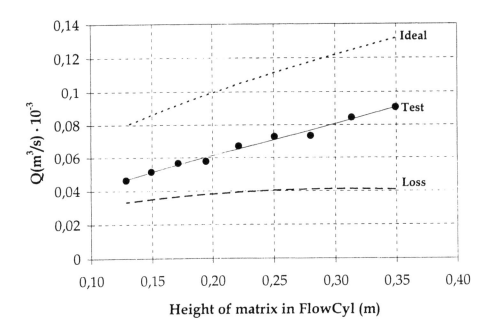

Fig. 2. Typical FlowCyl test data for a matrix showing curves for an ideal fluid, measured points for the actual matrix (fmpaste) and calculated curve for loss.

$$\lambda_Q = k \, (k_c \, c/v + k_s \, s/c + k_f \, f/c)^n \tag{2}$$

where k, k_c, k_s, k_f and n   are constants found by regression analysis of test data
       c/v   is cement water ratio by weight
       s/c   is silica fume cement ratio by weight
       f/c   is filler cement ratio by weight

Based on experience with the actual materials on a concrete mixing plant, all the constants in eq. 2 can be determined and for future changes of the matrix, λ_Q can be calculated from eq. 2. In the tests performed, different numbers for k, k_c, k_s, k_f and n were found depending on type of materials used. However, all the constants are independent parameters. As an example, this means that when changing type of cement, only the constant k_c will be changed. The correlation between calculated λ_Q from eq. 2 and measured values is very good as shown in Figure 3 for different types of cement and filler.

4   The air voids modulus (Hm)

The air voids modulus (Hm) is based on the air voids space ratio of the fine (0.125 - 4 mm) and coarse (> 4 mm) portions of the particle system. The air voids ratio depends on grading, angularity, mineralogi and surface texture. Hence, at least within limited

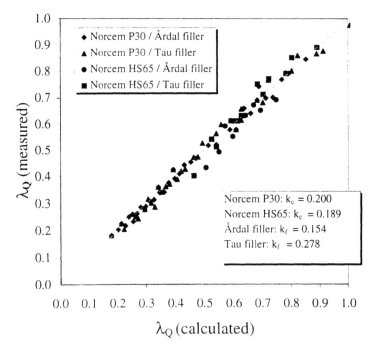

Fig. 3. Correlation between measured and calculated flow resistance ratio for pastes with different types of cement and filler

fractions of the aggregate, the air voids ratio represents the sum of the relevant properties with respect to concrete workability. However, the effects of these properties on concrete workability are more expressed for the fine particle fraction than the coarse. Hence, the fineness modulus (Fm) of the fine/coarse aggregates are introduced as correction factors when defining the air voids modulus.

A preliminary definition of the air voids modulus (Hm) is:

$$Hm = v_1(Hs/(Fm_s)^{0.5} + Ts) + v_2(Hp/(Fm_p)^{0.5} + Tp) \tag{3}$$

where v_1 is volume fraction of sand (0,125 - 4 mm) in the aggregate
v_2 is volume fraction of coarse aggregate (>4 mm) in the aggregate
Hs   is air voids ratio in the sand (0,125 - 4 mm) (volume %)
Hp   is air voids ratio in the coarse aggregate (>4 mm) (volume %)
Ts   is an aggregate parameter for the sand (0,125 - 4 mm)
Tp   is an aggregate parameter for the coarse aggregate (>4 mm)
Fm_s is the fineness modulus for the sand (0,125 - 4 mm)
Fm_p is the fineness modulus for the coarse aggregate (>4 mm)

The physical interpretation of Hm is that it is equal to the paste volume (vol%) when the mix is changing from no slump to a small but measurable slump.

The air voids ratio is determined by the method CAT. D 446 [4], but other methods may also be used. The fineness modulus is found from the grading curve for the actual material (without any particles < 0,125 mm). The aggregate parameters can be determined in different ways depending on the situation.

1. If the mix is a mortar ($v_2 = 0$), Ts is found in a mix with the actual materials by gradually increasing the water content until the consistency is changing from no slump to a small but measurable slump. Hm is set equal to the actual paste volume, and Ts is calculated.
2. If the aggregate parameter for the total aggregate in an actual concrete is to be found, the same method as described for mortar can be used. This will result in an aggregate parameter including both the parameters for sand and coarse aggregate.
3. Both Ts and Tp can be found in concrete mixes by regression analysis of results for different mixes.

Equation 3 is not valid for all combinations of v_1 and v_2. v_2 should not be higher than about 60%. By using method 3, Ts and Tp are found to be 8,7 and 7,0 respectively when v_1 and v_2 are approximately the same in the actual concretes.

The definition of Hm has to be further developed in such a way that it can be determined on the aggregate itself without testing concrete or mortar.

5   Workability function for concrete (Kp)

The workability of the concrete, characterized by the slump or the flow table test, is finally expressed as a function of the flow resistance of the matrix (λ_Q), the air voids modulus (Hm) of the particle phase, and the volume fraction of the matrix.

The correlation between workability and volume fraction of the fluid phase is assumed to be S-shaped. A convenient S-shaped function is the Tanh-function. The workability function Kp is defined as:

$$Kp = (n - m) \cdot (Tanh(x) + 1)/2 + m \tag{4}$$

where m and n   are lower and upper asymptote values respectively for the workability function

  x   is the argument of the Tanh-function given by:

$$x = \alpha \cdot (2\, Fp - 1 - \beta) \tag{5}$$

where Fp   is volume fraction of fluid phase (fmpaste)
  α is expressing the steepness of the Kp-function
  β is placing the Kp-function along the x-axis and given by:

$$\beta = 2 \cdot Hm - 1 + 1/\alpha \tag{6}$$

By introducing x from eq. 5 and β from eq. 6, the workability function may be written as:

$$Kp = (n - m) \cdot (Tanh(2\alpha(Fp - Hm)/100 - 1) + 1)/2 + m \tag{7}$$

where Fp and Hm now are given in volume%.
    The factor α is found to be a function of λ_Q alone:

$$\alpha = A \cdot e^{-B \cdot \lambda Q} \tag{8}$$

where A and B are constants found by regression analysis.

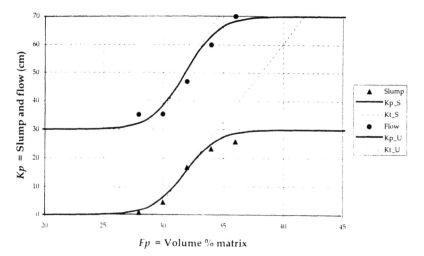

Fp = Volume % matrix

Fig. 4. Typical workability curves (slump and flow table test), showing measured points and calculated lines for a concrete series.

For the materials used in the actual concretes, A and B are found to be 115 and 4,45 respectively.

An example of the correlation between measured values for slump and flow table test and calculated values based on eq. 7 is shown in Figure 4 for a typical concrete series. The points are measured values and the lines are calculated. The S-shaped curve fits very well to the test data.

Correlation between calculated workability according to eq. 7 and test results is demonstrated in Figure 5 for the tested concrete mixes. The lower, left part and the upper right part of Figure 5 are representing slump and flow table test values respectively.

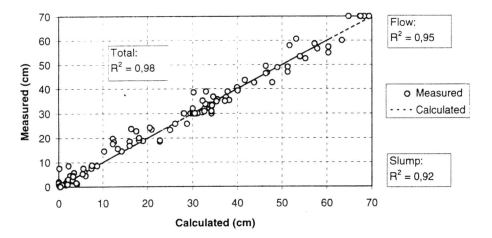

Fig. 5 Correlation between measured and calculated workability (slump and flow table test) for all the tested concretes

6 Practical application of the workability function

The main practical application of the model is in adjusting mix design in the ready mix concrete industry when properties or proportions of the constituants are changing. In such situations experience with the actual materials used have to be found by the concrete producer and used as input into the model.

As an example, the effect of changing cement quantity in mixes with three different aggregate gradings, will be demonstrated. The three grading curves are shown in Figure 6. According to the model, no finer particles than 0,125 mm are included in the aggregate.

Three mixes, one for each aggregate, are designed and Figure 7 shows how the slump is changing depending on cement content in the mixes. There are two changes involved, the increased quantity of cement and the reduced w/c-ratio. Increased quantity of cement will result in more paste and increased slump. Reduced w/c-ratio will result in a higher λ_Q and lower slump. The combined effect is demonstrated in Figure 7 for the three diffe-rent aggregate gradings. The effect is quite different depending on the aggregate and are in accordance with practical experiences. Within practical cement contents, it is also in accordance with results presented by Powers [5], saying that adding cement do not

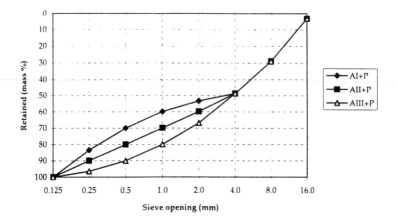

Fig. 6. Aggregate gradings used in demonstrating the workability function

change workability when all other parameters are unchanged.

Another example is shown in Figure 8 where the effect of increasing the quantity of three different types of filler in mixes with constant content of water and cement. The grading curve AII+P is used. As demonstrated in Figure 8, the effect of increasing filler content is the same for the three fillers up to around 50 kg/m³ because increased volume of fmpaste is dominating. For higher quantity of filler, the different properties of the three fillers become more pronounced. The natural filler (År) results in higher slump than the two crusher fillers (Ta and Va).

By comparing Figures 7 and 8, it is also possible see the different effects of cement and fillers on slump.

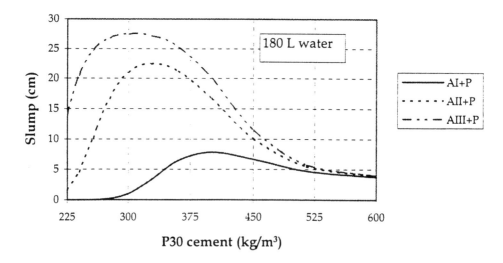

Fig. 7. Effect of cement content in mixes with different aggregate curves on slump

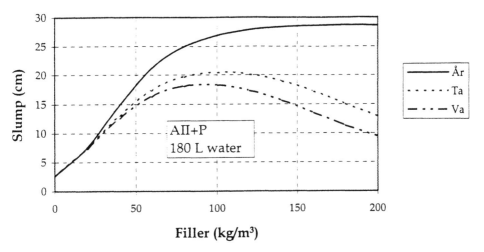

Fig. 8 Effect of different types of filler and quantities on slump

7 Conclusions

The simple model, where concrete is looked upon as a two component material (a fluid and a friction part), is suitable for predicting the workability of concrete, expressed by slump or flow table test.

The fluid part (matrix or fmpaste with all particles <0,125 mm) is described in a consistent way by the flow resistance ratio, λ_Q. λ_Q is determined in a new developed test method called FlowCyl.

The friction part (particle phase or aggregate >0,125 mm) is described by the air voids mudulus, Hm. Hm is calculated from volume fractions of sand (0,125< d >4 mm) and coarse aggregate (>4 mm), the fineness moduli for sand and coarse aggregate and an aggregate parameter for sand and coarse aggregate. Hm has to be further developed.

The workability function developed is describing the workability (slump or flow table test) of the concrete in an appropriate way for practical purposes. The main application of the function is to adjust concrete mix design, from a workability point of view, when properties or proportions of the constituent materials are changed.

Acknowledgement

This paper is based on a Dr. thesis [1] sponsored by the Research Council of Norway, Aker Betong A/S and the Norwegian University of Science and Technology.

8   References

1. Mørtsell E. (1996) *The effect of the constituent materials on the rheology of fresh concrete.* Dr. thesis at the Norwegian University of Science and Technology, Trondheim, Norway. (in Norwegian).
2. Mørtsell E., Smeplass S., Hammer T.A. and Maage M. (1996) *FLOWCYL - How to Determine the Flow Properties of the Matrix Phase of High Performance Concrete.* Proceedings. Fourth International Symposium on Utilization of High Strength/High Performance Concrete, Paris, 1996.
3. Mørtsell E., Smeplass S. and Maage M. (1995) *Characterization of the flow properties of the matrix phase in concrete.* BETONGindustrien no.3, Oslo, Norway, 1995 (in Norwegian).
4. Domone P. L. J. and Soutsos M. N. (1994) *An Approach to the Proportioning of High-Strength Concrete Mixes.* ACI, Concrete International, Oct. 1994.
5. Powers T. C. (1968) *The properties of Fresh Concrete.* Wilky & Sons.
6. Dewar J. D. (1983) *Computerized simulation of aggregate, mortar end concrete mixtures.* Proceedings, ERMCO, British Ready Mixed Concrete Association, London, 1983.

38 MIX DESIGN OF SELF-COMPACTING CONCRETE (SCC)

T. SEDRAN and F. de LARRARD
Laboratoire Central des Ponts et Chaussées, Paris, France
F. HOURST and C. CONTAMINES
Laboratoire Départemental de l'Herault, Montpellier, France

Abstract

Self-Compacting Concrete (SCC) has been spreading worldwide in recent years. However, few efficient and sound mix-design methods have been proposed that can take account of the actual geometry of the structure, including the reinforcement.

The Laboratoire Central des Ponts et Chaussées (LCPC) has recently developed two powerful tools, BTRHEOM™ and RENÉ-LCPC™, which lead to a new rational approach in concrete proportioning. The first is a new type of rheometer which can fully characterize the rheological behaviour of concrete. The second is software based upon a mathematical model aiming at optimizing the granular skeleton, while taking into account the degree of confinement in which the concrete has to be placed.

After a brief presentation of the two tools, the paper will highlight their application in the design of Self Compacting Concrete (SCC). The paper concludes by summarizing three full-scale experiments showing the effectiveness of the mixture proportioning method.
Keywords: granular optimization, mix-design, mixture-proportionning, plastic viscosity, self-compacting concrete, self-levelling concrete, slump flow, yield stress

1 Introduction

A new type of concrete is spreading worldwide in recent years, so-called Self-Compacting Concrete (SCC). This kind of concrete is very fluid and has a high filling ability even in the case of very dense reinforcement, with little segregation risk. As it does not need any vibration during casting, it is of great interest compared to classical concrete for productivity purposes and in situ quality of the hardened material.

Many technical papers (almost all Japanese) have shown that this kind of concrete can be successfully used on real projects [1] [2] [3] [4], but few of them are more

Production Methods and Workability of Concrete. Edited by P.J.M. Bartos, D.L. Marrs and D.J. Cleland.
Published in 1996 by E & FN Spon, 2–6 Boundary Row, London SE1 8HN. ISBN 0 419 22070 4.

specifically devoted to mix-proportioning [5] [6]. However, Self-Compacting Concretes (SCC) contain a paste volume higher than ordinary concretes, which leads to a more expensive material and promotes a tendency to higher shrinkage [7], creep and heat generation. So, it is of particular importance to optimize the granular skeleton to limit this paste volume while keeping the workability.

At present, a mix designer must proceed by trial and error to optimize SCC. A lot of devicess have been proposed in the literature to test the filling ability of SCC through narrow spaces [8], [9], such as the U-box or Grid-Box. However, they are cumbersome to perform (largesamples are required) and their use seems to be very limited for practical projects. More convenient tests such as funnel tests [8] have been developed, but more research is needed to make them representative of the real confinement of structures in the case of very dense reinforcements.

To reduce the number of tests, some authors have proposed guidelines [5], [6] for SCC proportioning taken from their own experience, which mainly consistof limiting the coarse aggregate volume and sand to powder ratio. As these guidelines are general, they are very safe and generally leads to paste volumes around 35% or more.

In this context, the « Laboratoire Central des Ponts et Chaussées » (LCPC, Paris) has started to study a new rational mix-design process based on different tools described in the following sections. The method presented hereafter is presently limited to concrete containing no viscosity agent but only aggregates, cement, mineral admixtures, superplasticizer and water.

2 The BTRHEOM™ rheometer

A new kind of rheometer has recently been developed at LCPC and extensively validated [10] [11] [12] to characterize the rheology of soft to fluid concrete in the fresh state. The BTRHEOM™ is a torsionnal rheometer where a 7-liter sample is sheared between a rotating top section and a fixed bottom one. This apparatus controlled by a portable computer can be used by a single operator on site as well as in the laboratory.

The Fig. 1 shows typical diagram obtained with rheometer. It can be seen that a soft to fluid concrete can be considered as a Bingham fluid and that BTRHEOM™ may provide up to five parameters to fully describe the concrete:

- the shear yield stress τ_0 (in Pa) and the plastic viscosity μ_0 (in Pa. s) of the concrete not submitted to any vibration;
- the shear yield stress $\tau_{v\,0}$ (in Pa) and the plastic viscosity $\mu_{0\,v}$ (in Pa. s) of the concrete under vibration; and
- the shear yield stress at rest $\tau_{r\,0}$ (in Pa), which describes the thixotropy of the concrete.

Of course, as far as SCC are concerned, only properties without vibration are of interest. Moreover, shear yield stress has proved to be strongly related to the slump of concrete [10] : the greater the shear yield stress, the lower the slump. Thus, SCC have very low shear yield stress between 0 to 500 Pa and are mainly characterized by their viscosity as for a newtonian fluid.

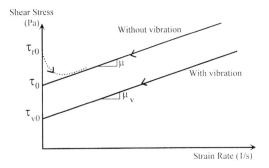

Fig. 1 Rheological parameters determined thanks to BTRHEOM™

Finally, according to LCPC's experience [11], it is suggested that the viscosity of a concrete should be less than 200 Pa.s in order to get:
- easy handling;
- easy pumping,
- easy finishing;
- acceptable appearance of concrete (little bubbling) after removal of forms.

For a given set of materials, the water content is the main parameter controlling the viscosity of the concrete. Thus, the mix-designer will have to find the right balance between the strength of concrete, which requires low water to binder ratio and a good workability on site.

3 The Solid Suspension Model

Part of the water in a fresh concrete is consumed to fill the porosity of the granular skeleton (binders and aggregates); only the other part may contribute to the workability of the mix. So, improving the packing density of the skeleton leads to using less water for the same workability, or to improving the workability at constant water content. In other words, optimizing the performances of hardened concrete is mainly a matter of improving the packing density of its skeleton.

Nowadays, more and more components are available to produce concrete: chemical and mineral admixtures, several kinds of cements and aggregates (crushed, rounded, etc.). But until now, engineers could only approximate the best packing of the grains using master curves, which cannot take the specific properties of each material precisely into account. On the other hand, iterative technological tests are of interest but quickly become burdensome to perform.

In this context, LCPC has developed, and implemented in RENÉ-LCPC™ software, a mathematical model called the Solid Suspension Model [11.13]. This model is based on the analysis of the granular interactions between the different components of a mixture. It can predict the packing density of all dry granular mixes with an accuracy better than 1% from a few properties of the different components which are:

- bulk density;
- grading curve;
- packing density; and
- mass proportion compared to total solid content of the mix.

The experimental methods used at LCPC to determine these values for aggregates as well as for binders are fully described in [13].

This model also provides a value called relative viscosity for a given concrete from the property of its solid skeleton and its water content. The workability of concrete can be assessed by this calculated relative viscosity: the lower the relative viscosity, the more workable the concrete. Though research is still in progress at LCPC to establish relationships between the rheological parameters of fresh concrete (i.e. shear yield stress and plastic viscosity) and this relative viscosity, it has proved to be efficient for mix-design [13].

Moreover the model can take into account the degree of confinement of the mixture. Hence, a packing of aggregate is loosened, due to the presence of the container boundary. This loosening may be evaluated through the Ben Aïm model [14], which assumes that the packing density β is lowered when the distance from the wall is less than d/2 (see Fig. 2).

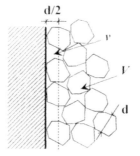

Fig 2: Boundary wall effect

Therefore, the the packing density β becomes:

$$\beta^* = \left(1 - \frac{v}{V}\right)\beta + \frac{v}{V}\, k\, \beta \qquad (1)$$

where V is the volume of the container, v the volume of the zone at a distance from the wall less than d/2, and k a coefficient. From experiments, the following values have been found:

k = 0.87   for rounded aggregate
k = 0.73   for crushed aggregate.

Thus, the wall effect due to a pumping pipe with an inside diameter of D can be described with the following equation:

$$\frac{v}{V} = 1 - \frac{(D-d)^2}{D^2} \tag{2}$$

In this context, the Solid Suspension Model is of great interest because it can account for the confinement effects exerted by the form and the reinforcement.

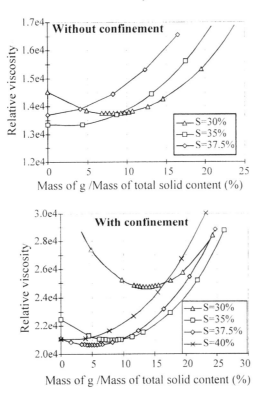

Fig. 3   Theorical influence of confinement

For example, consider a SCC designed to pass through parralel bars separated by 50 mm clear spaces. A conservative approach is to assume that the flowing of the concrete is the same as between two 50 mm-spaced parralel planes. In this case, the v/V ratio to be put in equation (1) related to the diameter of grains d (in mm) reads:

$$\frac{v}{V} = \frac{d}{50} \tag{3}$$

To evaluate the influence of such a confinement on a concrete optimization, calculations have been performed for a given set of materials with the Solid Suspension Model with and without confinement.

The paste volume and composition of the concrete were fixed . So percentages of sand (S), fine gravel (g) and coarse gravel (G) were optimized while S+g+G was kept constant at 78.7% of the total solid content mass of concrete.

To avoid segregation, gap-graded concretes must be excluded. Therefore, according to the experience, the percentage of fine gravel has to be greater than 10%. In this domain, Fig 3 shows that optimized concrete (with lower relative viscosity) contains 30% sand i.e. (G+g)/S=1.62 (by mass) when there is no confinement and 35% sand i.e. (G+g)/S=1.25 when the concrete is confined.

A well known feature of SCC - that is lower coarse aggregate to sand ratio than for normal concrete- is, here, rediscovered by the model.

More generally, the Solid Suspension Model is of interest because it can account quantitatively for the confinement and for the materials nature. For example, it is known [5] that with rounded gravel, SCC callows a larger volume of coarse aggregates. The model is in agreement with this fact as rounded aggregates have a greater packing density.

4 A rational mix design process for SCC

With the help of the two powerful tools presented in the previous sections, we propose the following rational mix design process for SCC (without any viscosity agent).

1. First of all, a set of specifications is written, which contains at least the two following general criteria:
    a) the slump flow should be comprised between 60 to 70 cm (or the shear yield stress measured with BTRHEOM should be less than 500 Pa);
    b) the plastic viscosity should be less than 200 Pa.s to ensure a good workability and a fast casting process on site (see section 2), but greater than 100 Pa.s to avoid segregation;
2. For each project, at least two other items are added:
    a) the mean compressive strength at 28 days;
    b) the most restricting confinement environment to be met on site (for example pipe when pumping or typical clear spaces between reinforcements), which gives a v/V ratio (see section 3);

3. A choice of relevant materials based on local experience is done:
    a) a set of aggregate fractions that permits to design continuous and well-graded aggregate skeleton;

b)   a compatible cement/superplasticizer couple;

c)   a compatible retarder; and

d)   mineral admixtures (limestone filler, fly ash, slag etc.) which are needed to limit the cement content, as SCCs require larger than normal powder volume...

4.   The mixture-proportioning can start according to the following procedure:

a)   A first combination of binders is a priori fixed (for example 70% of cement and 30% of limestone filler). When several mineral additions are available, the choice may be governed by local experience and/or particular specifications. For example, if a strength increase is aimed after 28 days, fly ash will be preferred.

b)   The saturation amount of superplasticizer [15] is determined for this combination of binders. As this amount may lead to too viscous a concrete, a half amount is tentatively chosen;

c)   The water demand of the binders combination is measured in presence of the superplasticizer. It is a kind of packing density measurement, which is needed to run the Solid   Suspension Model, as explained in section 3. The experimental method to measure the water demand is described precisely elsewhere [13];

d)   Calculations are performed with the Solid Suspension Model accounting for the confinement (with the v/V parameter in equ. 1). An arbitrary relative viscosity is fixed at $5 \ 10^4$ (which corresponds to typical SCC workability according to LCPC experience) and the water content is minimized. Here, the mineral admixture/cement ratio(s) are kept constant, at the levels chosen in step 4 a), while the mutual proportions of aggregates and binders combination are optimized.

e)   The concrete obtained is batched (a 10/15-litre sample is sufficient) and its water content is adjusted aiming to obtain the target viscosity;

f)   The superplasticizer dosage is adjusted to find a suitable slump flow (or yield stress). Note that this will not change significantly the plastic viscosity [12]. At this step, the concrete obtained corresponds to both criteria of yield stress (or slump flow) and plastic viscosity.

g)   The compressive strength of the concrete is measured (if time is let to do so) or estimated with an empirical formula. For this purpose, the classical Féret's formula has been recently generalized to apply for concretes containing mineral admixtures [16]:

$$Fc = \frac{Kg \ Rc}{\left(1+3.1 \dfrac{W+A}{C(1+K1+K2)+BFS}\right)^2} \tag{4}$$

where

- Kg is an aggregate coefficient (which can been fitted on a reference Portland cement concrete). Two typical values are Kg =5.4 (crushed aggregate) and Kg = 4.8 (rounded aggregate);
- Rc is the cement strength mesured on ISO mortar (sand:cement:water proportions: 3:1:0.5);
- W is the free water content of concrete (in kg/m3, including the one contained in admixtures),
- A is the volume of the entrapped air(in l/m3),
- C, PFA, SF, LF and BFS are the weights (in kg/m3) of cement, fly ash, silica fume, limestone powder and blast-furnace slag, respectively;
- $K1 = 0.4\ PFA/C + 3\ SF/C$ ($K1 \leq 0.5$) is a pozzolanic coefficient,
- $K2 = 0.2\ LF/C$ ($K2 \leq 0.07$) is a limestone filler activity coefficient.

If the strength is too high or too low, a new combination of binders is chosen and the mix design process restarted at step 4. b). However, the water content is not likely to change significantly as compared to the reference recipe. Thus, predictions made at this stage with the modified Féret's formula should be reliable enough to determine the final combination of binders. Therefore, the aimed concrete may be reached in few iterations.

h) Finally, the evolution of the rheological behavior must be studied [12] and retarder added in case of unacceptable stiffening of the concrete.

Of course a filling ability test (like U-box test for example) may be performed in laboratory prior to site utilization.

It is important to note that the Solid Suspension Model may converge, in some cases, on gap-graded concrete. This may happen if an intermediate fraction of grains has a low packing density (due to angular shape). Although such a concrete is optimized from the viewpoint of packing density, it will present a proneness to segregation. A more continuous skeleton must be preferred, even with a slightly higher water demand. For example, on Fig 3., in the confinement case, the best packing density is reached with g = 5%. Experience shows that segregation occurs, thus a larger amount of intermediate aggregate must be used (e.g. 15%). At this dosage, the optimum amount of sand becomes 35%. There is presently no mean to anticipate the segregation of a mixture, so that adjustments must be done between steps 4.d) to 4.f) by iterative trials.

5 Example of SCC mix design

An experimental project has been carried out in October 1995 in Montpellier (France) in which LCPC, in collaboration with the Herault District Laboratory, was in charge of the mix design of a SCC. The method used was close to the one presented herein.

This SCC was designed to be poured with a bucket in a funnel. The structure was a cross-shaped wall (see Fig. 4) where two horizontal reinforcement areas (on top and bottom) presented 50-mm clear spaces between rebars. The mixture-proportions are given in Tab. 1. The slump flow was between 60 and 70 cm, and the plastic viscosy was less than 150 Pa.s. The compressive strength at 28 days was about 50 MPa.

Table 1: SCC composition

Cement CEM I 52.5	Limestone 0/4 filler	Sand	4-10 Gravel	10-20 Gravel	Water	Superplasticizer*
350	134	852	363	571	168	7.1

all the aggregates are crushed limestone   *with a 20% solid content

The cross-shaped wall was successfully filled up without any vibration. Visual inspection after demolding showed that no honey-combing occured and that the quality of the surface was excellent (few bubbling).

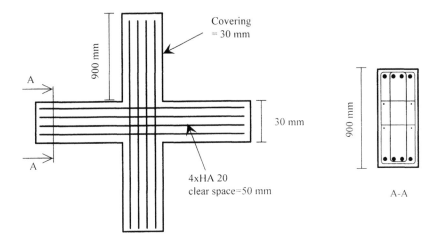

Fig. 4 The cross-shaped wall

The same concrete was used to fill 4-m3 non-reinforced accropode (see Fig. 5) without vibration and directly from the truck. In a previous project, these accropodes needed 15 min. or more to be cast with normal concrete and vibration. Here, only 2 min. were necessary to do the same job due to the low viscosity of the concrete. The surface quality was excellent with little bubbling and very sharp edges.

Finally two 5-metre-high walls were cast by pumping the concrete by the top and the bottom respectively. A very restricting pumping circuit with a length of 78 m. was chosen: the inside diameter of pipes was 125 mm with reductions to 100 mm, and many bends were introduced into the circuit.

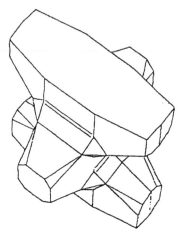

Fig. 5 One 4 m^3 accropode

As expected, the concrete was pumped into the first wall without any blocking into the pipes. Yet, some bubbling were detected on the concrete surface after demolding probably due to a too thin covering of reinforcement.

For the bottom-injected wall, the pumping became impossible while the concrete had only reached a level of 2.5 metres high. As pumping re-started when the pipe was disconnected from the form, it was concluded that blocking occured in the wall. This was confirmed by an important bubbling of the wall surface near the pipe connection. This problem was probably due to the fact that the slump flow of the concrete had drop to 50 cm at this moment.

6 Conclusions

Two powerful tools have been presented in this paper as following:

- the BTRHEOM rheometer which provides the rheological parameters of SCC (and particularly its plastic viscosity);
- the Solid Suspension Model which can optimize the granular skeleton of concrete. This model accounts quantitatively for the component nature (grading, packing density, bulk density) and the real confinement of the concrete (diameter of pumping pipes, density of reinforcement and so on).

Based on these tools, a rational mixture proportioning method for Self-Compacting Concrete has been proposed. The article concludes by summarizing a full scale experiment showing the effectiveness of the method.

Considering the multiplication of the materials available for concrete industry (binders and aggregates from several sources) such a mix design process is expected to provide a great help:

- to make the best use of available materials, in order to reach the desired properties of concrete at the lowest cost;
- to account for the specificity of a project (for example pumping or casting in very dense reinforcements);
- to save time during the mix design process by limiting the amount of tests;

Nevertheless more research is needed to understand and quantify the segregation risks of SCCs. Also, the method does not cover the use of viscosity agents. These products may be useful if substandards (poorly graded) aggregate are to be used, and/or where mineral admixtures are not available at a reasonnable price.

7 References

1. Tanaka K., Sato K.. Watanabe S.. Arima I. and Suenaga K. (1993) Development and Utilization of High Performance Concrete for the Construction of the Akashi Kaiskyo bridge. *American Concrete Institute SP140, High Performance Concrete in Severe Environments*, Detroit, pp 25-51.
2. Sakamoto J., Matsuoka Y. H., Shindoh T. and Tangtermsirikul S. (1993) Application of Super Workable Concrete to Actual Construction, Conference *Concrete 2000*, Dundee, 7-9 September.
3. Hayakawa M., Matsuoka Y. And Yokota K. (1995) Application of Superworkable Concrete in the Construction of 70-Story Building in Japan, *Second CANMET/ACI International Symposium on Advances in Concrete Technology, SP 154*, Las Vegas, pp. 381-398, june.
4. Izumi I., Yonezawa T., Ikeda Y. And Muta A. (1995) Placing 10, 000 m3 Super Workable Concrete for Guide Track Structure of Retractable Roof of Fukuoka Dome, *Second CANMET/ACI International Symposium on Advances in Concrete Technology, SP 154*, Las Vegas, Supplementary papers, pp. 171-186, june.
5. Okamura H. And Ozawa K. (1995). Mix design for Self-Compacting Concrete. *Concrete Library of JSCE, n°25*
6. Yurugi M., Sakata N., Iwai M. And Sakai G. (1993) Mix Proportion for Highly Workable Concrete, *Conference Concrete 2000*, Dundee, 7-9 September.
7. Ogawa A., Sakata K. & Tanaka (1995). A Study on Reducing Shrinkage of Highly-Flowable Concrete. *Second CANMET/ACI International Symposium on Advances in Concrete Technology*, SP 154, Las Vegas, pp 55-72.
8. Ozawa K., Sakata N. And Okamura H. (1995). Evaluation of Self Compactibility of Fresh Concrete using the Funnel Test. *Concrete Library of JSCE, n°25*.
9. Sedran T. (1995). Les bétons autonivelants (BAN). Synthèse bibliographique (Self Compacting Concretes: a litterature review), *Bulletin de Liaison des Laboratoires des Ponts et Chaussées n°196 (in French)*.

10. Hu C. de Larrard F. and Sedran T. (1996) A New Rheometer for High Performance Concrete, *BHP 96, fourth International Symposium on the Utilization of High Strength/High Performance Concrete*, Paris, 29-31 may.

11. F. de Larrard, C.Hu And T. Sedran (1995) Best Packing and Specified Rheology: Two Key Concepts in High-Performance Concrete Mix-Design, *Adam neville Symposium, Advances in Concrete Technology*, Las Vegas, June.

12. F. de Larrard, T. Sedran, C Hu, J. C. Szitkar, M. JOLY and F. Derkx (1996) Evolution of the Workability of Superplasticized Concretes :Assesment With BTRHEOM Rheometer, *RILEM International Conference on Production Methods and Workability of Concrete*, Paisley, june.

13. T. Sedran and F. de Larrard (1996) René-LCPC: a Software to Optimize the Mix Design of High Performance Concrete, *BHP 96, fourth International Symposium on the Utilization of High Strength/High Performance Concrete*, Paris, 29-31may.

14. Ben-Aïm R. (1970) *Etude de la texture des empilements de grains. Application à la détermination de la perméabilité des mélanges binaires en régime moléculaire, intermédiaire, laminaire (Study of the texture of granular packings. Application to the determination of the permeability of binary mixes, under molecular, intermediate and laminar regimes)*. Thèse d'Etat, University of Nancy (in French).

15. de Larrard F. (1990) A Method for Proportionning High Strength Concrete Mixtures, *Cement, Concrete and Aggregates*, vol 12, n°2, pp 47-52, Summer.

16. de Larrard F. (1993) Optimization of High Performance Concrete Mixtures, *Michromecanics of Concrete and Cementitious Composites*, edited by C. Huet, Presses Polytechniques et Universitaires romandes-Lausanne, pp 45-58.

39 INTERACTION OF PARTICLES IN FIBRE REINFORCED CONCRETE

P.J.M. BARTOS and C.W. HOY
Advanced Concrete Technology Group, Department of Civil, Structural and Environmental Engineering, University of Paisley, Paisley, Scotland, UK

Abstract

There has been a significant amount of work on the packing of aggregates, with regards to maximising the packing density. This has been done to improve durability and strength of the concrete and as such has gained popularity recently in the field of High Performance Concrete. However very little research has been done on the packing of mixtures containing non-spherical particles and even less on civil engineering materials containing non-spherical particles. It is believed that a greater understanding of the interactions between fibres and aggregate will lead to development of improved properties of fibre reinforced concrete, through optimisation of mix proportions and alternative methods of incorporating fibres into the mix.

This paper will discuss a laboratory study of the packing of mixtures of fibres and aggregates used in fibre reinforced concretes. Tests were carried out on a range of mixtures containing different aggregates and fibres. The fibres used include steel fibres ranging from very rigid and non-deformable fibres to very flexible fibres which are easily deformed. The aggregates included different size ranges and gradings.

Keywords: fibre-aggregate interaction, packing density, mix proportions.

1 Introduction

It has been said that concrete with higher aggregate contents can have improved strength and fatigue properties[1]. By increasing the aggregate content the drying shrinkage is reduced as is the creep and permeability and thus, the resistance to weathering. The strength may also be increased if good aggregates and a high quality paste are used. For these reasons it is important to optimise the packing of aggregates giving the highest feasible aggregate content and the lowest volume of voids. Most of

Production Methods and Workability of Concrete. Edited by P.J.M. Bartos, D.L. Marrs and D.J. Cleland. Published in 1996 by E & FN Spon, 2–6 Boundary Row, London SE1 8HN. ISBN 0 419 22070 4.

the voids between aggregate particles will later be filled with the cement paste and the remainder will remain filled with air.

Theoretical work at the beginning of this century indicated that the *best* aggregate grading was based on the packing characteristics, specifically, the reduction of voids. After considerable experimental work Fuller and Thompson[2] determined an *ideal* grading curve which closely followed the expression :-

$$P_t = \left(\frac{d}{D}\right)^{1/2} \tag{1}$$

where P_t is the mass of solids finer than diameter d, and D is the nominal maximum aggregate size. This resulted in a dense packing but the concrete it produced was very harsh.

Although there was significant interest in the aggregate grading earlier in the century, this has diminished as the benefits of an improved grading do not normally outweigh the additional cost of selectively blending the aggregates. Normally local aggregates are used in concrete production and the coarse and fine parts are blended in suitable proportions to achieve the required workability. The British Standards for concrete aggregates produce limits for the gradings allowable which are relatively broad, so as not to be too restrictive and costly.

In recent years interest has arisen in optimum aggregate gradings for High Performance Concretes. The development of High Performance Concretes (HPC) has followed two approaches[3]. The first is to reduce the quantity of non-hydrated water which in turn reduces the total water content. This has been achieved through the use of plasticisers and superplasticisers. The second method to improve performance focuses on forming a monolithic or solid rock like material, with the emphasis being on mix design. Effectively, the aim of the latter method is to increase the packing density. This is achieved by widening the range of grain sizes, for example by introducing ultra-fine particles such as silica fume and other fillers. Optimisation of aggregate grading is also used to improve the concrete performance. Dr. de Larrard's Solid Suspension Model[4] has been used to optimise particle proportions and gradings for HPC, including cement, silica fume and water. In developing HPC both of these approaches are combined to give some very high strength and durable concrete mixes.

It is thought that similar approaches to Fibre Reinforced Concrete (FRC) proportioning will allow increased fibre contents resulting in further improvements in properties. The packing of fibre reinforced concrete has not previously been studied. This study aims to investigate the interaction between fibres and aggregate, looking especially at limits to the aggregate size and at the requirements for additional cement paste in the mixture. It is also hoped that this will lead to an improved mixing procedure by identifying the best sequence in which to add each component.

2 Testing programme

Fibres with different physical properties were selected for use in these tests. The aim of the tests is to observe the packing characteristics of different fibre and aggregate combinations. This study focuses on fibre concentrations likely to be encountered in practice. The quantity of fibres used in FRC is normally less than 1% by volume($\%V_f$) and is rarely more than 2% by volume, due to workability constraints. The precise quantities depend on the end application of the concrete and, hence, the improvement in any property which is required. Table 1 below shows typical mixes used for the construction of airfield pavements[5]. The fibre contents are in the range of approximately 0.5 to 1 % of the concrete volume.

Table 1. Typical mix proportions for fibre reinforced concrete pavements[5].
(Values converted to metric equivalent)

Constituent	Mix proportions (kg/m^3)	
	Mix No. 1, 10 mm aggregate	Mix No. 2, 20 mm aggregate
Cement	295	310
Fly ash	140	150
Steel Fibres	50-85	50-85
Coarse aggregate, 10 mm max	870	-
Coarse aggregate, 20 mm max	-	790
sand	810	855
Water	150	170
Plasticiser	Per manufacturer's instructions	
Air-entrainment agent	Per manufacturer's instruction, for 6% air when subject to freezing and thawing conditions.	

Although the fibre content is only around 1% of the concrete, the solid volume of fibres takes up about 2.3% of the coarse aggregate volume. So, if a mixture containing just fibres and coarse aggregate had 2% fibres by volume, this would closely represent the interaction between fibres and coarse aggregate in a mixture with 1% fibre content. Around 5% fibres mixed with coarse aggregate would represent up to 2% by volume of concrete, with the fine aggregate, cement and water omitted. A study of the whole mixture of concrete would be of little use in itself as it would be impossible to isolate any one factor affecting packing so it is therefore important to observe the interaction between components individually.

3 Procedure

For each test, the quantity of aggregate and fibres were accurately weighed out. The fibres and aggregate were then mixed in the pan of a 14 litre mixer. Due to the excessive breakdown of coarse aggregate particles during mechanical mixing, it was decided that hand mixing would give better results. The breakdown of coarse aggregate would alter the bulk density of the aggregate alone and therefore give poor representation of the true results with fibres. Although mechanical mixing is used in concrete practice, the water and cement paste reduce the impact forces on the coarse aggregate. Once the mix appeared to be uniform, a sample of around 1kg was removed and the fibres were separated from the aggregate and weighed, to evaluate the uniformity of the mix. If there was a significant deviation from the expected values, the fibres and aggregate would be mixed for longer then a further sample would be taken. This was repeated until a uniform mix was obtained. Some mixes proved very difficult to mix evenly due to segregation of fibres. Once a uniform mixture was obtained the bulk density was calculated using a method based on the method describe in BS 812: Part 2[6].

The standard describes a test to measure the compacted bulk density of aggregates, an outline of which follows. The mixture (aggregate in the standard) is placed into a container of fixed volume in three equal layers. Each layer is tamped 30 times by a rod of 16 mm diameter being dropped from 50 mm above the surface of the mixture. When the container is filled, it is struck off level with the rim. Any obvious depressions in the surface are filled. The weight of the mixture in the vessel is taken so that the bulk density can be calculated using the containers volume. This should be carried out twice according to the standard, but for these tests it was carried out three times. A seven litre container was used for all of these tests.

From the bulk density the packing density was calculated. This is defined as the ratio of solids to total volume occupied by a mixture. It is found by simply dividing the bulk density by the solid density of the mixture.

4 Materials

4.1 Aggregate
Two aggregates sizes were used in the packing experiments, namely a rounded coarse aggregate and a sand. The rounded coarse aggregate was originally graded at 5 to 20 mm, but it was sieved to obtain a single sized aggregate of between 14 and 20 mm. The density of the aggregate was found to be 2600 kg/m^3. The aggregate was well

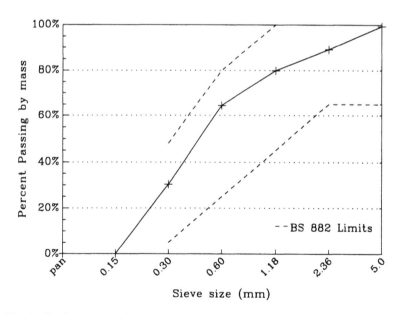

Fig. 1. Grading curve of the sand used, showing the limits for a medium sand.

rounded, like a river gravel. It consisted mainly of sandstone particles with a considerable amount of quartz particles. The largest size of coarse aggregate normally used in fibre concrete is 20 mm, so these are the largest particles which will interact with fibres in concrete. Tests with the largest size of aggregate will give an indication of the loosening effect of fibres. The sand was graded as 0 to 5 mm sand and was used with the fine particles (<0.15 mm) removed. The results of a sieve analysis are shown in figure 1. Figure 1 also shows the specified limits for a medium sand to BS 882 [7].

4.2 Fibres

Three different types of steel fibres were used in the tests. The fibre types were chosen for their varied properties. The fibres were very flexible, high strength Fibra-flex fibres, ductile hooked wire Dramix fibres and rigid machine chipped Harex fibres. This selection of fibres allowed a comparison between their rigidity. Fig. 2 shows a photograph of the three fibre types. A range of sizes were available for the fibres but the results described here were obtained using the following fibres:

- Fibra-flex fibres were 30 mm long, 1.6 mm wide and 30 μm thick.(FF30L6)
- Dramix fibres were 30 mm long and had a diameter of 0.5 mm.(ZL 30/.50)
- Harex fibres were 32 mm long with a varied cross-section. (SF 32.01)

Fibra-flex and Dramix fibres of other dimensions were used to investigate the effects of aspect ratio and size.

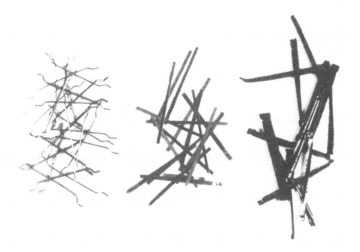

Fig. 2. Fibres used in these tests, from left to right they are :- Dramix ZL 30/.50, Fibra-flex FF30L6 and Harex SF 32.01.

5 Results

A selection of results are shown in figs. 3 to 5. As can be seen from fig. 3, addition of Fibra-flex fibres to aggregate causes a considerable reduction in the packing density, which becomes greater with an increase in fibre content. The reduction is slightly larger with the coarse aggregate than with the sand. With both Harex and Dramix fibres there is virtually no change in the packing density of sand with their addition, although there is a significant decrease in packing density when they were added to coarse aggregate.

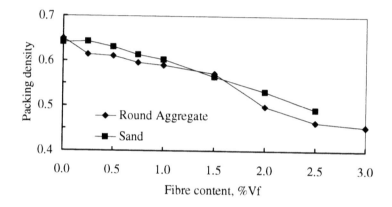

Fig. 3. Results of packing of Fibra-flex fibres and aggregate.

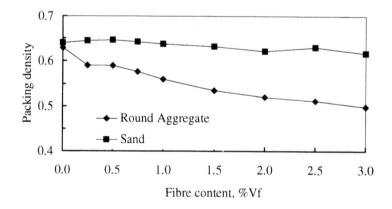

Fig. 4. Results of packing of Dramix fibres and aggregate.

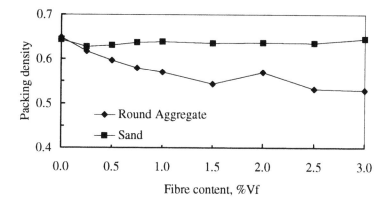

Fig. 5. Results of packing of Harex fibres and aggregate.

An explanation for these results is that the sand is able to pack tightly around a fibre, while the coarse aggregate is effectively pushed apart by the fibre's presence as shown in fig. 6. Inspection of fig. 6c. shows that the introduction of a single fibre can cause an increase in area of three to four times the fibre area. Things are much more complicated when there are many fibres in a three dimensional arrangement, but the figure serves to illustrate what can happen. From this explanation, the results are not unexpected, except for mixtures of sand and Fibra-flex. Observations of Fibra-flex fibres on their own shows that they have a very low packing density, approximately half of that for Harex and Dramix fibres. This means that Fibra-flex fibres have much more void space to be filled with sand. The problem is that these voids are very small, due to the fibre geometry, and as such are difficult to fill even with sand particles. It is often the case that the sand packs round a small clump of fibres (containing voids) rather than around individual fibres. This problem may be solved by using finer sands and this is currently under investigation.

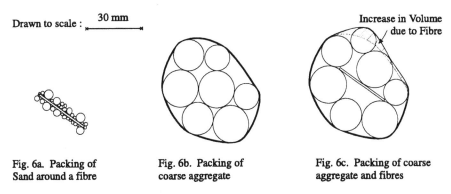

Fig. 6a. Packing of
Sand around a fibre

Fig. 6b. Packing of
coarse aggregate

Fig. 6c. Packing of coarse
aggregate and fibres

Fig. 6. Packing behaviour of sand and coarse aggregate containing fibres.

In total tests on seven fibres have been carried out, with up to 5%V_f in some cases. There have also been a number of tests carried out on angular aggregate which are not presented here. In general the results all followed trends similar to those shown here.

It was observed that the aspect ratio of the fibre has an influence on the packing characteristics. Fig. 7 shows the packing of Dramix fibres with aspect ratios of 37.5, 60 and 75 packed on their own and mixed with coarse aggregate at 3%V_f. As can be seen the lower the aspect ratio the higher the packing density. The dimensions of each fibre are noted on the figure.

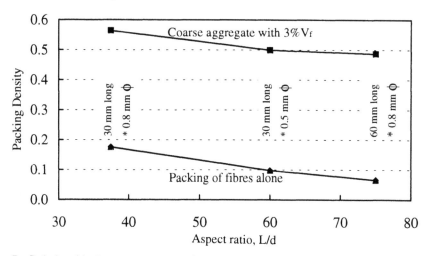

Fig. 7. Relationship between aspect ratio and packing density for Dramix fibres.

6 Discussion

6.1 Factors effecting packing

Results indicate that the aspect ratio of the fibres has an effect on the packing density of mixtures containing them (fig. 7). The aspect ratio also effects the packing of the fibres on their own. This indicates that the packing of the fibres in fact influences the packing of the mixture. The comparison between sand and coarse aggregate packing illustrates the effect that relative size of the particles has on packing. Therefore the main factors which appear to effect the packing of the mixtures is the packing density of the individual components involved, their relative sizes and the proportions. The packing of each individual component is in turn influenced primarily by their shape and also to a lesser extent on their surface characteristics and deformability. The aspect ratio describes the fibre shape and as such has an influence on the packing density of the fibres. The proportions and size of each of the components is normally known from the mix design and manufacturer's specifications (fibres) or grading curves (aggregate). The relative packing density of each component must be found by laboratory measurement. Any attempt to produce a model for packing of mixtures used in FRC must take account of all these factors to produce useful results.

6.2 Use of coarse aggregate

In terms of packing density, the best mixture appears to be one containing only sand and fibres. The detrimental particle interaction is minimised with only sand and, with the exception of Fibra-flex fibres, the sand appears to support the fibres in the mixture. The coarse aggregate tends to allow some settlement of fibres, as the voids are often large enough to allow this. However there are a number of other considerations which must be taken into account. Concrete (or mortar) mixtures with a high proportion of fine aggregate requires more cement paste to remain workable, due to the large surface area of the aggregate. In addition fibre concretes are generally less workable than conventional concrete mixes. Therefore from an economical viewpoint it is better to include some coarse aggregate.

Strength benefits of coarse aggregate inclusion are also worthy of note. Research on fibre reinforced concrete has shown that the maximum feasible fibre content is higher in sand only mixes[8]. The reason for this is that the fibres ball at a lower quantity with coarse aggregate. However the same research also noted that FRC with coarse aggregate gave higher strength benefits than those without. A mix containing $1.5\%V_f$ with only sand had about the same strength improvement as a mix containing $0.5\%V_f$ with a coarse aggregate:sand ratio of 2:1, with all other parts the same. The same mixes, both containing $1.5\%V_f$, yielded strength improvements of 23% and 15% respectively for the mixes with and without coarse aggregate. This would indicate that the benefits of addition of coarse aggregate is twofold - less cement paste will be required and there will be a more significant increase in strength. Of course the use of too much coarse aggregate will allow segregation of fibres.

6.3 Aggregate blending

In normal concrete the mixing of sand and coarse aggregate results in a mixture more dense than either component. It follows that if sand, coarse aggregate and fibres are blended the resulting density will lie between the worst case (coarse aggregate and fibres) and the best case (coarse aggregate and sand). This has been investigated with Harex fibres, producing the results shown in fig. 8. As can be seen, with $3\%V_f$, a ratio of 0.4:0.6 coarse aggregate to sand has a packing density greater than that of sand alone. The result of this blending is a mixture with a significant decrease in surface area, hence less cement paste will be required to coat all of the particles. This should also lead to improved strength based on previous results[8]. As can be seen the optimum ratio of sand to coarse aggregate is different with addition of fibres. Fig. 8 only shows a small number of results, more of which are necessary to find the optimum and how it varies with fibre content. The optimum points will also vary depending on the grading and packing density of each component being used.

The introduction of ultra-fine materials such as silica fume will fill some of the remaining voids so the packing density will further increase. This is to be investigated in the future.

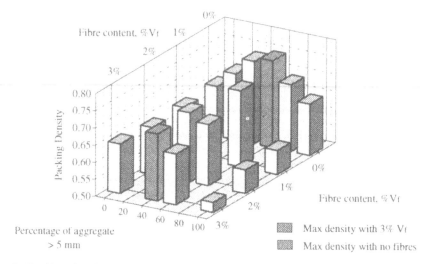

Fig. 8. Packing density of mixtures containing sand, coarse aggregate and Harex fibres

7   Conclusion

This research investigates packing of the particles involved in fibre reinforced concrete. The need for achieving a blend of coarse and fine aggregates, to improve strength and reduce the cement paste requirement is highlighted. The paper shows how this may be achieved by looking at particle interaction. Results show that fibres mixed with sands generally maintain good packing density while those with coarse aggregate decrease their packing density considerably. The parameters which have an influence on this behaviour are also isolated, namely :-

- Packing density of each component
- Volume proportion of each component
- Size of each component

A further consideration of this is can be applied to the mixing process by looking at alternative methods of mixing. More fibres can be added to a mortar mix than a concrete mix with a less detrimental effect, therefore it may be beneficial to add the coarse aggregate just before completion of the mixing.

8   Acknowledgements

This study was carried out in the Civil Engineering Department at the University of Paisley and was funded by the Engineering and Physical Science Research Council (EPSRC). It is part of an ongoing project into production of fibre reinforced concrete.

Thanks to David Wallace, an undergraduate student in the department, for assistance in the laboratory.

9 References

1. Palbøl, L., Goltermann, P. and Johansen, V. (1994) Optimization of Concrete Aggregates. *Betonwerk + Fertigteil-Technik*, Vol. 60, No. 11, pp. 58-62.
2. Mindess, S. and Young, J.F. (1981) *Concrete*, Prentice-Hall, New Jersey, p.239.
3. Malier, Y. (1992) *High Performance Concrete - From Material to Structures*, E & FN Spon, London. pp xiii - xv
4. de Larrard, F. and Sedran, T. (1994) Optimization of Ultra-High-Performance Concrete by the Use of a Packing Model. *Cement and Concrete Research*, Vol. 24, No. 6, pp. 997-1009.
5. ACI Committee 544 (1984) *Guide for specifying, mixing, placing, and finishing steel fiber reinforced concrete.* ACI, Detroit. ACI 544.3R-84.
6. British Standards Institute (1975) *Methods for Testing Aggregate. Part 2: Methods for determination of physical properties.* BSI, London. BS 812:Part 2.
7. British Standards Institute (1983) *Specification for Aggregates from Natural Sources for Concrete.* BSI, London. BS 882.
8. Narayanan, R. and Kareem-Palanjian, A.S. (1982/83) Factors influencing the workability of steel-fibre reinforced concrete. *Concrete*, Vol. 16, No. 10, pp. 45-48 and Vol. 17, No. 2, pp. 43-44.

40 STATISTICAL MODELS TO PREDICT FLOWABILITY, WASHOUT RESISTANCE AND STRENGTH OF UNDERWATER CONCRETE

K.H. KHAYAT, M. SONEBI and A. YAHIA
University of Sherbrooke, Quebec, Canada
C.B. SKAGGS
Kelco Co., Division of Monsanto, San Diego, California, USA

Abstract
Concrete used for underwater placement and repair should be stable, yet fluid enough to spread readily into place without consolidation. A factorial design was carried out to mathematically model the influence of key concrete mixture variables on fresh paste rheology and washout resistance of highly flowable underwater concrete.

Cementitious materials used for the concretes prepared to derive the models systematically included 8% silica fume and 20% fly ash. Crushed granite aggregate with 20-mm nominal size and a natural sand were used. In all, 28 mixtures were prepared to derive the models, and 10 mixtures were prepared to compare measured-to-predicted responses to evaluate the accuracies of the mathematical models. The proposed models are valid for concretes with cementitious materials contents ranging between 380 to 600 kg/m^3, water-to-cementitious materials ratios of 0.34 to 0.46, sand-to-total aggregate ratios of 0.42 to 0.50, as well as anti-washout admixture and high-range water reducer dosages varying between 0.005 and 0.265% and 0.05 and 2.65%, by mass of cementitious materials, respectively.

The resulting models show that slump flow is highly influenced, in order of significance, by the concentrations of anti-washout and cementitious materials, then by the water-to-cementitious materials ratio, dosage of high-range water reducer, and various coupled effects of these parameters. The washout mass loss is similarly affected, in order of importance, by the concentration of anti-washout admixture, cementitious materials content, water-to-cementitious materials ratio, high-range water reducer, and various interactions of these parameters. The sand-to-aggregate ratio had a secondary effect on test results. Comparisons of predicted to measured properties indicate that the models are reliable. Such models can be used to facilitate the test protocol required to proportion highly flowable, yet cohesive concrete for underwater placement.

Keywords: anti-washout admixture, concrete consistency, high-range water reducer, modeling, underwater placement, statistics, strength, washout resistance.

Production Methods and Workability of Concrete. Edited by P.J.M. Bartos, D.L. Marrs and D.J. Cleland.
Published in 1996 by E & FN Spon, 2–6 Boundary Row, London SE1 8HN. ISBN 0 419 22070 4.

1 Introduction

The underwater casting of flat concrete surfaces, such as foundation support of bridge piers, cofferdam seals, or structural elements, often presents a major challenge because of the need to ensure optimal balance between fluidity, cohesiveness, mechanical properties, and durability. Given the geometry and complexity of some underwater cast areas and the imposed placement conditions, including poor visibility and water currents, it may be required that the concrete be highly flowable to spread readily from the discharge point and around various obstacles and reinforcement without consolidation. The differential velocity at the interface between freshly cast concrete and surrounding water can erode some of cementitious materials and other fines. Such erosion can increase the turbidity and contamination of surrounding water and impair strength development, bond to reinforcing steel and existing surfaces, as well as durability of the underwater-cast concrete. Therefore, the concrete should also be highly resistant to water dilution, segregation, and bleeding. The objective of a mixture proportioning is to tailor design a mixture that is fluid enough to facilitate handling and casting given the selected placement method, yet cohesive enough to attain the required in-situ mechanical properties and durability.

Limited guidelines exist for selecting mixture proportioning of flowable, non-dispersive underwater concrete. The optimization of such concrete often necessitates several trial batches before achieving an optimal balance among the various mixture parameters that affect rheological properties, stability, strength, durability, and cost. This is due to the fact that several mixture parameters have contradicting influence on these properties. For example, the risk of water dilution can be reduced by decreasing the fluidity of the concrete or by maintaining a high fluidity and incorporating an antiwashout admixture (AWA). Such AWA increases the viscosity of the concrete and limits its self-consolidation and deformability. A high-range water reducer (HRWR) is therefore necessary to enhance the workability of an AWA concrete, unless more water is allowed in the mixture. However, high fluidity can again increase the risk of water dilution. Other measures necessary to enhance cohesiveness involve the reduction of water-to-cement ratio (W/C). For example, for underwater-cast concrete containing an AWA, standards of the Japan Society of Civil Engineers [1] recommend limiting the W/C to 0.50 and 0.55 for reinforced concrete cast in fresh and seawater, respectively, and 0.60 and 0.65 for non-reinforced concrete in fresh and seawater, respectively. Lower W/C can be used in mixtures with special performance characteristics, for example water-to-cementitious materials ratio (W/CM) of 0.38 to 0.42 for the repair of marine piles [2] and 0.38 for the repair of concrete damaged by abrasion-erosion [3]. The allowable dosage of water depends on the W/C and desired fluidity. For high slump and slump flow values of 250 and 500 mm, respectively, the water content can be on the order of 230 to 250 kg/m^3 [1]. The increase of cementitious materials content or addition of silica fume or sand with adequate content of fines can also improve stability. A high sand-to-total aggregate volume reduces segregation and water dilution and is often used for underwater concrete placement at ranges between 42 and 50%, by mass.

Engineers are frequently faced with the complex task of manipulating several variables to enhance concrete performance and reduce cost. Conventional design methods that vary one mixture parameter at a time are time consuming and produce limited information as to the various interactions that influence each measured property.

The main objective of this paper was to develop and evaluate the feasibility of using a statistical experimental design to optimize key mixture parameters that have significant effect on fresh paste rheology, washout resistance, and compressive strength (f'c) of highly flowable, yet washout resistant concrete. Such models can illustrate the relative significance of primary mixture parameters and the two-way interaction of such parameters on important concrete properties required to ensure successful underwater placement.

The models are based on a given set of cementitious materials, aggregate, and admixtures. Further studies are underway to modify these models in order to take into account different materials. The use of such models can improve the test protocol required to optimize a given mixture proportioning, hence reducing the effort necessary to optimize mixture proportions of a specified concrete.

2 Statistical design approach

The targeted concrete interest for this study is a flowable concrete that can spread readily into place without consolidation and with minimum risk of material separation, including water dilution. Such concrete should have a slump flow greater than 300 mm, a washout mass loss smaller than 10%, and a minimum 28-d f_c of 35 MPa when cast and consolidated above water.

Five key variables that could influence the above relevant properties of underwater concrete were identified to derive mathematical models for fluidity, washout resistance, and strength. The five variables included the concentrations of AWA and HRWR, the W/CM, the content of cementitious materials (CM), and the sand-to-total aggregate ratio (S/A), by mass.

It is important to note that the underlying factors that influence fresh concrete properties and strength development are too complicated to permit the development of an exact mathematical model. Therefore, in this case, an empirical statistical model was derived. Such model is a useful design tool provided the levels for each measured response are orthogonal and obtained over a reasonable working range.

A 2^{5-1} statistical experimental design was used to measure the influence of two different levels of each variable on fresh concrete fluidity, washout resistance, and f'c. Such two-level factorial design offers several advantages:

- It requires a minimum number of tests for each variable.
- Data analysis can identify major trends and predict the most promising direction for future studies.
- It can identify complimentary variables, or so called interactions between variables.
- The derived mathematical model describes the relative significance of each variable, thus providing key information required to optimize the design.
- Data analysis can be efficiently conducted using any number of statistical software packages for personal computers.
- The design may be used as a building block augmenting future studies.

For the development of models for fluidity, washout resistance, and strength of concrete, initial levels of the five selected mixture variables were carefully chosen after reviewing the demand constraints imposed by the raw materials and targeted concrete properties. Coded units of variables -1 and +1 considered in the factorial design corresponded to 430 and 530 kg/m^3 for the cementitious materials content, 0.37 and 0.43 for the W/CM, 0.07 and 0.20% for the AWA dosage, 0.7 and 2% for the HRWR concentration, and 0.42 and 0.50 for S/A.

The experimental design was expanded to include four replicate center points to estimate the degree of experimental error. The central points consisted of mixtures with the five variables set at coded values of zero corresponding to 480 kg/m^3, 0.40, 0.135%, 1.35%, and 0.42 for the cementitious materials content, W/CM, dosage of AWA, HRWR content, and S/A, respectively. Ten additional mixtures were used to expand the 2^{5-1} model to consider extreme values of four of the principal variables that proved to have important influence on the various measured responses. Given the extreme values, the regime of mixture proportioning covered by the model consisted of concretes with 380 to 600 kg/m³ of cementitious materials, W/CM of 0.34 to 0.46, AWA concentration of 0.005 to 0.265%, HRWR dosage of 0.05 to 2.65%, and S/A of 0.42 to 0.50.

The 28 mixture combinations, expressed in coded values, considered in the experimental design are listed in Table 1. The coded units of variables are calculated as follows:

coded AWA = (absolute AWA - 0.135) / 0.065
coded CM = (absolute CM - 470) / 50
coded W/CM = (absolute W/CM - 0.4) / 0.03
coded HRWR = (absolute HRWR - 1.35) / 0.65
coded S/A = (absolute S/A - 0.46) / 0.04

Table 1. Coded units of variables considered in the experimental design

	Mixture	AWA	CM	W/CM	HRWR	S/A
	1	-1	-1	-1	-1	1
	2	1	-1	-1	-1	-1
	3	-1	-1	-1	1	-1
	4	1	-1	-1	1	1
	5	-1	-1	1	-1	-1
Initial	6	1	-1	1	-1	1
2⁵⁻¹	7	-1	-1	1	1	1
model	8	1	-1	1	1	-1
	9	-1	1	-1	-1	-1
	10	1	1	-1	-1	1
	11	-1	1	-1	1	1
	12	1	1	-1	1	-1
	13	-1	1	1	-1	1
	14	1	1	1	-1	-1
	15	-1	1	1	1	-1
	16	1	1	1	1	1
Center	17	0	0	0	0	0
points for	18	0	0	0	0	0
experimental	19	0	0	0	0	0
error	20	0	0	0	0	0
	21	-2	0	0	0	-1
	22	2	0	0	0	-1
	23	0	0	0	-2	-1
Model	24	0	0	0	2	-1
expansion	25	0	-2	0	0	-1
	26	0	2	0	0	-1
	27	0	0	-2	0	-1
	28	0	0	2	0	-1

3   Material Properties

A blended silica fume cement containing 8% silica fume by mass complying with Canadian Standard CSA3-A5-M83 [4] was used. A Class F fly ash with a Blaine fineness of 360 kg/m^2 was added to all mixtures at a dosage of 20% of the mass of cementitious materials in order to enhance workability and reduce cement content. The silica fume cement was primary used to enhance mechanical properties and impermeability. The 20% replacement of cementitious materials by fly ash was adopted following trial batches to determine optimum dosage of fly ash on workability given the type of selected materials. The chemical analysis of the cement and fly ash are presented in Table 2.

A continuously graded, quarried granite coarse aggregate of two sizes (10-5 and 20-5 mm) was used. The coarse aggregate was washed to remove fine sandy particles that can increase water demand. The bulk specific gravity and absorption values of the 20-5 mm coarse aggregate are 2.73 and 0.68%, respectively, and for the 10-5 mm coarse aggregate 2.73, 0.68%, respectively. A well-graded, non-alkali reactive natural sand was used. Its fineness modulus, bulk specific gravity, and absorption values are 2.61, 2.66, and 0.65%, respectively. The grain-size distributions of the two coarse aggregate types and the sand are summarized in Table 3.

Table 2. Chemical analysis of cement and fly ash

	Silica fume cement	Class F fly ash
SiO_2	26.7%	45.0%
Al_2O_3	4.0%	24.6%
Fe_2O_3	3.0%	17.3%
CaO	58.7%	4.3%
SO_3	2.8%	1.3%
MgO	2.7%	0.8%
Na_2O eq.	0.83%	--
LOI	1.1%	3.0%

Table 3. Grain-size distribution of sand and coarse aggregate

	Sieve size (mm)										
	28	20	14	10	5	2.5	1.25	0.63	0.31	0.16	0.08
CA 20-5 mm	100	98.9	97.8	14.5	1.1	--	--	--	--	--	--
CA 10-5 mm	100	100	100	89.3	14.7	1.4	--	--	--	--	--
Sand	100	100	100	100	98.4	94.6	81.8	49.7	12.4	2.2	0.7

A naphthalene-based HRWR with a specific gravity of 1.21 and an active content of 42% was used. The HRWR meets the requirements of Canadian Standard CSA3-A266.6-M85 [4]. Welan gum was selected for the AWA. Welan gum is a high molecular-weight, water-soluble polysaccharide obtained through a controlled microbial fermentation. Welan gum is used to increase the viscosity of mixing water, and hence that of the cement paste. Solutions containing such AWA exhibit pseudo-plastic behavior where the resistance to flow decreases with the increase in shear rate. For example, the apparent viscosities of aqueous solutions containing deionized water, 0.01 Molar NaCl, and 0.3% welan gum, by mass, at shear rates of 0.1, 1, and 100 sec^{-1} are 16700, 2960, and 70 cP, respectively [5].

A hydroxyl carboxilic acid-based set retarder conforming with Canadian Standard CSA3-A266.2.M78 [4] was used to determine the effect of incorporating a set-retarding agent on the demand of HRWR. The dosage of HRWR required to secure various levels of slump flow of the concretes made with and without a set retarder are

non-consolidated concrete. The measurements of slump and slump flow were taken one minute following the removal of the cone. The washout resistance was evaluated using CRD C 61 [6]. The test consists of determining the mass loss of a fresh concrete sample weighing 2.0 ± 0.2 kg which is placed in a perforated basket and allowed to freely fall three times through a 1.7-m column of water. The air content in fresh concrete was measured using a pressure meter (ASTM C 231) [7].

5 Modeling fluidity, washout resistance, and compressive strength

Table 4 summarizes the mixture proportions and test results of the 28 concrete mixtures used in the factorial design. These results were used to develop prediction models of slump flow, slump, washout resistance, and 28-d f'c.

The first four mixtures in Table 4 are replicates at the center points of the experimental matrix made to verify the experimental error of the statistical models. Table 5 shows the mean measured responses of the four replicate mixtures, coefficients of variations, as well as the standard errors with 95% confidence limit for each of the four measured properties. The relative experimental errors for slump, slump flow, and 28-d f_c are shown to be limited to approximately 5%. On the other hand, the relative error for the washout mass loss response was 7.8% indicating the greater degree of experimental error for the washout model.

Table 5. Repeatability of test parameters

	Slump	Slump flow	Washout loss	28-d f_c
Mean (N = 4)	256 mm	484 mm	5.9%	62.0 MPa
Coefficient of variation	3.6%	5.2%	7.9%	2.9%
Estimated error (95% confidence limit)	10 mm	25 mm	0.46%	1.8 MPa
Relative error	3.5%	5.1%	7.8%	2.9%

It is important to note that the slump measurement is not as sensitive as the slump flow in reflecting the consistency of highly fluid concrete. Data obtained from the measurements of slump and slump flows of the 28 concrete mixtures in Table 4, as well as the 10 mixtures in Table 8 used to compare predicted-to-measured responses are used to develop a relationship between slump flow and slump (Fig. 2). For slump values lower than 200 mm, the slump measurement appears to be sensitive in reflecting the variation of fluidity. However, as the slump value approaches 225 mm, the measurement of slump flow becomes more useful in reflecting the level of concrete fluidity than it is with the slump test. The proposed best-fit relationship between slump flow and slump can be expressed as follows ($R^2 = 0.90$):

$$Slump\ flow = 161 + 0.086\ (slump)^2 - 0.0119\ (slump)^{2.5} + 0.00043\ (slump)^3 \quad (1)$$

$$Slump\ flow \geq 200\ mm$$

shown in Fig. 1. All concrete mixtures contained welan gum, a high content (590 kg/m³) of blended silica fume cement and had W/CMs of 0.41. The high content of cementitious materials was selected since the modeled region of concrete mixtures considered in this paper extended between a low of 380 kg/m³ to a high of 600 kg/m³. The set-retarding agent at 0.6 L/m³ (100 mL/100 kg of cementitious materials) significantly reduces the HRWR required to attain a given fluidity. Therefore, all concrete mixtures considered in the experimental design contained a set retarder to reduce HRWR dosage and enhance fluidity retention.

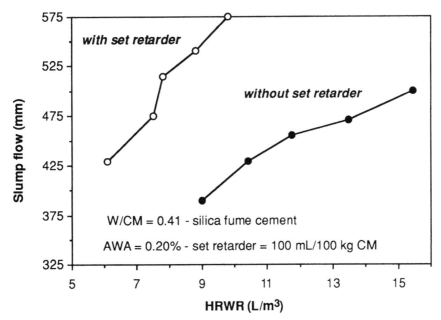

Fig. 1. Effect of set retarder on the demand of HRWR.

4 Experimental procedures

All concretes were prepared in 33-L batches with a rotating drum mixer. The mixture temperature was controlled to produce fresh concrete at 23 ± 2⁰C. The batching sequence consisted of homogenizing the sand and coarse aggregate for 40 sec., then adding half of the mixing water. Following 30 sec. of mixing, the cementitious materials were added, and the mixing was resumed for 80 sec. The remaining of the mix water was then added, and the concrete was mixed for 30 sec. The set retarder was diluted with part of the mixing water and added. Following 30 sec. of mixing, the AWA diluted with the HRWR was introduced. The concrete was then mixed for 3 min. After a minute of rest, the concrete was remixed for an additional minute.

The ambient and concrete temperatures were recorded. For each mixture, slump, slump flow (spread of concrete following the removal of the slump cone), washout mass loss after three drops in water, unit weight, and air content of fresh concrete were determined. Four 100 x 200 mm cylinders were cast and moist-cured to determine the 28-d f'c. The slump and slump flow were used to evaluate the ease of spreading of

Table 4. Mixture proportioning and test results of 28 concretes considered in the experimental design

Mixture	17	18	19	20	1	2	3	4	5	6
W/CM	0.40	0.40	0.40	0.40	0.37	0.37	0.37	0.37	0.43	0.43
Sand/sand + CA, by mass	0.46	0.46	0.46	0.46	0.50	0.42	0.42	0.50	0.42	0.50
HRWR (%)	1.35	1.35	1.35	1.35	0.70	0.70	2.00	2.00	0.70	0.70
AWA (%)	0.135	0.135	0.135	0.135	0.070	0.200	0.070	0.200	0.070	0.200
Silica fume cement (kg/m³)	389	388	389	388	342	345	343	342	349	344
Class F fly ash (kg/m³)	97	97	97	97	85	86	86	85	87	86
Total CM (kg/m³)	487	485	486	485	427	431	428	427	436	430
Extra mixing water (kg/m³)	186	185	185	185	154	156	146	146	183	181
Fine aggregate (kg/m³)	813	809	812	810	940	797	787	934	777	910
Coarse agg. 10/20 (kg/m³)	565	563	565	564	556	653	646	551	637	537
Coarse agg. 5/10 (kg/m³)	377	375	376	376	370	435	430	368	425	358
Volume of coarse agg. (L/m³)	346	344	345	345	339	399	394	337	389	328
Set retarder (mL/100 kg CM)	140	140	140	140	140	140	140	140	140	140
Slump flow (mm)	500	505	480	450	255	206	393	206	602	263
Slump (mm)	253	270	250	253	130	47	230	43	265	175
Washout after 3 drops (%)	6.0	6.2	6.15	5.2	7.1	1.0	11.7	2.4	18.3	2.0
Unit weight (kg/m³)	2437	2427	2433	2429	2452	2476	2450	2439	2463	2421
Air content (%)	2.0	1.0	2.2	2.3	3.4	2.6	3.2	3.4	1.4	3.5
Ambient temperature (C)	25.1	20.1	22.8	24.5	23.9	24.6	22.6	24.7	24.0	23.9
Concrete temperature (C)	24.3	22.8	21.7	22.2	23.2	23.4	23.2	23.9	23.2	23.9
28-d f'c (MPa)	61.7	63.4	59.5	63.3	64.3	64.9	63.3	61.3	58.5	53.7
C.O.V. (%)	2.5	2.1	4.8	1.8	2.0	2.0	0.9	1.5	2.9	2.6

Table 4. continued

Mixture	7	8	9	10	11	12	13	14	15
W/CM	0.43	0.43	0.37	0.37	0.37	0.37	0.43	0.43	0.43
Sand/sand + CA, by mass	0.50	0.42	0.42	0.50	0.50	0.42	0.50	0.42	0.42
HRWR (%)	2.00	2.00	0.70	0.70	2.00	2.00	0.70	0.70	2.00
AWA (%)	0.070	0.200	0.070	0.200	0.070	0.200	0.070	0.200	0.070
Silica fume cement (kg/m³)	344	346	430	429	438	428	426	434	438
Class F fly ash (kg/m³)	86	86	107	107	109	107	107	108	110
Total CM (kg/m³)	430	432	537	536	547	535	533	542	548
Extra mixing water (kg/m³)	173	174	194	193	187	183	224	228	220
Fine aggregate (kg/m³)	904	764	719	852	864	709	805	687	689
Coarse agg. 10/20 (kg/m³)	534	627	589	503	510	582	476	563	565
Coarse agg. 5/10 (kg/m³)	356	418	393	336	340	388	317	375	376
Volume of coarse agg. (L/m³)	326	383	360	307	311	355	290	344	345
Set retarder (mL/100 kg CM)	140	140	140	140	140	140	140	140	140
Slump flow (mm)	555	368	552	289	827	392	750	480	760
Slump (mm)	270	225	265	180	280	235	280	265	275
Washout after 3 drops (%)	20.9	2.5	12.2	1.6	59.8	3.3	61.1	4.9	33.0
Unit weight (kg/m³)	2410	2427	2437	2424	2463	2412	2359	2400	2414
Air content (%)	3.5	2.3	1.6	2.6	0.3	3.0	0.5	1.1	2.10
Ambient temperature (C)	22.8	24.2	25.0	25.8	24.5	25.3	24.5	27.3	24.6
Concrete temperature (C)	22.2	23.6	23.1	25.0	22.6	23.4	22.8	26.7	23.3
28-d f'c (MPa)	54.7	56.7	65.5	63.2	45.3	61.9	53.9	55.8	52.1
C.O.V. (%)	1.2	2.1	4.7	2.4	6.1	3.9	2.4	3.4	4.0

Table 4. continued

Mixture	16	21	22	23	24	25	26	27	28
W/CM	0.43	0.40	0.40	0.40	0.40	0.40	0.40	0.34	0.46
Sand/sand + CA, by mass	0.50	0.42	0.42	0.42	0.42	0.42	0.42	0.42	0.42
HRWR (%)	2.00	1.35	1.35	0.05	2.65	1.35	1.35	1.35	1.35
AWA (%)	0.200	0.005	0.265	0.135	0.135	0.135	0.135	0.135	0.135
Silica fume cement (kg/m^3)	434	394	391	389	391	304	481	391	397
Class F fly ash (kg/m^3)	109	99	98	97	98	76	120	98	99
Total CM (kg/m^3)	543	493	489	486	489	380	601	488	496
Extra mixing water (kg/m^3)	218	188	186	194	178	146	229	157	219
Fine aggregate (kg/m^3)	812	753	746	747	741	830	666	779	722
Coarse agg. 10/20 (kg/m^3)	479	616	610	613	607	680	546	638	592
Coarse agg. 5/10 (kg/m^3)	319	411	408	409	405	453	364	425	394
Volume of coarse agg. (L/m^3)	292	376	373	374	371	415	333	389	361
Set retarder (mL/100 kg CM)	140	140	140	140	140	140	140	140	140
Slump flow (mm)	563	792	340	200	510	210	612	273	622
Slump (mm)	267	273	216	0	260	57	280	163	277
Washout after 3 drops (%)	9.6	40.3	3.6	2.1	8.5	4.1	12.3	2.7	13.8
Unit weight (kg/m^3)	2387	2470	2448	2449	2437	2498	2417	2497	2433
Air content (%)	1.7	0.2	2.4	1.8	2.3	3.3	1.5	2.2	1.2
Ambient temperature (C)	23.2	23.4	24.5	25.6	24.2	24.5	22.4	25.8	25.5
Concrete temperature (C)	21.8	23.9	23.2	25.0	25.2	23.9	23.5	24.6	24.9
28-d f$'$c (MPa)	56.4	44.2	62.2	45.3	59.9	63.8	61.7	67.9	57.2
C.O.V. (%)	0.7	3.9	4.4	3.3	1.7	1.3	2.4	5.0	2.2

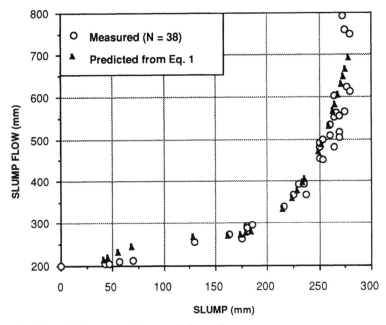

Fig. 2. Relationship between slump and slump flow measurements.

The derived statistical models for the slump flow, slump, washout mass loss, and 28-d f'_c along with correlation coefficients and Prob. >|t| values are shown in Table 6. The estimates for each parameter refer to the coefficients of the model found by least squares. The Prob. >|t| is the probability of getting an even greater t statistic, in absolute value, that tests whether the true parameter is zero. Probabilities less than 0.05 are often considered as significant evidence that the parameter is not zero, i.e. that the contribution of the proposed parameter has a highly significant influence on the measured response.

The presentation in Table 6 enables the comparison of various parameters as well as the interactions of the four measured responses. Except for a few interactions between some variables, the probabilities that the derived coefficients of the various parameters influencing each response are limited to 5%. This signifies that there is less than 5% chance, or 95% confidence limit, that the contribution of a given parameter to the tested response exceeds the value of the specified coefficient.

Unlike the other models, the strength model was divided using data generated in this paper which includes the 28 mixtures in Table 4 and the mixtures in Table 8. This was done to enhance the prediction accuracy of the strength model.

The correlation coefficients of the proposed models for slump flow, slump, washout mass loss, and 28-d f'c are 0.95, 0.94, 0.95, and 0.65, respectively. The high correlation coefficients of the first three responses demonstrates excellent correlations where it can be considered that at least 95% of the measured values can be accounted for with the proposed models. The low correlation coefficients of the strength model may be due to the fact that the range of measured 28-d f'c was narrow compared to the wide ranges of other measured responses. The spread between minimum and maximum values for all mixtures used to derive the slump flow, slump, and washout were 627 mm, 280 mm, and 60%, respectively, compared to 23.7 MPa for the strength results. The limited range of strength results given by the fact that the considered W/CM ranged from 0.34 to 0.46, can reduce the variability of the 28-d f'c results compared to the other models. This observation is confirmed by comparing the coefficients of variations of the slump flow, slump, washout, and 28-d f'c which were 41, 39, 126, and 11%, respectively.

Table 6. Parameter estimates of the four statistical models

Parameter	Slump flow $R^2 = 0.95$ Estimate	Prob. > |t|	Slump $R^2 = 0.94$ Estimate	Prob. > |t|	Ln washout $R^2 = 0.95$ Estimate	Prob. > |t|	28-d f'c $R^2 = 0.65$ Estimate	Prob. > |t|
Intercept	*460*		*246*		*1.841*		*62.1*	
W/CM	80.0	0	34.9	0	0.357	0	-2.6	0
CM	116.3	0	46.1	0	0.409	0	-1.08	0.163
AWA	-118.0	0	-27.9	0	-0.936	0	2.5	0
HRWR	53.7	0	11.5	0.039	0.294	0	NA	NA
S/A	NA	NA	-9.1	0.087	0.147	0.026	-1.16	0.141
AWA.AWA	28.2	0.012	NA	NA	0.202	0	-1.96	0
CM.AWA	-25.3	0.067	15.6	0.014	NA	NA	1.53	0.046
S/A.CM	27.5	0.025	NA	NA	0.115	0.075	NA	NA
HRWR.HRWR	-24.6	0.025	NA	NA	NA	NA	-2.28	0.005
W/CM.HRWR	-22.7	0.098	NA	NA	-0.214	0.007	NA	NA
W/CM.CM	NA	NA	-22.3	0	NA	NA	NA	NA
CM.CM	NA	NA	-21.8	0	NA	NA	NA	NA
W/CM.AWA	NA	NA	15.0	0.017	NA	NA	NA	NA
W/CM.W/CM	NA	NA	-8.9	0.064	NA	NA	NA	NA
S/A.AWA	NA	NA	NA	NA	-0.156	0.020	NA	NA
S/A.HRWR	NA	NA	NA	NA	NA	NA	-1.31	0.096

The contributions of the various parameters to the four measured responses are listed in order of significance in Table 7. A negative estimate signifies that an increase of the given parameter results in a reduction of the measured response. For each model, the effects of the primary parameter with the highest estimates vs. the contributions of other main parameters are given in Table 7. The increase in the content of cementitious materials within the range of the model can then be interpreted to have 1.5 times greater influence on enhancing the slump flow than the increase in W/CM, given that the AWA dosage and S/A are held constant. Similarly, the increase in AWA can be said to reduce washout mass loss by 3.2 times more than the reduction in HRWR. In comparing the relative effectiveness of each parameter, it is essential to consider the effect of independent variables. For example, the trade off between the effect of AWA and HRWR can be evaluated for mixtures containing a fixed content of cementitious materials since both variables are expressed in terms of cementitious materials content.

For any given response, the presence of parameters with coupled terms (HRWR.HRWR or AWA.AWA) indicates that the influence of the model is quadratic. As shown in Table 7, slump flow is influenced, in order of significance, by the concentrations of AWA and cementitious materials, the W/CM, dosage of HRWR, and various coupled effects of these parameters. The washout mass loss is similarly affected, in order of importance, by the concentration of AWA, content of cementitious materials, W/CM, HRWR, and various interactions of these parameters.   In both cases, the S/A had a secondary effect on test results.

Table 7. Statistical models and relative influence on various properties

Slump flow (mm)	Slump (mm)	Ln washout (%)	28-d f'c (MPa)
460	246	1.841	62.1
-118.0 AWA	46.1 CM	-0.936 AWA	-2.6 W/CM
116.3 CM (1.0)*	34.9 W/CM (1.3)	0.409 CM (2.3)	2.5 AWA (1.0)
80.0 W/CM (1.5)	-27.9 AWA (1.7)	0.357 W/CM (2.6)	-2.28 HRWR.HRWR
53.7 HRWR (2.2)	-22.3 W/CM.CM	0.294 HRWR (3.2)	-1.96 AWA.AWA
28.2 AWA.AWA	-21.8 CM.CM	-0.214 W/CM.HRWR	1.53 CM.AWA
27.5 S/A.CM	15.6 CM.AWA	0.202 AWA.AWA	-1.31 S/A.HRWR
25.3 CM.AWA	15.0 W/C.AWA	-0.156 S/A.AWA	-1.16 S/A (2.2)
-24.6 HRWR.HRWR	11.5 HRWR (4.0)	0.147 S/A (6.4)	-1.08 CM (2.4)
-22.7 W/CM.HRWR	-9.1 S/A (5.1)	0.115 S/A.CM	
	-8.9 W/CM.W/CM		

* relative effect of primary parameter vs. other tested primary parameter

6   Evaluation of accuracy of proposed models

To verify the accuracy of the proposed models of slump flow, slump, and washout mass loss, 10 concrete mixtures were prepared to measure various responses corresponding to very stiff to highly fluid concretes that lie within the modeled range of mixture proportioning. The mixture proportionings and properties of the 10 concretes are summarized in Table 8. The predicted-to-measured values for slump flow, slump, washout mass loss, and 28-d f'c are shown in Fig. 3, 4, 5, and 6, respectively. On each figure, the four duplicative mixtures used as central points in the prediction models are also included. The errors corresponding to 95% confidence limits (Table 5) of slump flow ($\pm$ 25 mm), slump ($\pm$ 10 mm), washout mass loss ($\pm$ 0.5%), and 28-d f'c ($\pm$ 2%) are indicated on Fig. 3, 4, 5, and 6, respectively.

The majority of measured slump flow, slump, and 28-d f'c values lie close to the predicted values. On the other hand, deviations up to 2% were obtained between the measured and predicted washout values. On the average, the ratios of predicted-to-measured slump flow, slump, washout mass loss, and 28-d f'c for the 14 mixtures were 0.95, 0.98, 1.04, and 0.99, respectively. The coefficient of variations of these ratios were 11.1, 8.3, 32.0, and 3%, respectively. In general, the proposed models for slump flow, slump, and 28-d f'c appear to be satisfactory in predicting fluidity and strength with low scattering between the measured and predicted values. On the average, the washout resistance model provides good prediction for the actual values. However, even when the estimated error in measuring the washout mass loss is taken into consideration, the scattering of predicted-to-measured ratio of the washout mass loss is considerably greater than similar ratios for the other three responses.

Table 8. Mixture proportioning used to compare predicted-to-measured properties

Mixture	29	30	31	32	33	34	35	36	37	38
W/CM	0.43	0.37	0.37	0.43	0.40	0.43	0.43	0.40	0.37	0.415
Sand/sand + CA, by mass	0.42	0.42	0.50	0.50	0.42	0.42	0.46	0.46	0.42	0.48
HRWR (%)	1.00	1.00	1.00	1.00	0.67	0.51	1.35	1.35	0.70	1.68
AWA (%)	0.255	0.255	0.255	0.255	0.260	0.200	0.200	0.135	0.070	0.168
Silica fume cement (kg/m^3)	436	347	429	343	390	436	439	436	435	415
Class F fly ash (kg/m^3)	109	87	107	86	97	109	110	109	109	104
Total CM (kg/m^3)	545	434	536	429	487	545	549	545	544	519
Extra mixing water (kg/m^3)	229	156	193	180	194	231	225	146	196	204
Fine aggregate (kg/m^3)	691	802	853	907	748	692	758	775	727	822
Coarse agg. 10/20 (kg/m^3)	566	657	503	536	613	567	527	538	596	526
Coarse agg. 5/10 (kg/m^3)	377	438	336	357	409	378	351	359	397	351
Volume of coarse agg. (L/m^3)	346	401	307	327	375	346	322	329	364	322
Set retarder (mL/100 kg CM)	140	140	140	140	140	140	140	140	140	140
Slump flow (mm)	490	212	295	293	367	433	535	555	565	517
Slump (mm)	250	70	185	180	237	258	260	270	275	270
Washout after 3 drops (%)	4.5	1.8	1.1	3.3	2.9	4.1	5.0	8.7	10.4	5.6
Unit weight (kg/m^3)	2414	2492	2426	2414	2452	2416	2420	2435	2464	2434
Air content (%)	1.8	2.6	2.6	3.0	1.6	1.3	1.3	1.1	1.2	1.3
Ambient temperature (C)	23.4	22.0	23.9	22.7	25.2	24.3	20.1	20.2	20.3	20.4
Concrete temperature (C)	22.8	22.6	22.8	22.4	24.8	24.1	22.3	23.2	23.0	22.7
28-d f'c (MPa)	59.1	59.7	62.3	56.9	59.3	59.0				
C.O.V. (%)	2.4	10.5	4.0	3.0	2.4	2.5				

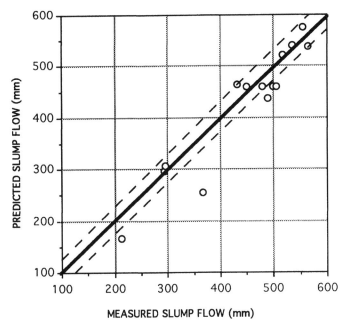

Fig. 3. Predicted vs. measured slump flow values.

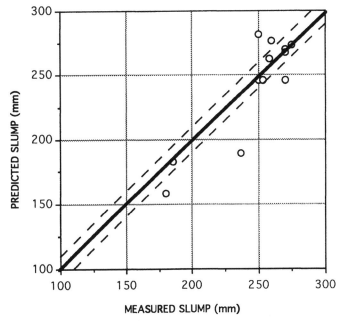

Fig. 4. Predicted vs. measured slump values.

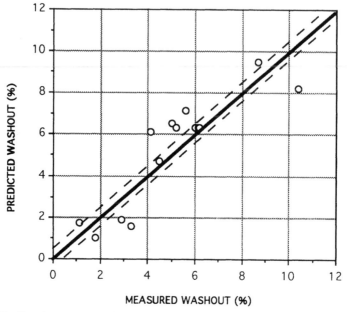

Fig. 5.   Predicted vs. measured washout mass loss values.

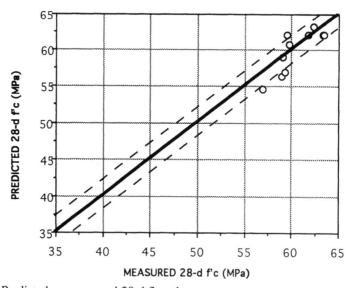

Fig. 6.   Predicted vs. measured 28-d f'c values.

7 Comparison of predicted fluidity and washout resistance

The derived statistical models can be used to compare fluidity to washout resistance. Such comparison is shown in Fig. 7 for 220 data points representing various possible combinations of the five key mixture parameters that lie within the range of the models. As the fluidity of concrete increases, the resistance to washout is reduced. However, for any given level of fluidity, it is shown that mixtures with relatively low risk of water dilution can be obtained. For example, at slump flow values below 300 mm, it is possible to proportion concretes with washout mass losses as low as 2%. For concrete mixtures with slump flow values between 500 and 550 mm, it is possible to proportion mixtures with washout mass losses on the order of 7%, although values as high as 15% can also be obtained.

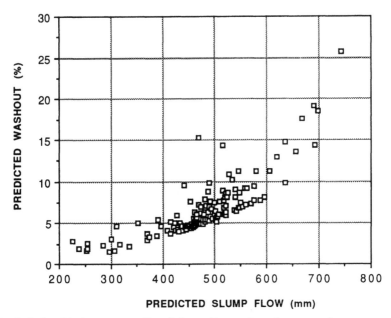

Fig. 7 Relationship between predicted slump flow and washout mass loss.

The proposed statistical models can also be used to test the effects of a group of variables on properties affecting the quality of underwater-cast concrete. For example, the effect of increasing the W/CM vs. that of the HRWR on fluidity and the impact on washout resistance and strength can be evaluated for mixtures with fixed contents of cementitious materials, concentrations of AWA, and S/A values. Furthermore, the models can be used along with material cost to obtain economical and high-performance concrete for underwater placement. This can reduce the effort often required in carrying out trial batches for mixture optimization.

8 Extension of existing models

Mathematical models can provide efficient design techniques, however, their limitations must be considered. A key benefit of using statistical experimental design is to employ the existing model as a building block to augment future studies. For example, a future mixture proportioning could involve different combinations and types of cementitious materials and/or different admixtures. The existing models can be used for mixture optimization even when mixture constituents are changed, providing that such changes do not significantly affect the prediction accuracy of the model. Assuming that for the proportioning of a flowable underwater concrete, the same cementitious materials and type of AWA and HRWR considered in deriving the existing models are available, however, a different kind of aggregate and sand would have to be used. Such aggregate is likely to influence the models in Table 6, but to what degree remains to be seen. A logical design approach would be to use the existing model to predict the optimal design, then carry out selected tests to quantify the influence of the new aggregate and sand on the model. A minimal estimate is obtained by repeating selected tests at levels within the desired range for the new job specifications. For example, if the specified slump flow, maximum washout mass loss, and minimum 28-d f'c are 370 $\pm$ 30 mm, 5%, and 40 MPa, mixtures 8, 12, and 22 of Table 4 can be duplicated. These mixtures are selected because the predicted properties from the existing model lie within the specified range of fluidity, washout resistance, and strength. The data obtained from the duplicated mixtures are then compared with the predicted values from the existing models.

The above approach can also be used to verify the reliability of the models for a greater range of concrete properties. For example, if the slump flow and washout mass loss should be limited to 620 mm and 15%, mixtures with predicted responses corresponding to extreme and central values can be duplicated. These mixtures can include mixtures 2, 17, 26, and 28 in Table 4. Again, the degree of variation between the new tests and predicted values can indicate the validity of the existing models to predict material properties when the source of sand and aggregate is changed. This can determine the additional test protocol required for the new materials.

When the variability between measured and predicted responses is unacceptably large, a pooled statistical design can be carried out to incorporate an extra variable to the experimental design matrix in Table 1. In the case of using a different sand and aggregate, the sixth variable will be S/A for the second aggregate type. A limited number of mixtures can then be prepared to adjust the existing model to take into consideration the influence of the newly considered sand and aggregate on concrete properties that are relevant to the quality of underwater-cast concrete.

When several material constituents are changed, and the existing models in Table 6 fail to accurately predict the various responses, a simplified experimental design can be used to derive modified models. With the understanding of the effect of key mixture variables on concrete properties elaborated by the models presented here, it is possible to construct a new experimental design using a limited number of mixtures to accommodate for any changes in responses resulting from the use of new types of mixture constituents. Instead of carrying out 28 mixtures that were necessary to derive the existing models, less than 10 mixtures may be necessary to obtain models for the newly considered materials.

9 Conclusion

The models established using a statistical design approach provide an efficient means to evaluate the influence of key variables on mixture proportioning. Such models can provide cost effective tool for evaluating the potential mixture proportions by weighing out the effects of key parameters on desired properties necessary to obtain a high-performance underwater concrete. The proposed models can reduce the extent of trial batches needed to achieve optimum balance among various mixture variables which affect flowability, washout resistance, and strength. The derived models include mixtures with 380 to 600 kg/m^3 of cementitious materials, 0.34 to 0.46 W/CM, 0.42 to 0.50 for the S/A, as well as AWA and HRWR dosages varying between 0.005 and 0.265% and 0.05 and 2.65%, respectively. Slump flow measurement is shown to be more sensitive in reflecting the deformability of highly flowable concrete than measuring the slump value.

The models indicate that the slump flow is highly influenced, in order of significance, by the concentrations of AWA and cementitious materials, then by the W/CM, dosage of HRWR, and various coupled effects of these parameters. The washout mass loss is similarly affected, in order of importance, by the AWA concentration, cementitious materials content, W/CM, HRWR content, and various interactions of these parameters. The S/A had a secondary effect on test results. The proposed model for washout mass loss offers a good prediction of measured values. However, the degree of scattering of the predicted-to-measured values is high. On the other hand, the slump, slump flow, and f'c models appear to be accurate in predicting responses and have low scattering between predicted and measured values.

10 References

1. Japan Society of Civil Engineers. (1991) *Recommendations for Design and Construction of Antiwashout Underwater Concrete*, Concrete Library of JSCE, No. 67, 89 p.
2. Khayat, K.H. (1992) In-situ Properties of Concrete Piles Repaired Under Water. *ACI Concrete International*, Vol. 14, No. 3, pp. 42-49.
3. Hester, W.T., Khayat, K.H., Gerwick, B.C. (1989) Properties of Concretes for Thin Underwater Placements and Repairs. *ACI SP114*, Vol. 1, pp. 713-731.
4. CAN/CSA-A23.5-M90, Standard
5. Khayat, K.H. (1995) Effects of Anti-Washout Admixtures on Fresh Concrete Properties. *ACI Materials Journal*, Vol. 92, No. 2, pp. 164-171.
6. U.S. Army Engineer Waterways Experiment Station. (1949) *Handbook for Concrete and Cement*, U.S. Army Engineer Waterways Experiment Station, Vicksburg, Mississippi, (with quarterly supplements).
7. *Annual Book of ASTM Standards* (1995) *Concrete and Aggregates*, ASTM, Philadelphia, V. 04.02.

41 A MODEL FOR SELF-COMPACTING CONCRETE

Ö. PETERSSON and P. BILLBERG
Swedish Cement and Concrete Research Institute, Stockholm, Sweden
B.K. VAN
Department of Civil Engineering, International Institute of Technology,
Thammasat University, Bangkok, Thailand

Abstract

Self compacting concrete is a highly flowable concrete. The concrete is without consolidation capable of filling into corners of formwork through congested reinforcement without causing segregation of coarse aggregate. Mortar for self compacting concrete has been tested in a paste viscometer. The Bingham model has been used to express the concrete's rheology in yield stress and plastic viscosity. By this method different types of filler have been investigated. The amount of and type of superplasticizers have been tested. Testing has also been done on some types of viscosity agents. The evaluation of the concrete has been done with the so-called slump flow and the L-box. The investigation has given a method and criteria to achieve self compacting concrete. Using the model, different binder, filler and admixture can be evaluated in the process to achieve a mix composition for self compacting concrete.

Keywords: Mix design, self compacting concrete, super workable concrete, workability.

1 Introduction

In recent years research and development of self compacting concrete has been carried out by many research organisations in Japan. The concept "High Performance Concrete" was proposed by Ozawa et al [1],[2] in 1989. High Performance Concrete is one of several names of this type of concrete. Other names are "Self Compacting Concrete", "Super Workable Concrete", "Self-placeable Concrete". In this paper the name "Self Compacting Concrete" SCC will be used.

The main interest for using SCC has been to improve the quality of concrete work and to give ability for automation of construction work. It will also improve the

Production Methods and Workability of Concrete. Edited by P.J.M. Bartos, D.L. Marrs and D.J. Cleland.
Published in 1996 by E & FN Spon, 2–6 Boundary Row, London SE1 8HN. ISBN 0 419 22070 4.

working conditions on construction sites.

SCC is without consolidation capable of filling into corners of formwork through congested reinforcement without causing segregation of coarse aggregate. To produce SCC it is necessary to add superplasticizers to create high mobility and eliminate segregation by adding viscosity admixture or a large volume of powdered material. The ability of concrete to pass between the steel reinforcing bars is controlled by the rheological properties of mortar and volume of coarse aggregate. A model for mix design of SCC is outlined in Fig. 1.

The different steps in the process will be discussed; research experiments together with background of the different steps will be given below.

2 Construction criteria

Construction criteria are given by the special demands that exist for each project. Some of them can be:

- Concrete strength.
- Concrete strength by time.
- Durability
- Gap between reinforcement

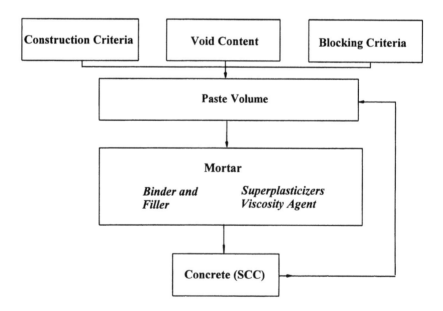

Fig. 1 Simplified process for mix design of SCC

The given concrete strength for the project will automatically specify the water to cement ratio (or water to binder ratio) accordingly to normal relationship between concrete strength and water to cement ratio. Concrete strength by time will also inflict the W/C ratio and/or specify a special type of cement. Durability demands will probably require a special type of cement , air entraining agent and other type of specification. The construction criteria step will give the "normal" specification (except from gap between reinforcement) for mix design without the special task to achieve self compacting concrete.

3   Void content

The first step is to find the minimal paste volume from the mixture between coarse and fine aggregate. This is done by measuring the void content for different coarse to fine aggregate ratio. The coarse to fine aggregate ratio affects not only the void content but also the total aggregate surface area. The method used to find the void content is based on a slightly modified ASTM C 29/C29M, *Van* [3].

The minimum paste volume (V_{pw}) should fill all voids between the aggregate and also cover all surfaces of the aggregate particles. Two different mixtures of fine and coarse aggregate can have different surface area even if they have the same solid volume. Larger surface area of the aggregate requires larger covering paste volume to give the same deformability.

In the investigation two types of aggregate were used, both with maximum aggregate size of 16 mm. The grading curve for sand and gravel is given in Fig. 2. Both the sand and gravel are river type aggregate. The gravel is partly crushed.

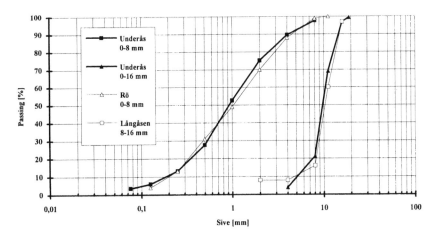

Fig. 2.   Particle size distribution curves for two types of fine and coarse aggregate.

4   Blocking criteria

Ozawa et al [4] have studied the mechanism of mortar flowing through round holes and the role of sand in fresh mortar on blocking. They found that the risk of blocking can be computed by the linear summation of the effect of each single size of sand. They proposed an equation for computing the risk for blocking according to Eq. 1.

$$\text{Risk of blocking} = \sum (n_{si} / n_{sbi}) \leq 1 \tag{1}$$

n_{si} = Volume ratio of an aggregate of single-size group i (to total volume of concrete).
n_{sbi} = Blocking volume ratio of an aggregate of single-size group i (to total volume of concrete).

Notice the similarity to Miner particial-damage hypothesis used in construction for calculation of fatigue risk.

Sand is defined as particles greater than approximately 1/10 of holes. Particles smaller than this size, including powder, have different roles on blocking than sand. Ozawa et. al. [4] also found that the ratio between size of opening and the size of sand particles is also an important factor for blocking of flowing mortar. Van [3], Tangtermsirikul and Van [5] has been studying the effect of total aggregate, which means coarse and fine aggregate on blocking. At CBI we have studied the blocking risk for aggregate type Underås (river coarse aggregate). We found a relationship accordingly to Fig. 3 between blocking volume aggregate ratio (n_{abi}) and clear spacing to particles fraction ratio (c/D_{af}). In Fig. 3. a curve from reference [5] is also shown. The aggregate type is crushed coarse limestone aggregate.

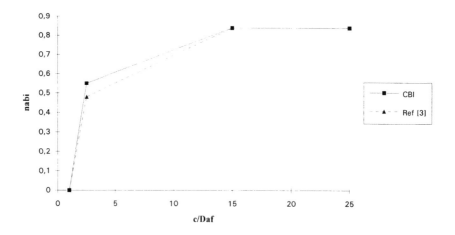

Fig. 3   Blocking volume ratio to clear spacing to fraction diameter of particle

$n_{abi} = V_{abi}/V_t$

V_{abi} = Blocking volume of aggregate group i

V_t = Total volume of the concrete mix

c = Clear spacing between reinforcement

$D_{af} = M_{i-1} + 3/4 \, (M_i - M_{i-1})$,

M_i and M_{i-1}= Upper and lower sieve dimension of aggregate

If we compare working with Miners rule for fatigue design then Fig. 3. express the fatigue model curve for the used material.

With the model, accordingly to Fig. 3, we can calculate the maximum total aggregate content for not causing blocking. The following Eq. 2 can be used for calculation of maximum allowable aggregate volume (means minimum paste volume) corresponding to ratio between coarse and total aggregate according to blocking criteria.

$$\text{Risk of blocking} = \sum_{i=1}^{n}(n_{ai}/n_{abi}) = \sum_{i=1}^{n}\frac{(V_{ai}/V_t)}{(V_{abi}/V_t)} = \sum_{i=1}^{n}(V_{ai}/V_{abi}) = 1 \qquad (2)$$

V_{ai} = Volume of aggregate group i

V_{abi} = Blocking volume of aggregate group i

4   Paste volume

First the void content for the mixture between coarse aggregate and fine aggregate is investigated accordingly to method mentioned in chapter 2. Ten different ratios between amount of coarse aggregate (kg) to total aggregate (S_a) has been investigated for aggregate type "Underås". The void content of 100 % fine or 100 % coarse aggregate is always measured. Obtained results for the aggregate is given in Fig. 4. As can be seen from Fig. 4. the minimum paste volume is given by approximately 58 % coarse aggregate to total aggregate.

By using equation 2 together with the blocking criteria accordingly to Fig. 3. we can calculate the minimum paste content for different gravel to total aggregate ratios, see Fig. 4. We can now decide the amount of paste depending on the choosen free spacing between reinforcement (gap).

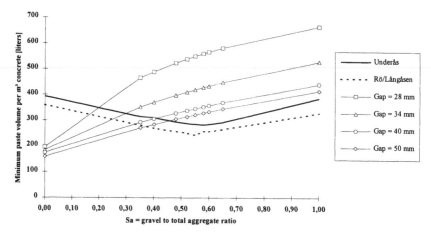

Fig. 4.   Relation between minimum paste volume for different gap between reinforcement. The paste volume curve for different ratio between gravel to total aggregate is also shown in the figure.

5   Mortar

Different types of binder and filler have been investigated. The investigation has been carried out in a paste viscometer of type Haake rotovisco Rv 20 rheometer (measuring system CV 20). Mortar with maximum aggregate size of 0,25 mm taken from the aggregate type "Underås" has been used together with different binder and filler, see Table 1. In an investigation at CBI [6] we found that the use of aggregate smaller than 0,25 mm gave good correlation between mortar and concrete and the repeatability was better with mortar than with only paste in the reometer.

Table 1   Different type of fillers.

Type	Mineral	Blaine (m²/kg)	Average diameter (µm)	Density (kg/m³)
Cement	Slite (S)	360	12	3,12
Cement	Skövde (SH)	500	18	3,13
Myanit C	Dolomit		60	2,85
Myanit E	Dolomit		36	2,85
Myanit 0-10	Dolomit		4	2,85
Myanit 0-30	Dolomit		14	2,85
I100	Limestone		-	2,85
I200	Limestone		10	2,85
K500	Limestone		20	2,85
Silica fume	SiO₂		-	2,70
Merit 5000	Blast furnace slag		12	2,94
Flyash	Flyash		25	2,20
Microfiller©	Glass		50	2,50

Table 2 shows the different mixes that was used in the investigation of mortar. The mixes with silica slurry are not really comparable to the others due to different amount of superplasticizers

Table 2   Mix design for tests.

Test nr	C type	C kg/m³	vbt	S_a	Filler type	Filler kg/m³	SP type	SP %*	Visco type	Visco %**
60	Slite	525	0,345	0,520	-	-	Sik12	2,5	-	-
61	Slite	325	0,345	0,520	Merit 5000	200	Sik12	2,5	-	-
62	Slite	325	0,345	0,520	Myanit C	200	Sik12	2,5	-	-
63	Slite	325	0,345	0,520	Myanit 0-10	200	Sik12	2,5	-	-
64	Slite	325	0,345	0,520	Myanit E	200	Sik12	2,5	-	-
66	Slite	325	0,345	0,520	Glass	200	Sik12	2,5	-	-
67	Slite	325	0,345	0,520	Fly ash	200	Sik12	2,5	-	-
68	Slite	325	0,345	0,520	I 200	200	Sik12	2,5	-	-
69	Slite	325	0,345	0,520	K 500	200	Sik12	2,5	-	-
70	Slite/SH	325	0,345	0,520	Myanit C	200	Sik12	2,5	-	-
71	Slite	325	0,345	0,520	Myanit C	180	Sik12	2,5	-	-
72	Slite	400	0,550	0,475	Silica slurry	10	92M	0,7	1	2,66
73	Slite	400	0,550	0,470	Silica slurry	0	92M	0	2	9,09
80	Slite	325	0,345	0,520	Myanit C	100	Sik12	2,5	3	0,01
81	Slite	325	0,345	0,520	Myanit C	150	Sik12	2,5	3	0,008

*	Superplasticizer dose [% of (filler+cement) weight]
**	Viscosity agent [% of waterweight]

In Fig. 5. the result with different binders and fillers is shown. The "Bingham model" has been used to calculate the results into yield shear stress (τ_0) and plastic viscosity (μ .

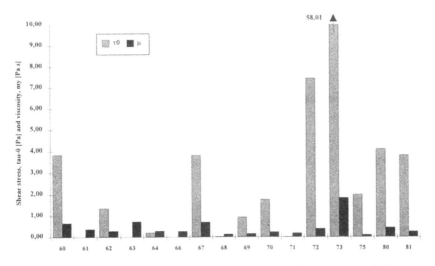

Fig. 5.   Result from viscosity tests on mortar with different types of filler.

As can be seen from Fig 5. it is important that the binder and filler fit together. If we use only cement (60) we find a high yield shear stress and high plastic viscosity compared to tests 61 and 62 where we have replaced some of the cement with fillers. From test 80 and 81 with less amount of fillers compared to 62 we can also find that we need a powder amount in the region of 500-525 kg/m^3 to achieve good rheological properties without any segregation.

8   Concrete

Slump flow is a method to test the workability by using the normal slump-cone without compaction and measure the attained diameter of concrete. The L-box is a method to investigate the concrete´s ability to flow through gap between reinforcement without blocking. It consists of a L-shaped box with three reinforcement bars. Between the bars there is a specified gap . A certain amount of concrete is allowed to flow through the gap between the bars and the ability of not coursing blocking and self levelling ability is measured , see Fig. 6.

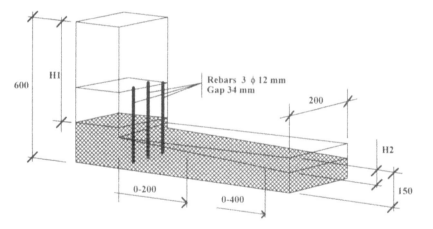

Fig. 6.   L-shaped box for measuring the ability for concrete to flow through reinforcement.

We have tested different types of mixes based on the mix design method. In Table 3 the mixes and results are presented using the slump flow and the L-box.

Table 3. Mix proportion and experimental results for aggregate type "Underås".

Mix	Cement kg/m^3	Dolomit kg/m^3	W/P	Nga	Vpw l/m^3	Va l/m^3	SP %	T50 sec	SF cm	H1 cm	Blocking OK/--
F18	330	203	0,345	0,40	400	600	2,00	4	700	51	Ok
F26	310	191	0,345	0,40	369	631	2,90	2	760	51	Ok
F7	332	204	0,345	0,48	400	600	1,60	2	670	50	Ok
F25	318	195	0,345	0,48	378	622	2,10	2	690	48	----
F23	374	230	0,345	0,52	427	573	1,30	2	695	50	Ok
F24	365	224	0,345	0,52	417	583	1,33	2	693	50	Ok
F9	332	205	0,345	0,52	400	600	1,50	3	680	49	----
VF3	331	204	0,345	0,55	400	600	1,8*	2	700	49	----
F37	355	219	0,345	0,58	423	577	1,00	1	660	50	Ok
F12	333	205	0,345	0,58	400	600	1,35	1	680	47	----

* 0,07 % viscosity agent Sa = Gravel to total aggregate ratio Wpw = Paste volume
Va = Aggregate volume SP= Superplasticizers % of cement h1 = See Fig. 6
W/P = Water to total powder ratio.

In Fig. 7. the results from the tests are presented (blocking or no blocking) together with void content curve and curve for blocking criteria for reinforcement with 34 mm gap. These result have been used to create the curve for blocking volume ratio to clear spacing to fraction diameter (c/D$_{af}$) in Fig. 3.

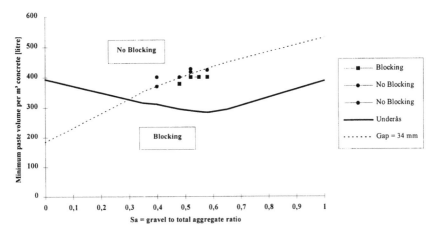

Fig. 7. Tests results on blocking (background for criteria in Fig. 3).

We have used the model to achieve a first approach to mixes for full scale testing with good results. We make the final adjustment with the slumpflow and the L-box. Our criteria for slump flow are between 670 to 720 mm. In the L-box we want to achieve a h1 greater than 490 mm.

9   Conclusions

A design model for self compacting concrete has been presented. The model can be used for different types of aggregate. The model use the least amount of paste based on void model and blocking model.

Superplasticizers and viscosity agents can be adjusted to the right level by testing mortar in an paste viscosimeter.

The L-box method has been used to evaluate the properties of self compacting concrete.

Future research will concentrate on modelling the filler and binder to optimise the paste. Also the role of viscosity agent to minimise the filler content will be investigated.

Research will also be carried out to optimise the economical properties for self compacting concrete.

10   References

1   K. Ozawa, K. Maekawa, H. Okamura, *Development of High Performance Concrete*, Proceedings of JCI, Vol 11, No 1, 1989.

2   K. Ozawa, K. Maekawa, H. Okamura, *High performance concrete with high filling capacity*, University of Tokyo, Department of Civil Engineering, Rilem Symposium, Admixtures for Concrete, Barcelona May 14-17 1990.

3   Bui K. Van, *A method for the optimum proportioning of the aggregate phase of highly durable vibrations-free concrete*, A Master thesis submitted to AIT, Bangkok 1994.

4   K. Ozawa, Tangtermsirikul, Maekawa K, *Role of Powder Materials on the Filling Capacity of Fresh Concrete,* Fourth CANMET/ACI International Conference on Fly Ash, Silica Fume Slag ..in Concrete, Page 121- 137, Istanbul, Turkey, 1992.

5   Tangtermsirikul. S, Bui. K. Van, *Blocking criteria for aggregate phase of self compacting high performance concrete,* Proceedings of Regional Symposium on Infrastructure Development in Civil Engineering, SC-4, pp 58-69, December 19-20, 1995, Bangkok, Thailand

6   Ö Petersson, P Billberg, J Norberg, A Larsson, *Effects of the second generation of superplasticizers on concrete properties*, Swedish Cement and Concrete Research Institute, CBI report 2:95, Stockholm 1995

PART NINE

SPECIAL CEMENTS AND CONCRETES

42 SOME ASPECTS OF FRESH CLAY–CEMENT MIXES FOAMED BY A BIOLOGICAL AGENT

M. RUZICKA, D. LORIN and M. QUÉNEUDEC
Université de Rennes I, I.U.T. Département Génie Civil, Rennes, France

Abstract
The studied material is a clay-cement mixture foamed by an air-entraining agent. The clayey part is presently an unused by-product of the exploitation of clay-containing sand. The air-entraining agent, a recycled ox blood, is also an industrial waste-product.

The presence of small bubbles in and the colloidal character of this material in its fresh state make its rheological behaviour very complex. The knowledge and control of this behaviour are absolutely essential to the workability and efficient performance of the hardened concrete. Experimental results, including problems encountered, are herein discussed.

In the first part, some characteristics of the mixture components, the experimental process and the technical equipment are reviewed. The second part treats the general rheological behaviour of the fresh clay-cement and foamed mixes. In the third part we present the influence of several factors in relation to the apparent viscosity. The factors considered are firstly the composition of the mixture and secondly the various procedures of the experimental process.

The raw experimental results themselves provide considerable information about the rheological behaviour of this new cellular clayey concrete and the possibilities of its use in the design of insulating building units.
Keywords: Clay, clay-cement mixes, air-entraining agent, rheology, workability.

1 Introduction

Le matériau étudié est un composite argile-ciment, allégé par un entraîneur d'air dont la mise oeuvre dans l'industrie du bâtiment est motivée par plusieurs raisons:

Production Methods and Workability of Concrete. Edited by P.J.M. Bartos, D.L. Marrs and D.J. Cleland.
Published in 1996 by E & FN Spon, 2–6 Boundary Row, London SE1 8HN. ISBN 0 419 22070 4.

La première est la revalorisation de sous-produits d'exploitation de granulats, actuellement inutilisés. L'entraîneur d'air est un déchet recyclé de l'industrie agro-alimentaire. De plus la fabrication s'effectuant sans consommation importante d'énergie (pas de cuisson, pas d'autoclavage, etc.) il s'ensuit un coût global intéressant.

A l'état frais, le caractère colloïdal de la matrice argile-ciment et l'aspect mousseux rendent très complexe le comportement rhéologique de ce matériau. Mais, il s'avère absolument indispensable de maîtriser ce comportement, en vue d'une bonne ouvrabilité (la mise en moule ou le pompage éventuel) mais aussi pour assurer des performances optimales à l'état durci. Les résultats développés dans ce travail concernent les premières déterminations de certains caractéristiques rhéologiques.

2 Constituants

* *Matériau argileux:* il est constitué en quasi-totalité de kaolinites (silico-aluminates hydratés dont la structure cristalline est en feuillets), [1], [2]. Sur le plan granulométrique, [3], on distingue une fraction d'*Argile fine* dont le diamètre est inférieur à 2 µm, représentant 55% de passants cumulés et une fraction d'*Argile-limon* dont le diamètre est compris entre 2 et 70 µm, représentant 45% de passants cumulés (fig.1). Rappelons toutefois que l'extraction de la fraction argileuse se fait avec ajout d'un floculant. Ceci revient très probablement à surestimer la fraction limoneuse par rapport à la distribution granulaire du matériau d'origine. Les limites d'Atterberg de l'argile employée sont: $W_l = 48\%$; $W_p = 36\%$; $I_p = 12\%$.
* *Liant:* la stabilisation chimique du matériau utilise un liant hydraulique, le ciment portland artificiel, CPA-CEM I 52.5 , (EN 196-1).
* *Entraîneur d'air:* sous forme d'une poudre, on utilise de l'hémoglobine de boeuf, traitée suivant un procédé protégé (EP 0-171-480-A1) par la société *ABC Bioindustries* et commercialisé sous le nom d'*ABC Proteolith*.
* *Eau:* selon NF P15-401, on emploie une eau potable à une température 20 ± 2° C.

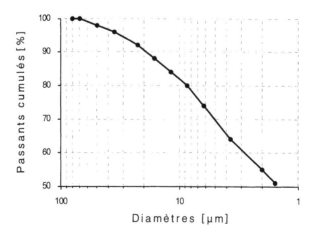

Fig. 1. Analyse granulométrique de l'argile exploitée, [1].

3 Processus expérimental

3.1 Dosage des constituants

La quantité de la partie argileuse (A) étant fixée, le *dosage standard* des autres constituants se fait selon les formules : E/A=0.75 ; C/A=0.25 , H/A=0.025 (E=eau, C=ciment, H=hémoglobine). Pour l'étude comparative de la viscosité (voir §5) ces rapports varient: C/A=0 à 0.55 ; E/A=0.65 à 0.85 ; H/A=0 à 0.050 .

3.2 Mise en oeuvre des échantillons

L'argile et le ciment sont d'abord malaxés à sec (2 min) dans un malaxeur pour mortier normalisé (EN 196-1). L'eau est ensuite ajoutée progressivement durant le malaxage jusqu'à homogénéisation de la pâte (2 min). Le malaxeur étant arrêté, l'hémoglobine est alors ajoutée. L'ensemble est malaxé pendant 30 minutes à la vitesse de 120 tours/min. La présence de l'hémoglobine crée dans la matrice argile-ciment un grand nombre de petites bulles, provoquant ainsi le moussage et l'allégement.

3.3 Appareillage viscosimètrique

L'appareil employé est *Rhéomètre Brookfield, modèle DV-III*, [4]. Il appartient à la famille des viscosimètres rotatifs. Le principe de son fonctionnement consiste à mesurer l'effort nécessaire (un couple) pour faire tourner un mobile immergé dans un fluide étudié. Le mobile est mû par un moteur synchrone à travers un ressort calibré. Le couple C, exprimée en pourcents de la déformation du ressort est l'une des deux valeurs principales des mesures. C'est la vitesse ω du mobile en tours/min, maintenue constante, qui représente la deuxième valeur. A partir des deux grandeurs et une constante d'échelle K on déduit la viscosité apparente:

$$\mu \;=\; K\,\frac{C}{\omega} \quad [\text{mPa.s}]$$

4 Comportement général

Dans le comportement rhéologique nous nous limitons ici à la relation entre la grandeur dynamique (le couple) et la grandeur cinématique (la vitesse du mobile), ce que l'on exprime graphiquement dans un rhéogramme. Il est ainsi possible de définir si le matériau étudié se comporte comme un corps linéaire (newtonien) ou non-linéaire (rhéofluidifiant ou rhéoépaississant), plastique (avec un seuil d'écoulement) ou non et s'il est dépendant du temps (thixotrope ou antithixotrope).

Il faut d'abord souligner que le comportement des échantillons testés manifeste une forte dépendance du temps. Le segment «A» sur la fig. 2 se montre typiquement thixotrope, la viscosité apparente diminue avec le temps de cisaillement. Ceci peut être dû à la destruction progressive des liaisons entre les particules ou l'alignement des éléments non-sphériques au sens de cisaillement. Mais la cause probablement la plus importante est la modification de la répartition des bulles dans l'espace cisaillé. Notons que le phénomène observé n'est pas après un long repos entièrement réversible, il s'agit donc plutôt d'une fausse-thixotropie.

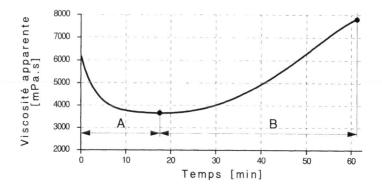

Fig. 2. Viscosité apparente en fonction du temps de cisaillement. Vitesse du mobile ω=75 t/min.

La phase «B», à la différence de la phase «A», montre une augmentation de la viscosité apparente. Ce comportement est vraisemblablement provoqué par les réactions chimiques qui prennent naissance au contact de l'eau et du ciment.

Pour cerner l'influence de la thixotropie dans la construction du rhéogramme, les différents échantillons ont été soumis à des vitesses variables mais constantes lors de chaque essai. Ensuite, pour un instant donné le rhéogramme a été tracé à partir de l'ensemble de ces mesures. La fig. 3 montre un rhéogramme, et le modèle de Casson, qui se prête le mieux à la modélisation de ce matériau.

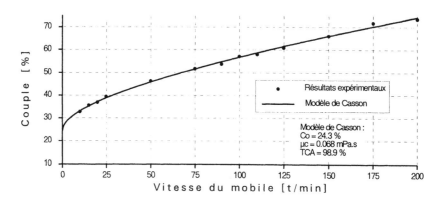

Fig. 3. Rhéogramme pour l'instant 0.5 min de cisaillement. Composition standard.

Modèle de Casson, [4]:

$$\omega = 0 \qquad \text{pour } C < C_0$$
$$\sqrt{C} = \sqrt{C_0} + \sqrt{\mu_c \cdot \omega} \qquad \text{pour } C > C_0$$

C_0 est le seuil d'écoulement [%]
μ_c est la viscosité plastique [mPa.s]
TCA est le taux de confiance d'ajustement du modèle [%]

En conclusion, le matériau étudié devrait être considéré comme un corps plastique et légèrement non-linéaire (rhéofluidifiant), manifestant une thixotropie assez importante.

5 Etude comparative de la viscosité

5.1 Influence du ciment

On constate que l'échantillon sans ciment est, contrairement à ce que l'on pourrait attendre en raison de la présence moins importante de fines, nettement plus visqueux qu'avec du ciment, quelle qu'en soit la teneur (fig. 4). De même, l'introduction de l'hémoglobine dans le mélange A+E produit une rigidification de celui-ci tandis que l'ajout dans le mélange A+E+C provoque une fluidification, (voir aussi §5.3). L'augmentation de la teneur en ciment cause en principe une légère diminution de la viscosité, jusqu'à la limite C/A=0.45 . La variation du rapport C/A modifie aussi la formation de la mousse, à savoir que, le volume de mousse obtenu est en effet proportionnel à ce rapport.

L'une des causes de ces phénomènes peut résider notamment dans le domaine de pH et sa variation. A savoir, le ciment en solution aqueuse a un pH de l'ordre de 13.5 tandis que l'argile est légèrement acide (pH=6.2), [3]. L'adjuvant utilisé est aussi stabilisé à un pH voisin de 6. D'après [3] le matériau étudié atteint un pH d'environ 12. Sans ciment on se trouve donc à un pH voisin de la neutralité, ce qui pourrait expliquer la modification de la fluidité. En effet selon [5], les mélanges eau-kaolinite aux pH élevés subissent une forte répulsion entre les particules et il y a une défloculation.

En revanche, la tendance à la croissance de la viscosité pour C/A>0.45 est probablement due à une consommation d'eau plus importante.

Notons encore que le ciment seul ne diminue pas la viscosité. La fig. 5 montre qu'en absence d'hémoglobine ce phénomène ne se manifeste pas.

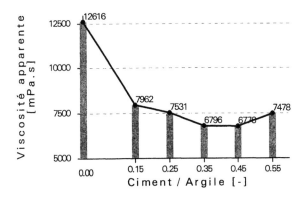

Fig. 4. Influence du dosage en ciment sur la viscosité apparente pour 0.5 min de cisaillement; ω=50 t/min; composition: E/A=0.75 , H/A=0.025 .

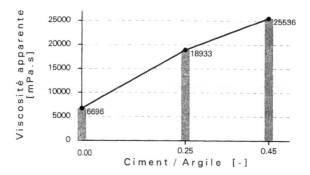

Fig. 5. Influence du ciment sur la viscosité apparente en absence de l'hémoglobine pour 0.5 min de cisaillement; ω=50 t/min; composition: E/A=0.75 , H/A=0.0 .

5.2 Influence de l'eau de gâchage

Malgré une allure relativement simple du phénomène observé dans le cas de la viscosité (fig. 6), l'influence de l'eau de gâchage sur le comportement rhéologique est plus complexe. Plus on augmente la quantité d'eau, plus on diminue la résistance intrinsèque du matériau (la viscosité), simultanément avec l'élévation de la production (le volume) de mousse. Toutefois, nous avons pu constater que l'étalement qui est la propriété primordiale pour l'optimisation de la mise en moule était peu ou pas amélioré. Une étude plus approfondie est en cours actuellement visant à expliciter ce phénomène.

5.3 Influence de l'hémoglobine

En absence de l'hémoglobine la pâte ciment-argile reste inchangée et donc très visqueuse (fig. 7). En revanche, l'augmentation de la teneur en hémoglobine jusqu'à la limite H/A=0.033 fait diminuer la viscosité apparente. Au-delà, la structure mousseuse est défavorable à l'écoulement et se manifeste par une légère hausse de viscosité.

Rappelons toutefois, comme dans le cas précédent (§5.2), qu'une faible viscosité affichée par le viscosimètre ne signifie pour autant que ce matériau mousseux s'étale.

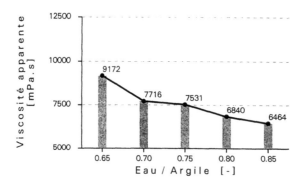

Fig. 6. Influence du dosage en eau de gâchage sur la viscosité apparente pour 0.5 min de cisaillement; ω=50 t/min; composition: C/A=0.25 , H/A=0.025 .

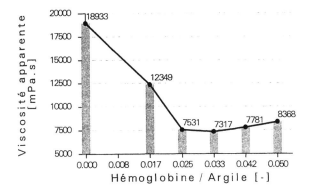

Fig. 7. Influence du dosage en hémoglobine sur la viscosité apparente pour 0.5 min de cisaillement; ω=50 t/min; composition: C/A=0.25 , E/A=0.75 .

Dans le chapitre 5.1 nous avons déjà signalé que l'introduction de l'hémoglobine dans le mélange en présence de ciment produit une fluidification tandis que sans ciment on rencontre une rigidification. Ceci est confirmé sur la fig. 8, la gâchée composée uniquement d'argile et d'eau est relativement fluide alors que l'ajout d'une petite quantité d'hémoglobine (H/A=0.017) augmente brutalement la viscosité. Il est difficilement concevable qu'une augmentation de la partie fine de 1.7% puisse quasiment faire tripler la viscosité du mélange même si la surface spécifique de la fraction ajoutée est très importante. On peut donc supposer que l'origine de ce phénomène est liée à un mécanisme d'interaction argile-hémoglobine qui n'apparaît pas en milieu basique, c'est-à-dire en présence du ciment. Pour le dosage H/A=0.025 et 0.033 la viscosité diminue, et même pour ce dernier dosage très fortement. Il y a aussi une légère production de mousse. En revanche, le dernier échantillon, H/A=0.050 montre une faible remonté de la viscosité explicable par l'augmentation de la partie fine du mélange sec.

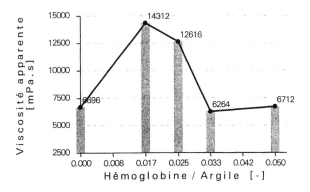

Fig. 8. Influence de l'hémoglobine sur la viscosité apparente en absence du ciment pour 0.5 min de cisaillement; ω=50 t/min; composition: E/A=0.75 , C/A=0.0 .

5.4 Influence de la granulométrie du matériau argileux

A l'état sec, le matériau argileux est traité sur différents tamis, d'ouvertures: d=2.5 ; 1.25 ; 0.63 ; 0.315 ; 0.16 [mm]. La fig. 9 montre qu'une finesse d'argile plus élevée peut faciliter l'écoulement. Les éléments les plus grossiers représentent probablement les obstacles à l'écoulement. Il est à noter que l'échantillon traité sur le tamis d=0.315 mm et notamment d=0.16 mm ont manifesté aussi une hausse en volume de la mousse obtenue.

5.5 Influence de la durée de malaxage

La viscosité ou cours de malaxage a été évaluée pour les durées 0.5 ; 2 ; 10 ; 20 ; 30 ; 40 ; 50 [min].

Sur la fig. 10, les deux premiers échantillons (0.5 et 2 min) montrent la plus petite viscosité donnant ainsi l'image d'efficacité de l'hémoglobine en tant que fluidifiant. A savoir que, si l'on compare les valeurs de la viscosité avant et juste après l'ajout de l'hémoglobine on rencontre une différence de facteur 10. Notons, qu'à ce stade le mélange n'est pas encore allégé (moussé).

En revanche, au bout de 10 minutes de malaxage, on retrouve une viscosité nettement plus élevée bien que l'échantillon manifeste encore un bon étalement. La mousse commence à se former.

Pour la durée de malaxage de 20 minutes on obtient la plus grande viscosité. A l'oeil nu on remarque que pour cette phase la mousse semble la plus volumineuse et que l'étalement est pratiquement nul.

A partir de 20 min la viscosité commence à descendre. Pour 30 min on constate une mousse encore très riche mais elle sera détruite progressivement au cours du malaxage pour finir par presque disparaître au bout de 50 minutes.

5.6 Influence de la vitesse de malaxage

Les échantillons ont été malaxés aux vitesses 60, 120 et 180 tours/min. La plus petite vitesse ne produit d'après l'observation pratiquement aucune bulle, la viscosité apparente sur la fig. 11 est ainsi très faible (il y a uniquement fluidification par l'hémoglobine).

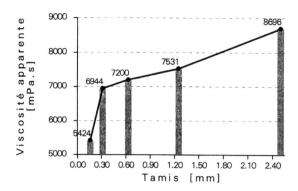

Fig. 9. Influence de tamisage de l'argile sur la viscosité apparente pour 0.5 min de cisaillement; ω=50 t/min; composition standard.

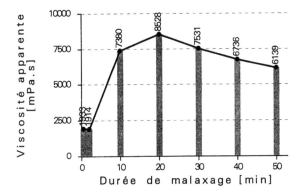

Fig. 10. Influence de la durée de malaxage sur la viscosité apparente pour 0.5 min de cisaillement; ω=50 t/min; composition standard.

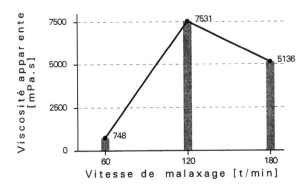

Fig. 11. Influence de la vitesse de malaxage sur la viscosité apparente pour 0.5 min de cisaillement; ω=50 t/min; composition standard.

En revanche l'échantillon allégé et mousseux (120 t/min) montre une viscosité environ dix fois plus grande que celui sans mousse. Les bulles entraînent alors une résistance à l'écoulement très importante.

La préparation du matériau à la vitesse 180 t/min n'est pas très avantageuse. D'une part la mousse ne se produit pas ou faiblement (ou elle est tout de suite détruite), d'autre part l'échantillon obtenu est quand même assez visqueux.

6 Conclusion

Le travail effectué a permis d'évaluer expérimentalement certaines propriétés rhéologiques du béton argileux allégé par l'hémoglobine pulvérulente. D'une manière générale, on peut constater des propriétés thixotropes assez importantes ainsi que le caractère plastique (la présence d'un seuil d'écoulement) avec une tendance non-linéaire (rhéofluidification).

En vue d'atteindre une bonne ouvrabilité et un entraînement d'air optimal on peut résumer l'influence des facteurs principaux comme suit:

- *Ciment:* le dosage faible en ciment modifie ou élimine assez brutalement la fluidification et l'entraînement d'air par l'hémoglobine. En revanche, l'augmentation de la teneur en ciment en présence d'une quantité normale d'hémoglobine provoque une légère amélioration en étalement ainsi qu'une hausse en production de la mousse.
- *Eau:* en augmentant le dosage en eau de gâchage on fait diminuer la viscosité apparente, simultanément avec l'élévation du volume de mousse obtenue. Toutefois, l'ouvrabilité ne semble pas être effectivement améliorée.
- *Hémoglobine:* elle est en présence du ciment très efficace au niveau de l'entraînement d'air mais en dépit de ses performances fluidifiantes dans le matériau sans mousse, faciliter la mise en moule par cette voie est quelque peu délicat.
- *Granulométrie:* l'argile plus finement tamisée favorise l'écoulement ainsi que l'entraînement d'air occlus dans le matériau résultant.

7 Références bibliographiques

1. Estéoule-Choux, J. (1970) *Contribution à l'étude des Argiles du Massif Armoricain.* Thèse de Doctorat, Rennes, France.
2. Jigorel, A. (1990) *Connaissance et Restauration de l'Habitat en Terre.* Comptes rendus du colloque, 10 et 11 Avril 1990, Montfort sur Mer, Ille et Vilaine, France.
3. Al-Rim, K. (1995) *Le Béton Argileux Léger, Généralisation à d'autres fines de roche et Applications à la conception d'éléments de construction préfabriqués.* Thèse de Doctorat, Université de Rennes I, France.
4. Brookfield Engineering Laboratoires. (1992) *More solutions to sticky problems.* Documentation technique, Stoughton, Massachusetts, U.S.A.
5. Coussot, P. (1993) *Rhéologie des boues et laves torrentielles. Etude de dispersions et suspensions concentrées.* Thèse de Doctorat, Camagref, N°5, Laboratoire de rhéologie, UJF, INPG, CNRS, Grenoble, France.

43 PRODUCTION AND TESTING OF CONCRETE WITH HIGH SPECIFIC HEAT

J. LINDGÅRD and K. JOHANSEN
Department of Cement and Concrete, SINTEF Civil and Environmental
Engineering, Trondheim, Norway

Abstract
Heat development in mass concrete can be decreased in many different ways, like use
of a low heat cement, a low cement content or a low fresh concrete temperature. It is
also common to cool the hardening concrete by use of cooling pipes with cold water.
An alternative method to decrease the concrete temperature during the cement
hydration is to use an aggregate with high specific heat.

SINTEF Civil and Environmental Engineering have since 1985 been working with
the mineral Olivine (in the rock type Dunite) as concrete aggregate. Olivine is
produced by A/S Olivine situated at Åheim in the western part of Norway.

The laboratory tests have shown that the Olivine aggregate is very interesting for
making concretes with moderate temperature development. The rather high specific
heat (0.95 kJ/kg·°C) and density (3230 kg/m^3) of the Olivine can reduce the maximum
temperature developed due to the cement hydration up to 20% as compared to other
Norwegian aggregates. This paper presents the main results from the performed
laboratory tests with the Olivine aggregate, with emphasis on the temperature reduction
effect in the hardening concrete.

The Olivine aggregate is crushed. The paper also comments the technology of
blending crushed and natural sand, especially with respect to workability properties.

Another very interesting field of use for the Olivine aggregate is concrete exposed
to high temperatures. Some comments on this theme are also included in the paper.

During the summer 1996 a full scale test is planned using Olivine aggregate in the
concrete for a pedestrian subway. The intention is to lower the maximum temperature
developed in the concrete to meet the requirements from the builder.
Keywords: Aggregate, compressive strength, durability, heat capacity, heat conductivity,
heat development, workability.

Production Methods and Workability of Concrete. Edited by P.J.M. Bartos, D.L. Marrs and D.J. Cleland.
Published in 1996 by E & FN Spon, 2–6 Boundary Row, London SE1 8HN. ISBN 0 419 22070 4.

1 Introduction

According to ENV 206 of March 1990, section 10.7, the average maximum temperature of the concrete shall not exceed 60°C (individual value less than 65°C) during the hardening process. For mass concrete with relative low water/binder ratios it may be difficult to satisfy this requirement, without taking any precautions.

Heat development in mass concrete can be decreased in many different ways, like use of a low heat cement, a low cement content or a low fresh concrete temperature (gained by cooling the aggregates or the fresh concrete by use of cold water, ice or liquid nitrogen). It is also common to cool the hardening concrete by use of cooling pipes with cold water. An alternative method to decrease the concrete temperature during the cement hydration is to use an aggregate with high specific heat. It is also possible to combine several of the methods mentioned.

SINTEF have since 1985 been working with the mineral Olivine (in the rock type Dunite) as concrete aggregate. Olivine is produced by A/S Olivine situated at Åheim in the western part of Norway. This paper presents the main results from the laboratory tests, with emphasis on the temperature rise in the concrete.

2 What is Olivine?

Olivine is a water free magnesium-iron silicate with chemical formula $[Mg,Fe]_2SiO_4$. Olivine may therefore be regarded as a solid solution of the two minerals Forsterite $[Mg_2SiO_4]$ and Fayalite $[Fe_2SiO_4]$. The pureness of Olivine may be described by the ratio between magnesium and iron. The high quality Olivine produced at Åheim in the South Western part of Norway is characterised by a $Mg_2SiO_4 : Fe_2SiO_4$ ratio 93:7 on molar basis. Olivine contains no free Quartz. The silicon oxide is chemical bound to magnesium and iron.

The typical main chemical composition of the Olivine from Åheim is; 49.0% MgO, 41.0% SiO_2 and 7.0% Fe_2O_3.

Below we have given some of the beneficial characteristic properties of Olivine, which makes it an industrial very applicable rock type;
· high specific density (3220 kg/m³)
· high specific heat capacity (0.95 kJ/kg·°C)
· high heat conductivity (4.76 W/m·°C at temperature 20°C)
· high fire resistance (melting point about 1760°C)
· linear thermal expansion (about 1.1 % up to 1200°C)
· environmental friendly, weak basic material (pH about 9.0-9.5).

Olivine deposits in Norway: Olivine is a common mineral in several rocks, but it is seldom the dominant mineral in large deposits. However, in Norway we find several such rock deposits. Geologically these deposits are some of the oldest rock beds in Norway, with an age of approximately 1500 million years. The biggest known deposit of high quality Olivine is situated at Åheim in the South Western part of Norway. It contains about 2000 million tons of the Olivine rock (Dunite). The deposit is controlled by A/S OLIVINE.

Olivine deposits in the world: in a global perspective there are several commercial utilized Olivine deposits. However, the quality of these vary with respect to the ratio between magnesium and iron and degree of serpentination. The most important deposits are situated in Norway, Russia, Mexico, Austria, USA, Spain, Japan and Pakistan.

3 Production and use of Olivine

3.1 Production of Olivine at A/S OLIVINE

Today A/S OLIVINE at Åheim produce about 2.5 million tons of Olivine sand and -stone each year. This makes A/S OLIVINE to the market leader and the biggest producer of Olivine products in the world. With this production the Olivine deposit will last for about 800 years. In addition, the A/S OLIVINE daughter company INDUSTRI-MINERALER A/S in Raudbergvik produce about 400.000 tons of Olivine sand and -stone each year.

A/S OLIVINE was established in 1948. The company has about 220 employees. The Norwegian State owns 100% of the share capital. The open pit with Olivine is situated about 4 kilometres from the sea. The Olivine is transported via conveyors in an underground tunnel to the purifying plant at the quay. Here the Olivine is crushed and sifted, before it is transported to silo storage or to drying out, dependant of type of Olivine product. The production equipment has a great flexibility, which allows production of several types of Olivine products. It is possible for ships up to about 80.000 tons dead weight to enter the quay. Today about 95% of the Olivine products are exported. Germany is one of several big markets.

In 1994 A/S OLIVINE had a turnover on about 290 millions NOK. The economy result before taxes was about 55 millions NOK.

3.2 Field of application for Olivine

Olivine is today used as an industrial product within the following areas;

- *Slag incrustation in blast furnaces:* Olivine has displaced Dolomite for slag incrustation in the production of pig iron. The high MgO content of about 50% gives advantages both with respect to the process and to the energy economy. Olivine is used in most of the pig iron production in Europe, and the use for slag incrustation is the main marked for Olivine products today.
- *Ballasting material:* the relative high density and chemical stability of the Olivine makes the mineral very well suited as ballasting material in concrete oil platforms and in pillars for bridges. More than 1 million tons of Olivine is delivered for this purpose. The projects include both the Norwegian Coondeep platforms: "Statfjord A" and "Statfjord B", "Gullfaks B", "Oseberg A" and "Sleipner A"; and the bridges "the West Bridge" and "the East bridge" at the Great Belt Link in Denmark.
- *Protection for oil pipes:* several under water oil pipes are covered with Olivine materials for protection. The relative high density makes it well suited for this purpose. The projects include "Statpipe Karmøy", "Zeepipe" and "Stena Strathpey".
- *Moulding sand:* Olivine is common used as moulding sand in moulds and cores in steel mills. The properties as no health risk, fire resistant, low and linear thermal expansion and a resulting smooth surface of the cast iron goods, make Olivine to a

quality sand within this area.

· *Blast sand:* Olivine used as blast sand has environmental advantages as no risk for silicosis and no leakage of toxic metals to the environment by depositing. Coming guidelines for disposal of such rest products will give a very advantagous market situation for Olivine sand.

· *Use in fireproof concrete:* due to the high fire resistance of Olivine, it is used as raw material for production of fireproof concrete. This concrete is used in floors in smelting plants, as fire protection, in heat absorbing materials etc. The concrete is based on high quality Olivine and aluminate cement. The concrete withstand temperatures up to 1600°C.

3.3 Why concrete is an interesting market

About 20 years ago preliminary laboratory tests were performed at SINTEF with concrete containing crushed Olivine sand. The concrete achieved acceptable properties both with respect to workability and compressive strength. Due to the relatively high production costs of crushed Olivine sand at that time compared to natural sand, the plans of using Olivine as concrete aggregate were put aside.

Today, however, A/S OLIVINE has a much more cost effective purifying plant. Further is the company more aware of the mineralogical preference of the Olivine, such as the high specific heat, the fireproofness and the high density. Due to these characteristic properties the concrete may achieve properties which normally is impossible to obtain by natural aggregates.

4 Olivine as aggregate for concrete production

4.1 Performed tests in the laboratory

In 1993-95 SINTEF performed further tests and documentation of the crushed Olivine both from Åheim [1] and Raudbergvik [2] with the intention of using it as concrete aggregate. The following aggregate investigations were performed; *grading, water absorption* and *petrographical analysis* including *thin section analysis* (including evaluation of the risk of alkali aggregate reactions).

The Olivine aggregates were also used in trial mixes with three different concrete types with water/binder ratios 0.60, 0.45 and 0.41, respectively. In addition to the Olivine aggregates, also a reference Granite aggregate from Aker NorRock were used in some of the trial mixes. The following concrete parameters were investigated in the tests; *density, workability, consistency, air content, water requirement* and *compressive strength.*

In addition, the following tests were performed on the concrete with the lowest water/binder ratio (0.41); measurement of *temperature development* during hydration (including calculation of the *specific heat capacity* of the Olivine aggregate) and determination of *thermal conductivity.*

The following durability properties of concrete containing Olivine aggregate from Åheim were also investigated; *capillary absorption* and *frost resistance.* In addition the possibility for *oxidation* of the bivalent iron to trivalent iron was investigated.

4.2 Results from tests with Olivine aggregate

4.2.1 Material properties of Olivine aggregate

According to the *mineral composition* the Olivine aggregate from Åheim is classified as the rock type Dunite (contains > 90% of the mineral Olivine). The Olivine aggregate from Raudbergvik is classified as the rock type Peridotite (contains 86% of the mineral Olivine).

The measured *water absorption* of the Olivine aggregate from Åheim was 0.85%, i.e. at the same level as a normal Norwegian aggregate.

The *particle shape* of the aggregate fractions from Åheim (0-3 mm, 1-5 mm, 6-10 mm and 8-25 mm, respectively) was rather good. The content of cubical grains was as high as 80-85%. Half of the grains were even partly rounded. These fractions are crushed in several steps. The particle shape of the aggregate fractions from Raudberg-vik (0-4 mm, 2-8 mm and 4-16 mm, respectively) was less favourable. In all the fractions the content of cubical grains was 60-70%. This material, however, is only crushed in one step. By crushing the aggregate in several steps, it is assumed that the particle shape will be as good as the shape of the aggregate from Åheim.

The *grading* of all the investigated aggregate fractions, except the sand fraction from Åheim, were rather favourable with respect to workability properties of concrete. The sand fraction 0-3 mm from Åheim had a "sand hunch". The total Olivine aggregate grading therefore had a too small content of particles in the fraction 0.5-2 mm. However, today A/S OLIVINE has purifying plants which makes it possible to adjust the aggregate grading according to the desired concrete properties.

The content of *physical weak aggregate grains* was low (0-5%). The grain surfaces were fresh and unweathered.

4.2.2 Fresh concrete properties

All the performed trial mixes show that the received aggregates both from Åheim and Raudbergvik are suitable for producing stable and workable concrete.

The type of coarse aggregate had no significant influence on the water requirement or the workability properties when using identical sand type.

By use of Olivine sand, however, both the water requirement [1 and 2] and the requirement of superplasticizer [3] increased (see table 1). The reason for this is the "sand hunch" of the Olivine sand from Åheim [1 and 3] and the relatively high content of flaky and elongated particles in the Olivine sand from Raudbergvik [2]. As earlier mentioned, it is possible to improve both these aggregate properties during the aggregate production. By optimizing the grading and the particle shape and/or blending natural - and crushed sand, one may thereby obtain a better concrete with respect to workability and need for water and superplasticizer [4]. In addition the concrete economy will be improved.

Table 1 shows the test results from the measurement of fresh concrete properties of the three trial mixes with the lowest water/binder ratio (0.41) [3]. The calculated real concrete compositions and the measured compressive strengths are also included in Table 1.

510   *Lindgård and Johansen*

Table 1. Concrete mixes, real concrete composition and test results [3]

Materials (kg per m^3 of concrete)				
Trial mix no.		1	2	3
Cement (Norcem HS65[1)])		385	381	397
Silica fume (CSF)		15.4	15.2	15.9
Crushed Olivine from Åheim,	0-3 mm	---	---	725
fraction marked	6-10 mm	---	381	397
	8-25 mm	---	838	998
Natural Granite aggregate	0-8 mm	683	676	-
from Aker NorRock Årdal,	8-11 mm	192	-	-
fraction marked	11-16 mm	414	-	-
	16-25 mm	419	-	-
Plastiment BV-40[2)]		2.7	2.7	2.8
Sikament 110[2)]		0.8	0.1	3.7
Sika air (AEA)[3)]		0.10	0.10	0.10
Water (60% water in admixtures included)		166.3	164.4	172.1
Water/binder ratio (including 4% CSF)		0.41	0.41	0.41
Slump (mm)		190	150	150
Apparent density (k/m^3)		2290	2470	2723
Air content (%, measured)		5.9	6.9	3.2
Air content (%, corrected)[4)]		6.5	7.4	3.2
Compressive strength at age	7 days (MPa)	43.5	40.2	45.2
	28 days (MPa)	58.3	53.0	62.0
	90 days (MPa)	66.7	60.3	65.4
	365 days (MPa)	69.3	61.3	65.1
Compressive strength (MPa)	2 % air	71.4	67.3	65.7
at 28 days adjusted[5)] to	4 % air	64.3	60.6	59.1

[1)] The Norcem HS65 cement fulfils the requirements of ENV 197-CEM I-52,5 LA
[2)] Including 60% water in the plasticizer
[3)] Including 90% water in the superplasticizer
[4)] The recipes were adjusted to 1 m^3 by correcting the measured air content
[5)] Under the presumption that 1 % increase in air content leads to a 5 % reduction in the compressive strength

Due to the higher dosage of SP in mix no. 3 in Table 1, the effect of the AEA was reduced.

The concrete density increased about 8.5% by exchanging the coarse Granite aggregate (mix no. 1) with coarse Olivine aggregate (mix no. 2) and 17% by exchanging the total Granite aggregate with Olivine aggregate (mix no. 3).

4.2.3 Strength properties in concrete

Table 1 shows the compressive strengths of the three concretes from the trial mixes with the lowest water/binder ratio (0.41) [3]. For similar air content, the compressive strength of the concrete with only Olivine aggregate was about 5-8% lower than for the concrete with only Granite aggregate. For the concrete with higher water/binder ratios [1 and 2], the difference in compressive strength was less pronounced.

The Olivine aggregate is well suited for production of concretes at least up to strength class C65 (according to NS 3420). In laboratory tests SINTEF have obtained compressive strengths of about 75 MPa without any signal of reaching the "strength limit" of the Olivine aggregate.

4.2.4 Thermal conductivity for concrete containing Olivine aggregate

The results from the measurement of thermal conductivity of the three concretes in table 1 are given in Table 2 [3].

Table 2. Measured thermal conductivity of the concretes in Table 1 (kJ/m·h·°C)

Mix no.	Moisture conditions	Temperature level at testing		
		approx. 20°C	approx. 55°C	approx. 150°C
1 (Granite)	dry	5.4	---	---
	wet	7.3	---	---
2 (Granite/ Olivine)	dry	6.4	---	---
	wet	8.9	---	---
3 (Olivine)	dry	7.9	8.1	6.2
	wet	9.3	9.5	---

In a wet condition, the thermal conductivity is approximately 30 % higher for the Olivine aggregate concrete than for the Granite aggregate concrete. In a dry condition, the difference is approximately 45 %.

4.2.5 Heat development of concrete containing Olivine aggregate

In Fig. 1 the calculated adiabatic temperature rise of the three mixes in Table 1 are shown [3]. The final adiabatic temperature rise was 58°C for the concrete containing Granite aggregates only (mix no. 1), 51°C for the concrete with combined Granite and Olivine aggregates (mix no. 2), and 46°C for the concrete containing Olivine aggregates only (mix no. 3). By only exchanging aggregate type, the maximum adiabatic temperature was lowered as much as 12°C. A reduction of the cement content of the Olivine concrete (to a level corresponding to the Granite concrete) would have reduced

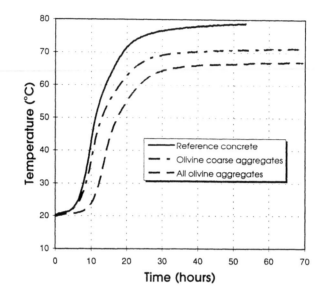

Fig. 1. Calculated adiabatic temperature rise [3]

the adiabatic temperature rise with an additional 2-3°C. By use of Granite sand and Olivine stone, the maximum temperature was lowered 7°C.

Similar tests have been performed by the Norwegian contractor Selmer A/S, by use of the same aggregate types and the same water/binder ratio (see Table 1). In these tests they obtained almost the same differences in temperature rise as in the present study.

The specific heat capacity of the Olivine aggregate was determined to 0.95 kJ/kg·°C, assuming that the specific heat capacity of the Granite aggregate is 0.80 kJ/kg·°C. This figure is somewhat higher than reported earlier /5/. A possible explanation may be that the specific heat capacity of the Granite aggregate in this case is overestimated. However, no earlier results have indicated such a problem. The most reasonable explanation is therefore that the specific heat capacity of the Olivine from Åheim is higher than the specific heat capacity of the Olivine reported earlier /5/. The mix water absorption of the aggregates may not have caused an increased specific heat capacity of the Olivine aggregate, since the water absorption of the Granite aggregate is quite similar (see section 4.2.1).

The high density of the Olivine aggregate also contributes to lower the maximum temperature rise in the concrete.

4.2.6 Durability properties in concrete

Both the Olivine aggregates, i.e. Dunite and Peridotite (see section 4.2.1), are classified as *non-alkali-reactive* on the basis of the thin section analysis.

Capillary absorption tests showed that the aggregate type, i.e. Granite or Olivine, had no significant influence on the *porosity/permeability* of concrete with water/binder ratio 0.45. Testing of *frost resistance* of the same concretes according to [6] showed that the concrete containing Olivine aggregates was not frost resistance when the air content was as low as about 1%. In this case also some of the aggregate particles were broken. When the air content in the hardened concrete was raised to 3.0%, however, the concrete had good frost resistance. Provided use of a sufficient frost resistance cement paste, use of Olivine aggregate is therefore not expected to give any problems with respect to frost resistance, even in a wet and cold environment.

In the investigation of the possibility for *oxidation* of the bivalent iron to trivalent iron, mortar prisms (with w/c=0,80 to provide easy access of oxygen) containing Olivine aggregates were stored for 1 year in 100% RH and 38°C. The results from measurement of the possible expansion of the prisms and SEM-analyses of the mortar showed that no such oxidation of the Olivine aggregate in concrete occurred.

Due to the firm bond of the iron in the Olivine aggregate, no problem with dis-coloration of the concrete surface when exposed in a humid environment, is expected.

5 Potential benefits by use of Olivine in concrete

5.1 Heat development in mass concrete
The main reasons for the requirement not to exceed an average maximum temperature of 60°C in the concrete during the hardening process, are the risk of formation of temperature cracks during the cooling period and a potential delayed ettringite for-mation. A normal supplement requirement is that the difference between the maximum temperature inside the concrete section and the surface shall not exceed 20°C during the cooling period.

For mass concrete with relative low water/binder ratios it may be difficult to satisfy these requirements, without taking any precautions such as use of a low heat cement, a low cement content, a low fresh concrete temperature or by use of cooling pipes with cold water. If some of these precautions fails, there is still a risk of exceeding the maximum temperature allowed. By use of Olivine aggregate the maximum temperature rise in the concrete during the cement hydration will be considerable lowered. This will reduce or avoid the use of some of the mentioned expensive cooling precautions.

The combination of low maximum temperature rise and high thermal conductivity of concrete containing Olivine aggregate, is special favourable for cold joints. The risk of cracking due to temperature differences between the "old" and the "new" concrete may be reduced both due to the lower starting temperature difference and improved heat exchange between the sections. In general, the significance of such effects can only be thoroughly evaluated by case studies.

For a lot of structures, especially with mass concrete, use of Olivine aggregates will be favourable both for the contractor (less cooling precautions and less risk of crack formation) and the builder (probably less expensive).

5.2 Fire resistance
Due to the high fire resistance of Olivine, it is used as raw material for production of fireproof concrete (see the section 2 and 3.2). The high fireproofness and the linear

thermal expansion are also assumed to be favourable for the fire resistance of normal concrete.

SINTEF have therefore in co-operation with A/S OLIVIN made a proposal with the goal to develop a fireproof concrete with Olivine aggregate. The concrete is planned to stand very high temperature loads.

5.3 Concrete with high density
The high density of the Olivine aggregate may be favourable for production of concrete with requirement to high density, for instance heavy foundations.

6 References

1.   Haugen, M. and Dahl, P.A. *Crushed Olivine as concrete aggregate. Petrographical analysis and influence upon basic concrete properties.* SINTEF-rapport nr STF70 F93026, Trondheim, March 1993.

2.   Haugen, M. and Dahl, P.A. *Crushed Olivine as concrete aggregate. Petrographical analysis and influence upon basic concrete properties.* SINTEF-rapport nr STF70 F93122, Trondheim, December 1993.

3.   Johansen, K, Lindgard, J. and Smeplass, S. *Crushed Olivine as aggregate in Norwegian bridge concrete. Workability analysis and heat of hydration.* SINTEF-rapport nr STF70 F95036, Trondheim, March 1993.

4.   Johansen, K., Laanke, B., Smeplass, S.: *Crushed sand as a Complementary Aggregate.* Nordic Concrete Research no.2 1995

5.   Mason, B., Berry, L.G.: *Element of Mineralogy.* Freeman 1968

6.   Swedish Standard SS 13 72 44, *Concrete Testing - Hardened Concrete - Frost Resistance*, Second Edition, 1988

44 FLOWABLE CONCRETE FOR STRUCTURAL REPAIRS

A. McLEISH
W.S. Atkins Structural Engineering, Epsom, Surrey, UK

Abstract
For large repairs to reinforced concrete structures the use of a flowable concrete is appropriate. This paper looks at the required properties of flowable repair concretes, methods of testing both in the laboratory and in the field, to ensure suitability and compliance, the method of placement and potential problems that may be encountered.

The specification and testing regime adopted for a major repair contract are used as illustrations.
Keywords: concrete repair, fluidity, compliance tests, self-compacting mix, site tests.

1 Introduction

Concrete for use in repair often demands properties not generally considered in detail in concrete for new works. In particular the flowable repair concrete for many applications must:

- have small maximum aggregate size to allow flow in the restricted spaces between parent concrete, reinforcement and formwork;
- be self levelling and self compacting as access for vibrators is often limited;
- be placed such that no air or water is trapped;
- have a high rate of strength gain, particularly where a lot of small localised repairs have to be undertaken;
- be dense and have a low permeability as often repairs result from a combination of low cover and chloride/carbonation attack of the original construction which must be overcome;

Production Methods and Workability of Concrete. Edited by P.J.M. Bartos, D.L. Marrs and D.J. Cleland. Published in 1996 by E & FN Spon, 2–6 Boundary Row, London SE1 8HN. ISBN 0 419 22070 4.

- not suffer from any bleed that would destroy the bond of the repair concrete to the parent concrete, thus reducing the structure effectiveness and presenting a leakage path into the reinforcement.

The specifications and test requirements for a specific repair contract recently undertaken are now outlined and used to illustrate the use of flowable concrete.

2 Requirements for repair concrete

A problem had arisen on a motorway flyover because of deteriorated joints which allowed the penetration of de-icing salts from the road deck above on to the supporting reinforced concrete beams.

Repair to these beams was necessary because of extensive cracking and spalling particularly on the soffit and top deck due to chloride induced corrosion.

The loading conditions applied to the beams required that, to avoid extensive and complex temporary support, the repairs had to be carried out piecemeal such that only small areas could be repaired at any one time. This necessitated concrete with a high rate of strength gain to minimise delays between successive repairs. In many other repair situations, however, this rapid strength gain would not be essential thus allowing greater flexibility in the design of the flowable concrete mix.

Access difficulties precluded the use of poured concrete and resulted in the choice of a flowable, self levelling, self compacting concrete. Congested reinforcement (Fig. 1) and low cover in some areas dictated a maximum aggregate size of 8 mm.

Fig. 1 Congested reinforcement in soffit

The concrete used for the repair was thus essentially a cementitious grout with the addition of 8 mm nominal size aggregate.

The key elements of the specification for the flowable repair concrete can be summarised as follows:

Minimum cement content	450 kg/m^3
Maximum aggregate size	8 mm
Maximum water/cement ratio	0.40
Minimum compressive strength	30 N/mm^2 at 3 days at 20°C
	30 N/mm^2 at 10 days at 5°C
Maximum compressive strength	60 N/mm^2 at 7 days at 20°C
Flow along test trough	750 mm in 30 secs at 20°C
(after mixing and after further 30 mins)	
No shrinkage at 7 days	

To ensure good quality control it was decided to use proprietary factory dry batched repair concrete. Details of the various repair concrete mixes used remain confidential to the manufacturers. However in general the mix constituents were as follows:

	OPC in range 480 to 550 kg/m^3
	Pfa typically 20% of total cement content
or	GGBS typically 30% of total cement content
	Microsilica 5% or less of total cement content
	Aggregate (8 mm maximum size) in range 1220 to 1320 kg/m^3
	W/C ratio around 0.35 to 0.40
	Plasticizer and other undefined admixtures

3 Testing

Early in the development of the specification it was realised that the required properties of the flowable concrete were highly sensitive to small changes in the constituents and in any case difficult to achieve. A three stage testing regime was therefore adopted to ensure compliance:

1. Initial compliance tests: these consisted of extensive laboratory and field trials of the concrete to ensure full compliance with the specification.
2. Production control tests: each batch of concrete manufactured was tested for flow and compressive strength.
3. Site tests: routine site tests on flow and strength to compare with previous initial compliance and production control test results.

Additionally two types of placement trials were required:

- a simulated soffit test in which the flow of the concrete around reinforcement was checked through a perspex sheet representing the broken back soffit of the beam being repaired;
- a full sized repair trial in which a specially cast section of beam was repaired and then cored to check on air voids and debonding at the repair/parent concrete outerface. Inspection after 56 days was carried out to check for shrinkage cracking.

Parameters such as shrinkage, bleed, segregation and "placeability" were therefore checked by realistic trial repairs rather then placing total reliance on laboratory tests.

3.1 Initial compliance tests

Laboratory tests were undertaken on a range of proprietary mixes and designed mixes. The tests, which were generally carried out at 5°C, 12°C and 20°C to simulate the range of site conditions, are listed below. In all cases no vibration or other means of compaction was applied as the concrete was intended to self compact.

3.1.1 Flow trough tests

The flow characteristics were assessed using the flow trough equipment shown in Fig. 2. Each test consisted of six readings, three taken immediately after completion of mixing and three taken 30 minutes later. For compliance none of the flow tests was to exceed the specified time.

Fig. 2 Flow trough test

3.1.2 Simulated soffit tests

The flow characteristics were also assessed by simulated soffit tests. The general arrangement of this test is shown in Fig. 3. The layout of the reinforcement for this test was selected to be the most onerous in terms of achieving placement of the concrete that was likely to be encountered. After the concrete had set the specimen was cored (Fig. 4) or saw-cut into two sections which were examined to assess the amount of voidage around the reinforcement and at the repair/substrate interface, bleed at the interface, cracks and any other defects.

3.1.3 Air content

The air content of the fresh concrete was measured.

3.1.4 Compressive strengths

These were measured at 3, 7 and 28 days. In addition to meeting the minimum and maximum strength criteria the difference between the minimum and maximum test results were restricted to 20% of the mean strength.

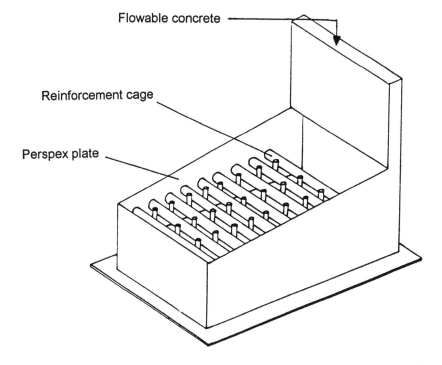

Fig. 3 Simulated soffit test

Fig. 4 Cores taken into trial repair

The maximum strength requirement was applied as an indirect means of controlling the brittleness of the repair concrete. Previous trials had used very high strength concrete which, during break-out to repair adjacent areas, had fractured.

3.1.5 Chemical tests

Chemical tests were undertaken to determine the cement, chloride and alkali contents of the concrete.

The alkali contents were to be maintained below 3 kg/m^2 to minimise any risk of ASR in the parent concrete. This was not always easy to achieve due to the high cement contents required to ensure high early strengths.

3.1.6 Expansion and shrinkage

During development work on the flowable concrete short term expansion and drying shrinkage tests were undertaken to check on the performance of the concrete admixtures. These were subsequently omitted from routine approval testing because no correlation was found between the results obtained from the laboratory testing and the actual performance in the full scale mock-up crossbeam.

3.2 Production control tests
Following the initial compliance tests various concrete formulations were accepted as being suitable for use in repairs to the structure. During the course of the repair works production control tests were carried out on each new batch of concrete supplied by the manufacturer. Production control tests were restricted to flow trough and compressive strength testing both carried out at 20°C only. These tests proved to be a

very valuable method of detecting poor quality or defective batches of material and providing a check on the manufacturer's quality control.

3.3 Site tests
Whilst the production control tests demonstrated the suitability of each batch of repair concrete further site tests were carried out on site on every pour.

For each pour of concrete a flow test was carried out and the compressive strength gain was monitored by testing cubes stored alongside the repair areas at ambient temperature.

4 Method of placement

The consistency of flowable concrete makes it ideal for pumping and this is generally the most efficient method of placement for all but very small quantities. The inclusion of only small sized aggregate allows a smaller diameter pipeline to be used. The concrete can either be pumped into the top of the formwork, or introduced through a valve in or near the soffit. This latter approach is particularly appropriate for repairs where the introduction of the concrete at the bottom of the lift minimises the risk of trapped air which could reduce bond to the substrate and provide a path for water penetration. It is also the best method of placement for underwater application.

A convenient method of placement of small quantities of flowable concrete is by means of a funnel and tube (Fig. 5). A transparent perspex face to the formwork allows the placement of the concrete to be monitored and provides a check on compaction and absence of voids (Fig. 6).

Fig. 5 Placement by funnel and tube

PLACEMENT TEST

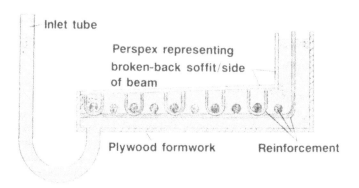

Fig. 6 Perspex shutter

5 Problems

One problem that can occur with flowable concrete is cracking due to drying shrinkage. For repairs the concrete generally has a high cement content and, even where a shrinkage compensating admixture is used, is susceptible to long term drying shrinkage. It is cast against a substrate which is often much larger in mass than the repair, which is therefore effectively totally restrained against shrinkage. Crazing and fine cracks have been recorded in several instances using some repair concretes, although this was not serious enough to cause a durability problem.

Another problem encountered is brittleness of the repair concrete. The requirement for a high early strength can lead to extremely high 28 day strength and brittleness of the concrete.

For repairs requiring small batches of concrete, quality control of strengths and flow properties can be difficult to achieve.

When used for soffit repairs, bleed of repair concrete can result in poor bond with the substrate and an increased risk of water penetration to reinforcement. Flowable concrete used for repairs must be designed, and use constituents (particularly sand) which minimise the amount of bleed that occurs.

6 Acknowledgement

This paper was taken from a chapter in Structural Grouts, published by Blackie Academic and Professional, 1994.

45 INFLUENCE OF FINE MINERAL ADDITIONS ON WORKABILITY AND MECHANICAL PROPERTIES OF CEMENT PASTES AND MORTARS

J.L. GALLIAS, S. AGGOUN, R. CABRILLAC and R. KARA-ALI
IUP Génie Civil, Université de Cergy-Pontoise, France

Abstract
Twelve fine minerals, extracted from natural deposits and from industrial processes are studied with the aim of comparing their effectiveness as additions in concrete They cover a large range of textural properties and mineral compositions.

The comparison of the fine minerals is based, at first, on the study of normal consistency pastes with various proportions and with and without a superplasticizer. The granular arrangement of the particles in the pastes and the effectiveness of the superplasticizer are appraised by the variation in the water requirement of the pastes. The results show that no correlation exists between the water requirement of the pastes and the nature or the textural parameters of the fine minerals.

The results obtained on pastes are confirmed by the rheological and mechanical characterisation of mortars with various formulations and with constant workability. It is demonstrated that the capacity of the fine particles to fill the solid skeleton of the mortars, to improve their packing and consequently their mechanical strengths, is closely correlated with the water/solid ratio of the normal consistency pastes.
Keywords: Fine mineral, cement, superplasticizer, paste, mortar, consistency, workability, packing of particles, mechanical strengths.

1 Introduction

The abundant literature dedicated on the silica fumes in high performance concrete indicates that, in a general way, their efficiency is based on a double action: the increase of the packing of particles through the filling of the solid skeleton of the concrete and the reinforcement of the binding action of the cement through pozzolanic reactions with free lime.

Production Methods and Workability of Concrete. Edited by P.J.M. Bartos, D.L. Marrs and D.J. Cleland. Published in 1996 by E & FN Spon, 2–6 Boundary Row, London SE1 8HN. ISBN 0 419 22070 4.

On the other hand, the action of other fine additions in the concrete, natural fine materials particularly, less studied than silica fumes, reveals results more mitigated [1], [2], [3], [4], [5]. Every fine material shows a specific behaviour in the concrete related to its actual textural and physico-chemical characteristics and without a connection to those of other fine materials with possibly similar characteristics.

Nevertheless, some natural deposits can be easily micronized and provide a good supply of fine materials in quantities and at a cheap price. The use of these natural fine materials as additions into the concrete could improve the concrete performance as well as open the market of the construction in some local materials. It is with this objective of local development in France that the present study has been carried out.

2 The fine materials

2.1 Nature and origin
Eight fine materials obtained by the crushing of natural deposits and four fine materials coming from industrial processes have been selected for the study. The principal physico-chemical and textural characteristics of the fine materials are indicated in Table 1. They can be distinguished in four groups according to their nature and origin.

2.1.1 Natural carbonates

Four fine materials belong to this group, composed of crystallised calcite (>99%) micronized industrially. They are comparable to calcareous fillers with fineness superior than the cements.:

- Two limestones, LM1 and LM2, from the same deposit micronized at different fineness.

Table 1 Mineral, granular and textural characteristics of fine minerals

Fine	Mineral Composition	Density kg/m3	Mean part. size μm	Specific area m²/g	Particles shape
LM1	Limestone	2690	1.8	5.2	Angular
LM2	Limestone	2690	3.8	2.5	Angular
CK1	Chalk	2710	2.1	2.6	Sub-spherical
CK2	Chalk	2710	1.5		Sub-spherical
CP1	Carbonate precipitate	2620	0.1	20.0	Sub-angular
CP2	Carbonate precipitate	2620	0.2	11.0	Needles
QZ1	Quartz	2630	6.2	1.6	Angular
QZ2	Quartz	2630	3.1	3.7	Angular
DT1	Diatomite	2140	20.1	38.4	Very irregular and porous
DT2	Diatomite	2230	10.7	8.7	Very irregular and porous
SF1	Silica fume	2340	0.2	14.9	Spherical
SF2	Silica fume	2340	0.1	21.1	Spherical
CM1	Cement CEM I	3150	6.6	0.3	Angular
CM2	Cement CEM II	3100	5.1	0.5	Angular

- Two micronized chalks, CK1 and CK2, coming from different deposits; CK2 has been submitted at a hydrophobe treatment.

2.1.2 Carbonate precipitates

Two fine materials CP1 and CP2 belong to this group, resulting from a controlled chemical reaction between carbonic gas and lime milk, allowing us to obtain particles of a great fineness, comparable to the silica fumes and of a great chemical purity ($CaCO_3$>99.9%). CP1 contains calcite and CP2 aragonite.

2.1.3 Natural silicates

Four fine materials belong to this group:

- Two quartzes of great purity (98.8%), QZ1 and QZ2, coming from the same natural deposit of fine sand micronized in different finenesses.
- Two diatomites, DT1 and DT2, containing amorphous silica, coming from the same deposit exploited industrially, but treated in a different way. DT1 is a brute dried and crushed diatomite; DT2 is calcinated.

2.1.4 Condensed silica fumes

Two condensed silica fumes, SF1 and SF2, obtained from two ferro-silicium industries, used currently as additions in high strength concrete, have been selected in order to be compared with the natural fine materials. They contain more than 80% of amorphous silica.

2.2 Granular and textural characteristics
The fine materials possess a very big range of granular and textural characteristics. According to their fineness, the fine materials can be distinguished in three groups:

- The fine materials with fineness, sub-micronic as the silica fumes (SF1 and SF2) and carbonate precipitates (CP1 and CP2).
- The fine materials with middle fineness, superior than that of cements, as the natural carbonates (LM1 and LM2, CK1 and CK2) and quartzes (QZ1 and QZ2).
- The diatomites DT1 and DT2 with rough fineness inferior to the cement and comparable to the common carbonate fillers.

  Concerning the texture we can distinguish two groups:

- For fine materials with particles which are weakly porous the specific surface depends essentially on the particles' size and shape. In this group belong the silica fumes SF1, SF2 with spherical shaped particles; the chalks CK1, CK2 with sub-spherical particles; the carbonate precipitate CP1 with sub-angular particles; the limestones CA1 and CA2; the quartzes QZ1, QZ2, QZ3, QZ4 and the pumice stone PM with strongly angular particles; and the carbonate precipitate CP2 with needle shape particles.

- Fine materials with strongly porous particles distinguishing those of open porosity characterised by a very important specific surface without connection to the particles' size or shape, as the pulverised silica PS and the volcanic tufa VT and, on the other hand, those of close porosity as the diatomites DT2 and DT3 whose specific surface is comparable to the non-porous fine materials of the first group.

3 The tests

3.1 Sampling

Most of the tests were executed on a main sample of each fine material. For some tests on pastes, concerning the fine minerals LM1, LM2, CK1, QZ2, DT2 and SF1, a second sample has been used. The results obtained on this second sampling show low deviation in connection to the first, due to the variation of the granular characteristics from one sample to another.

3.2 Tests on pastes

With the first sampling of fine minerals, four mixtures have been prepared with a CPA-CEM I 52.5 cement (according to NF P 15 301 standard), noted CM1; the granular and textural characteristics of CM1 cement are similar to the quartz QZ1 (Table 1):

- Mixtures with 100% of fine and 0% of cement.
- Mixtures with 25% of fine and 75% of cement.
- Mixtures with 10% of fine and 90% of cement.
- Mixtures with 0% of fine and 100% of cement.

All mixtures (first and second sampling, with and without superplasticizer) have been mixed with the required quantity of water to obtain a paste with normal consistency, determined with Vicat apparatus, according to the EN 196-3 standard (Table 2). A second series of mixtures (Table 3), with the same proportioning of fine material and cement was mixed under the same conditions (required water for normal consistency) with addition of 2.5% of a superplasticizer of the polyvinyl-naphthalenes sulphonates family. The superplasticizer is miscible with water and its dry content is equal to 28%. The proportioning recommended by the manufacturer is between 1.0 and 3.5% of the mass of the binder.

With the second sampling of fine materials, four mixtures have been prepared with a CPJ-CEM II 32.5 cement (according to NF P 15 301 standard), noted CM2 and without superplasticizer (Table 2). The CM2 cement contains 25% of calcareous filler and its fineness is higher than CM1's (Table 1):

- Mixtures with 100% of fine and 0% of cement.
- Mixtures with 20% of fine and 80% of cement.
- Mixtures with 10% of fine and 90% of cement.
- Mixtures with 0% of fine and 100% of cement.

Table 2 (upper part – CM1)

	Measured W/(F+C) (kg/kg)				Calc. W/(F+C)		Calc-Meas.(%)	
CM1	0%	75%	90%	100%	75%	90%	75%	90%
Fine	100%	25%	10%	0%	25%	10%	25%	10%
LM1	0.386	0.294	0.280		0.301	0.283	2.2	1.2
LM2	0.306	0.276	0.267		0.281	0.275	1.6	3.1
CK1	0.284	0.267	0.272		0.275	0.273	3.0	0.4
CK2	0.358	0.245	0.257		0.294	0.281	19.8	9.2
CP1	0.838	0.400	0.311		0.414	0.329	3.4	5.7
CP2	0.922	0.376	0.304		0.435	0.337	15.6	10.9
QZ1	0.320	0.296	0.282		0.284	0.277	-4.1	-1.8
QZ2	0.400	0.308	0.284		0.304	0.285	-1.3	0.3
DT1	1.340	0.560	0.390		0.539	0.379	-3.8	-2.9
DT2	1.984	0.770	0.460		0.700	0.443	-9.1	-3.7
SF1	0.638	0.330	0.293		0.364	0.309	10.2	5.3
SF2	0.338	0.318	0.280		0.289	0.279	-9.3	-0.5
CM1	0.27							

Table 2 (lower part – CM2)

	Measured W/(F+C) (kg/kg)				Calc. W/(F+C)		Calc-Meas(%)	
CM2	0%	80%	90%	100%	80%	90%	80%	90%
Fine	100%	20%	10%	0%	20%	10%	20%	10%
LM1	0.330	0.268	0.265		0.270	0.263	0.7	-0.9
LM2	0.294	0.271	0.259		0.263	0.259	-3.0	-0.0
CK1	0.268	0.266	0.258		0.258	0.256	-3.2	-0.7
QZ2	0.456	0.305	0.266		0.295	0.275	-3.2	3.4
DT2	1.960	0.640	0.447		0.596	0.426	-6.9	-4.8
SF1	0.592	0.328	0.282		0.322	0.289	-1.7	2.4
CM2	0.26							

Table 2. Measured and calculated water/solid ratio of normal consistency fine mineral - cement pastes without superplasticizer (upper part: first sampling of fine minerals and CM1 cement, lower part: second sampling and CM2 cement)

Table 3 (upper part – CM1)

	Measured Wt/(F+C) (kg/kg)				Calc. Wt/(F+C)		Calc-Meas.(%)	
CM1	0%	75%	90%	100%	75%	90%	75%	90%
Fine	100%	25%	10%	0%	25%	10%	25%	10%
LM1	0.224	0.176	0.188		0.210	0.207	19.2	10.1
LM2	0.198							
CK1	0.198	0.172	0.190		0.203	0.204	18.2	7.5
CK2	0.284							
CP1	0.651	0.223	0.192		0.317	0.250	42.0	30.0
CP2	0.523							
QZ1	0.318	0.218	0.214		0.233	0.216	7.0	1.1
QZ2	0.408							
DT1	1.195	0.446	0.294		0.452	0.304	1.4	3.4
DT2	1.826							
SF1	0.536	0.290	0.238		0.288	0.238	-0.8	0.0
SF2	0.290	0.182	0.182		0.226	0.214	24.3	17.3
CM1	0.21							

Table 3 (lower part – CM2)

	Measured Wt/(F+C) (kg/kg)			
CM2	0%	80%	90%	100%
Fine	100%	20%	10%	0%
LM1	0.188			
LM2	0.210			
CK1	0.178			
QZ2	0.200			
DT2	0.500			
SF1	0.244			
CM2	0.19			

Table 3. Measured and calculated water/solid ratio of normal consistency pastes with addition of superplasticizer (upper part: with 2.5% of A superplasticizer, lower part: with 2.0% of B superplasticizer)

Table 4 (upper part – CM1)

	(W-Wt)/W (%)			
	0%	75%	90%	100%
	100%	25%	10%	0%
LM1	42.0	40.1	32.9	
LM2	35.3			
CK1	30.3	35.6	30.1	
CK2	20.7			
CP1	22.3	44.3	38.3	
CP2	43.3			
QZ1	0.6	26.4	24.1	
QZ2	-2.0			
DT1	10.8	20.4	24.6	
DT2	8.0			
SF1	16.0	12.1	18.8	
SF2	14.2	42.8	35.0	
	24.6			

Table 4 (lower part – CM2)

	(W-Wt)/W (%)			
	0%	80%	90%	100%
	100%	20%	10%	0%
LM1		29.9		
LM2		22.5		
CK1		33.1		
QZ2		34.4		
DT2		21.9		
SF1		25.6		
	24.7			

Table 4. Water reduction in the paste obtained by the superplasticizer

For the second sampling of fine minerals, only the mixtures with 20% of fine and 80% of cement were mixed with 2% of a second superplasticizer of the -naphthalenes sulphonates family (Table 3). This superplasticizer is miscible with water and its dry content is equal to 40%. The proportioning recommended by the manufacturer is between 0.3 and 2% of the mass of the binder.

The capacity of particles to stand closer together into the mixtures without superplasticizer has been judged through the W/(F+C) ratio, where W is the mass of the water, F the mass of the fine materials and C the mass of the cement (Table 2) In the same way, the Wt/(F+C) ratio, where Wt is the total mass of the water (including the quantity of the added water and that brought by the superplasticizer), has been used when superplasticizer was added in the mixtures (Table 3).

3.2 Tests on mortars

Three reference mortars, referred R500, R450 and R400 according to the cement content (Table 5), have been used for the tests on the mortars, using the CM1 cement and the first superplasticizer of tests on pastes. The water quantity of the reference mortars has been adjusted in order to obtain a workability corresponding to 9.5 ± 1.5 s (measured with the workability-meter « LCL » according to the NF P 18 452 standard). The presence of the calcareous filler 0/0.6 mm (mean particle size 60 μm) was judged necessary to obtain a high density of packing with a limited amount of cement.

A first series of mortars with addition of fine minerals (of the first sampling) have been obtained by substitution of 100 kg/m^3 of cement in the reference mortar R500 by an equivalent mass of fine material preserving the proportioning of the other solid components and of the superplasticizer. Only the water quantity has been adjusted in order to maintain the same workability as the reference formulations (Table 5).

A second series of mortars was prepared under the same conditions as in the first series (water required for constant workability) by substitution of 50 kg/m^3 of cement in the reference mortar R500 by an equivalent mass of fine mineral, preserving the proportion of the components other than water (Table 5).

4 Results and discussion

4.1 Study of pastes

The values of the W/F ratio of pastes at 100% of fine material (Table 2) vary substantially with the capacity of the fine particles to stand close together in presence of water:

- The carbonate precipitates CP1 and CP2, the diatomites DT1 and DT2 and one of the silica fumes SF1 show higher W/F ratios between 2.0 and 0.5.
- The limestones LM1 and LM2, the chalks CK1 and CK2, the quartzes QZ1 and QZ2 and the second silica fume SF2 show lower W/F ratios, less than 0.5, comparable to the W/C ratio of the paste at 100% of cement (0.272 for CM1 and 0.255 for CM2).

Formulation	Cement kg/m3	Sand kg/m3	Fine kg/m3	Superplast. kg/m3	Water kg/m3	W/C Kg/kg	W/(C+F) Kg/kg	Workability s	Compacity %	Compr. str. MPa	Flexural str. MPa
LM1	400	1485	100	10	146.0	0.383	0.306	9.5	77.3	82.8	10.7
LM2	400	1485	100	10	153.0	0.401	0.320	10.9	77.0	82.3	10.9
CK1	400	1485	100	10	145.0	0.381	0.304	9.3	77.9	86.0	11.6
CK2	400	1485	100	10	160.0	0.418	0.334	8.7	75.4	71.6	10.3
CP1	400	1485	100	10	202.5	0.524	0.419	11.2	70.4	47.2	7.9
CP2	400	1485	100	10	168.0	0.438	0.350	8.5	74.8	65.7	9.9
QZ1	400	1485	100	10	162.5	0.424	0.339	9.2	76.7	75.0	9.5
QZ3	400	1485	100	10	157.0	0.411	0.328	9.2	76.7	80.4	10.1
DT1	400	1485	100	10	288.0	0.738	0.590	9.5	68.5	49.8	6.7
DT2	400	1485	100	10	255.0	0.656	0.524	10.9	70.2	59.0	8.6
SF1	400	1485	100	10	193.0	0.501	0.400	9.8	70.6	56.7	9.6
SF2	400	1485	100	10	173.0	0.451	0.360	10.1	73.1	67.9	10.7
LM1	450	1535	50	10	153.0	0.356	0.320	8.0	77.0	83.8	9.9
CK1	450	1535	50	10	152.0	0.354	0.318	8.5	77.5	81.8	10.5
CP1	450	1535	50	10	177.0	0.409	0.368	8.2	73.6	65.2	8.6
QZ1	450	1535	50	10	160.0	0.372	0.334	9.5	76.1	80.1	10.3
DT1	450	1535	50	10	220.0	0.505	0.454	8.4	73.4	68.6	9.5
SF1	450	1535	50	10	173.0	0.400	0.360	8.5	75.3	75.8	10.2
SF2	450	1535	50	10	165.0	0.383	0.344	8.1	74.1	72.2	11.2
R500	500	1485	0	10	160.0	0.334	0.334	9.2	76.5	84.2	11.3
R450	450	1535	0	10	158.0	0.367	0.367	9.2	76.2	82.7	10.6
R400	400	1585	0	10	157.0	0.411	0.411	9.0	76.3	76.4	10.0

Table 5. Composition, physical and mechanical characteristics of mortars

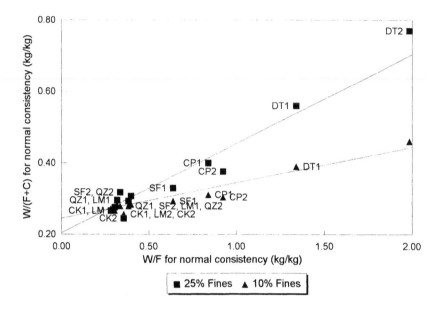

Figure 1. Relation between the water/fine ratio of 100% fine pastes (first sampling) and the water/solid ratio of binary, fine mineral-cement, mixtures (without superpl.)

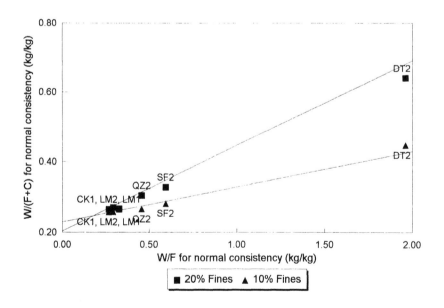

Figure 2. Relation between the water/fine ratio of 100% fine pastes (second sample) and the water/solid ratio of binary, fine mineral-cement, mixtures (without superpl.)

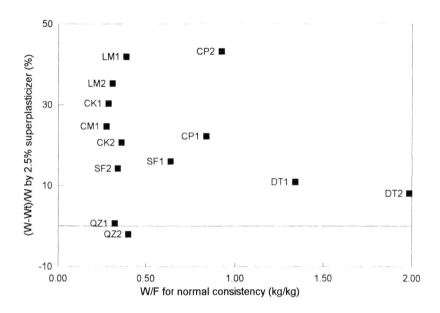

Figure 3. Relation between the water/fine ratio of 100% fine pastes (without super-plasticizer) and the water reduction obtained by the superplasticizer on the paste

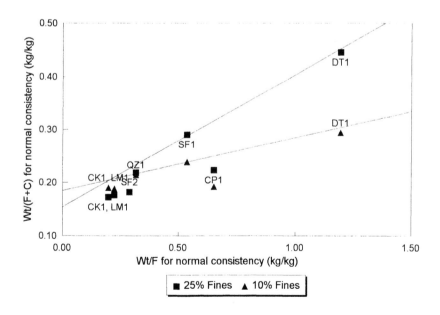

Figure 4. Relation between the water/fine ratio of 100% fine pastes (first sampling) and the water/solid ratio of binary, fine mineral-cement, mixtures with superplast.

The difference on the W/F ratio between the two samplings of fine minerals is weak and insufficient to upset the classification of fine minerals into the two previous groups.

The W/F ratio is independent of the nature (siliceous or calcareous), the origin (natural deposit or industrial product) and the fineness (specific area and mean particle size) of the fine material. However, the W/F ratio is important for the fine materials with irregular particle shape and with an important infra-granular porosity (DT1, DT2).

The measured $W/(F+C)$ ratios in the binary (fine materials - cement) mixtures with normal consistency depend at first on the W/F ratio of mixtures at 100% of fine material and the proportion of fines in the mixture (Figs. 1 and 2). Nevertheless, the relation between $W/(F+C)$ ratio and W/F ratio could deviate from linear low up to $\pm 20\%$ (Figs. 1 and 2, Table 2). For instance, the CK2, CP2 and SF1 mixtures show a $W/(F+C)$ ratio between 5% and 20% less than calculated, showing that the granular arrangement into the binary mixtures is closer than the corresponding to the juxaposition of the single phase mixtures. All happens as a part of the particles of the fine addition fills the spaces between cement particles, reducing the total porous volume of the binary mixture. On the contrary, the DT1, DT2 and SF2 binary mixtures present a volume expansion up to 9% in comparison with the single phase mixtures, as the fine particles push away the cement particles.

The presence of superplasticizer in the mixtures leads generally to a better dispersal and lubrication of the particles and consequently to an increase in the fluidity of paste. In our case, as the consistency of pastes remains constant (Table 3), the gain of fluidity corresponds to a water reduction calculated by the difference between the $W/(F+C)$ ratio of mixtures without superplasticizer and the $Wt/(F+C)$ ratio of the same mixtures in the presence of superplasticizer (Table 4).

In the case of pastes at 100% of fine material, the water reduction by the first superplasticizer varies greatly and depends essentially on the nature of the fine materials (Fig. 3). The calcareous fine materials show important water reduction from 21% to 43%, comparable to that of cement alone (24%), while siliceous fine materials have a small reduction between 0% and 18%.

In the case of the binary, fine material and cement mixtures, the effect of the superplasticizer is more important. The water reductions exceed, in a great majority of cases, those measured in the mixture at 100% of cement (remaining, nevertheless, more favourable for the calcareous fine materials in the case of the first superplasticizer). In this manner the presence of superplasticizer amplifies the capacity of the fine particles to fill the porosity of the cement particles in the mixture. In fact, the deviation between calculated and measured $Wt/(F+C)$ ratio (Fig. 4, table 3) is, with one exception near to zero, systematically positive.

4.2 Study of mortars

As in the case of the pastes, the evaluation of the fine mineral additions in mortars with constant workability (equal to the reference mortars) can be done of the W_t/C or $W_t/(F+C)$ ratios where W_t includes the quantity of added water and the water amount brought by the superplasticizer.

It is worth stating at this point that the W_t/C or $W_t/(F+C)$ ratios of mortars with fine materials vary in the same manner as the W/F, W_t/F, $W/(F+C)$ and $W_t/(F+C)$ ratios

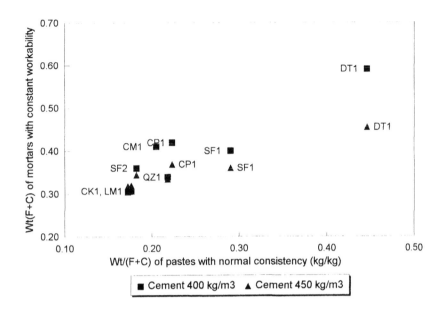

Figure 5. Relation between the water/solid ratio of pastes (25%fine, 75%cement and 2.5% superplastic.) and the water/solid ratio of mortars for various cement contents

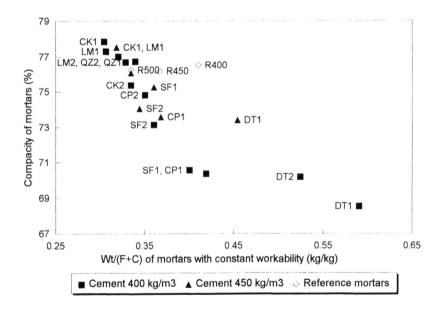

Figure 6. Effect of the water/solid ratio of mortars with constant workability on the compacity for various fine and cement content

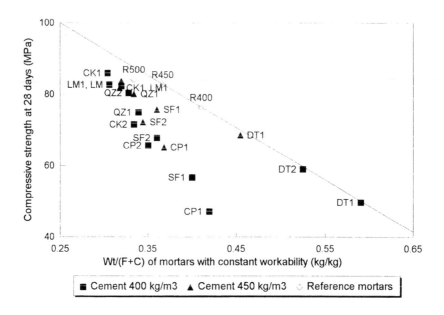

Figure 7. Effect of water/solid ratio of mortars with constant workability on the compressive strength at 28 days for various fine and cement contents

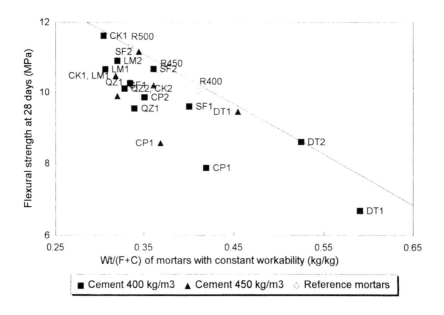

Figure 8. Effect of water/solid ratio of mortars with constant workability on the flexural strength at 28 days for various fine and cement contents

of the pastes. A statistical analysis of results has shown that the best correlation of values is obtained between the $W_t/(F+C)$ ratios of mortars and the $W_t/(F+C)$ ratios of pastes at 75% of cement and 25% of fine materials with superplasticizer. This correlation has been confirmed equally in the second series of mortars containing 450 kg/m^3 of cement (Fig. 5). Two groups of fine minerals can be distinguished in the same way as these observed for the pastes:

- The carbonate precipitate CP1, the diatomites DT1 and DT2, and one of the silica fumes FS1 lead to the most important Wt/(F+C) ratios between 0.590 and 0.400.
- The limestones LM1 and LM2, the chalks CK1 and CK2, the second carbonate precipitate CP2 and the second silica fume SF2 lead to low Wt/(F+C) ratios between 0.360 and 0.304. Particularly the LM1, LM2, CK1 and QZ2 mortars show a Wt/(F+C) ratio less than that of the R500 reference mortar, where the fine materials have been substituted by the cement.

All this proves that the granular arrangements in the pastes are completely representative of those produced in the more complicated mixtures, as if the effect of the finest size range particles has not been significantly influenced by the presence of the rough size range particles. It remains also that the Wt/(F+C) ratio of pastes, and consequently that of the mortars, is not directly correlated either with the mean particles size or with the specific area of fine materials but, rather, with the shape of the particles and their porosity.

On the other hand, the Wt/(F+C) ratio is an essential characteristic of the mortars and influences directly the packing density of the mixtures (Fig. 6), and consequently the mechanical strengths. The relation between the Wt/(F+C) ratio and the compressive and flexural strength of mortars (Figs. 7 and 8) is limited by an upper line corresponding at a binding activity of the fine material equal to that of the cement. Otherwise the values of reference mortars and the mortars with strongly pozzolanic fine materials (DT1, DT2, SF2) take place on this upper line.

5 Conclusion

The previous results induce the conclusion that the $W_t/(F+C)$ ratio of binary, fine material and cement mixtures with a normal consistency in presence of superplasticizer express the capacity of spatial arrangement of the particles of the mixture which is not a priori dependent either on the physico-chemical nature or on the fineness of the fine material.

The $W_t/(F+C)$ ratio of pastes is also representative of granular arrangements in presence of aggregates and can constitute a total index of the aptitude of the fine material to be incorporated at any cementitious matrix material (mortar or concrete). The fine materials with low $W_t/(F+C)$ ratio in the binary mixtures, incorporated into the mortars, lead to a water reduction when the workability remains constant, and increase the packing density and ameliorate the physico-mechanical performances.

Otherwise the fine materials with low $W_t/(F+C)$ ratio in the binary mixtures can be added with a constant amount of water in order to improve the workability of the formulation. On the contrary, the fine materials with a high $W_t/(F+C)$ ratio will systematically demand an adaptation of the formulation (increasing the proportioning of ad-mixture or of cement) in order to cover the loss of performances led by the excessive water requirement and the corresponding loss of the packing density.

6 References

1.   Neto, C. and Campiteli, V. (1990) The Influence of Limestone Additions on the Rheological Properties and Water Retention Value of Portland Cement Slurries, in *Carbonate Additions to Cement,* (ed. Klieger/Hooton) ASTM, USA, STP 1064, pp. 24-29.
2.   Kronlof, A. (1994) Effect of Very Fine Aggregate on Concrete Strength, *Materials and Structures* No. 27, pp. 15-25.
3.   Bombled, J.P. (1982) Influence des Fillers sur les Propriétés des Mortiers et des Bétons, *Ciments Bétons Plâtres et Chaux,* Vol. 5, No. 73, pp. 282-290.
4.   De Larrard F. (1989) Ultrafine Particles for the Making of Very High Strength Concrete, *Cement and Concrete Research,* Vol. 19, pp. 189-172.
5.   Harada, H., Shizawa, M. and Kotani H. (1986) Influence of Super Fine Granualted Blast Furnace Slag and Limestone on the Properties of Blast Furnace Slag Cement, *Concrete Association of Japan Revue,* Vol. 34, pp. 66-69.

Author index

Subject index

This index has been compiled from the keywords assigned to the papers by the authors, edited and extended as appropriate. The numbers refer to the first page number of the relevant paper.

RILEM

RILEM, The International Union of Testing and Research Laboratories for Materials and Structures, is an international, non-governmental technical association whose vocation is to contribute to progress in the construction sciences, techniques and industries, essentially by means of the communication it fosters between research and practice. RILEM activity therefore aims at developing the knowledge of properties of materials and performance of structures, at defining the means for their assessment in laboratory and service conditions and at unifying measurement and testing methods used with this objective. RILEM was founded in 1947, and has a membership of over 900 in some 80 countries. It forms an institutional framework for cooperation by experts to:

- optimise and harmonise test methods for measuring properties and performance of building and civil engineering materials and structures under laboratory and service environments;
- prepare technical recommendations for testing methods;
- prepare state-of-the-art reports to identify further research needs.

RILEM members include the leading building research and testing laboratories around the world, industrial research, manufacturing and contracting interests as well as a significant number of individual members, from industry and universities. RILEM's focus is on construction materials and their use in buildings and civil engineering structures, covering all phases of the building process from manufacture to use and recycling of materials.

RILEM meets these objectives though the work of its technical committees. Symposia, workshops and seminars are organised to facilitate the exchange of information and dissemination of knowledge. RILEM's primary output are technical recommendations. RILEM also publishes the journal *Materials and Structures* which provides a further avenue for reporting the work of its committees. Details are given below. Many other publications, in the form of reports, monographs, symposia and workshop proceedings, are produced.

Details of RILEM membership and the journal *Materials and Structures* may be obtained from RILEM, École Normale Supérieure, Pavillon des Jardins, 61, avenue du Pdt Wilson, 94235 Cachan Cedex, France. RILEM Reports, Proceedings and other publications are listed below and details may be obtained from E & F N Spon, 2-6 Boundary Row, London, SE1 8HN, UK. Tel: (0)171-865 0066, Fax: (0)171-522 9623.

RILEM Reports

1   **Soiling and Cleaning of Building Facades**
    Report of Technical Committee 62-SCF. *Edited by L.G.W. Verhoef*
2   **Corrosion of Steel in Concrete**
    Report of Technical Committee 60-CSC. *Edited by P. Schiessl*
3   **Fracture Mechanics of Concrete Structures - From Theory to Applications**
    Report of Technical Committee 90-FMA. *Edited by L. Elfgren*
4   **Geomembranes - Identification and Performance Testing**
    Report of Technical Committee 103-MGH. *Edited by A. Rollin and J.M. Rigo*
5   **Fracture Mechanics Test Methods for Concrete**
    Report of Technical Committee 89-FMT. *Edited by S.P. Shah and A. Carpinteri*